WHATSAPP IN THE WORLD

WhatsApp in the World

Disinformation, Encryption, and Extreme Speech

Edited by Sahana Udupa and Herman Wasserman

NEW YORK UNIVERSITY PRESS
New York

NEW YORK UNIVERSITY PRESS
New York
www.nyupress.org

CIP Data (Work)Library of Congress Cataloging-in-Publication Data

Names: Udupa, Sahana, 1977– editor. | Wasserman, Herman, 1969– editor.
Title: WhatsApp in the world : disinformation, encryption and extreme
    speech / edited by Sahana Udupa and Herman Wasserman.
Description: New York : New York University Press, [2025] | Includes
    bibliographical references and index. | Summary: ""WhatsApp in
    the World" presents a groundbreaking global analysis of the vastly
    popular instant messaging service, delving into the complex interplay
    between encryption and extreme speech"— Provided by publisher.
Identifiers: LCCN 2024040426 (print) | LCCN 2024040427 (ebook) |
    ISBN 9781479833269 (hardback) | ISBN 9781479833276 (paperback) |
    ISBN 9781479833283 (ebook) | ISBN 9781479833306 (ebook other)
Subjects: LCSH: WhatsApp (Application software) | Instant messaging—Social
    aspects. | Misinformation. | Hate speech. | Data encryption (Computer science)
Classification: LCC HM742 .W496 2025 (print) | LCC HM742 (ebook) |
    DDC 302.23/11—dc23/eng/20250213
LC record available at https://lccn.loc.gov/2024040426
LC ebook record available at https://lccn.loc.gov/2024040427

New York University Press books are printed on acid-free paper, and their binding materials are chosen for strength and durability. We strive to use environmentally responsible suppliers and materials to the greatest extent possible in publishing our books.

The manufacturer's authorized representative in the EU for product safety is Mare Nostrum Group B.V., Mauritskade 21D, 1091 GC Amsterdam, The Netherlands. Email: gpsr@mare-nostrum.co.uk.

Manufactured in the United States of America

10 9 8 7 6 5 4 3 2 1

Also available as an ebook

CONTENTS

Introduction: Lived Encryptions: WhatsApp,
Disinformation, and Extreme Speech   1
Sahana Udupa and Herman Wasserman

PART I: POLITICS OF POLARIZATION

1. Extreme Speech, Community Resonance, and Moralities:
   Ethnographic Notes on the Use of WhatsApp in
   Brazilian Favelas   29
   Carolina Parreiras

2. Exclusionary Politics and Its Contradictions: Peddling
   Anti-Immigrant Sentiments through WhatsApp
   in South Africa   50
   Nkululeko Sibiya and Iginio Gagliardone

3. Deep Extreme Speech: Intimate Networks for Inflamed
   Rhetoric on WhatsApp   68
   Sahana Udupa

4. Misinformation behind the Scenes: Political
   Misinformation in WhatsApp Public Groups
   ahead of the 2022 Constitutional Referendum in Chile   88
   *Marcelo Santos, Jorge Ortiz Fuentes, and
   João Guilherme Bastos dos Santos*

PART II: (UN)SAFE SPACES

5. Delete This Message: Media Practices of Anglophone
   Cameroonian WhatsApp Users in the Face of
   Counterterrorism   111
   Kim Schumann

6. Engaging and Disengaging with Political Disinformation
on WhatsApp: A Study of Young Adults in South Africa   126
Herman Wasserman and Dani Madrid-Morales

7. Discourses of Misinformation in the Russian Diaspora:
Building Trust across Instant Messaging Channels   141
Yulia Belinskaya and Joan Ramon Rodriguez-Amat

8. WhatsApp in the United States: The Political Relevance
of Subversive Platforms   159
Inga Kristina Trauthig

PART III: INFRASTRUCTURE

9. Contextualizing WhatsApp as Reporting Infrastructure   175
Ruth Moon

10. Beyond Algorithms: How Politicians Use Human
Infrastructure to Spread Disinformation and Hate Speech
on WhatsApp in Nigeria   189
Samuel Olaniran

11. Dis/Misinformation, WhatsApp Groups, and Informal
Fact-Checking Practices in Namibia, South Africa,
and Zimbabwe   204
Admire Mare and Allen Munoriyarwa

12. How to Approach Speech Regulation on WhatsApp:
Lessons from Regulatory Experiments in India   223
Amber Sinha

PART IV: METHOD

13. Methodological Challenges in Researching Disinformation
on WhatsApp in Turkey   241
Erkan Saka

14. Researching Political Communication on WhatsApp:
Reflections on Method   254
Tanja Bosch

15. Collecting WhatsApp Data for Social Science Research:
Challenges and a Proposed Solution   267
Simon Chauchard and Kiran Garimella

16. Automating Data Collection from Public WhatsApp
    Groups: Challenges and Solutions   286
    Nicholas Micallef, Mustaque Ahamad, Nasir Memon,
    and Sameer Patil

PART V: REFLECTIONS ON POLICY
17. Fact-Checking on WhatsApp in Africa: Challenges
    and Opportunities   303
    Cayley Clifford

18. Challenges of Fact-Checking WhatsApp Messages in India   309
    Jency Jacob

19. The Policy Problems of Coordinated Harm on WhatsApp
    in Africa: From Calculation to Observation   313
    Scott Timcke

    Acknowledgments   319

    Bibliography   321

    About the Contributors   357

    About the Editors   360

    Index   361

Introduction

Lived Encryptions

WhatsApp, Disinformation, and Extreme Speech

SAHANA UDUPA AND HERMAN WASSERMAN

In the months leading up to Chile's historic 2022 referendum on its proposed new constitution, WhatsApp groups in the country saw a significant surge in conspiracy theories about election fraud, as misinformation messages that were predominant within right-leaning groups in the beginning soon spread and "contaminated" left-leaning groups that were in favor of the proposal. During the fiercely contested presidential elections in Brazil in 2018 and 2022, several women in the favelas received messages in the WhatsApp church group, hailing right-wing conservative leader Jair Bolsonaro as the "man of God." There was little that could be disputed, in their mind, that "God, homeland and family" are deeply connected, and amid waves of WhatsApp messages that echoed such sentiments, they switched their loyalties from progressive parties to Bolsonaro's conservatism.

Anti-Black far-right ideologies are common within WhatsApp groups of "Operation Dudula," an anti-immigration group in South Africa, although a majority of its users are Black South Africans. Contrary to assumptions that hateful narratives ride on brazen falsehoods, members of this group spend much time to offer "accurate" information as a rhetorical ploy to ensure their groups are free from "inauthentic" and "criminal" groups who could spoil their "brand image" and upset their activities to prevent immigration from neighboring African countries into South Africa.

At traffic checkpoints in the Anglophone regions in Cameroon, amidst an escalating conflict between separatist and state forces, it is a routine practice for government authorities to stop passengers and demand to see not only their national identity cards but also their smartphones to manually check if they carry any "incendiary" messages. WhatsApp's encryption is overruled in such swift everyday acts of surveillance.

On WhatsApp groups in India, Hindu nationalists have transformed encryption from a technical feature of privacy to one that can foster obligatory and affective chambers for ideological talk. Disinformation proliferates within WhatsApp groups modeled as kin-like groups. Similarly, in South Africa and Kenya, convivial bonds within communities breed disinformation on WhatsApp groups since any act of correcting dubious messages passed on by known members of family or church members is viewed as impolite or disrespectful of one's elders.

As these examples from different contributions in this volume illustrate, ironies, contradictions, and thick social norms and community affect suffuse WhatsApp discourse, while their ramifications have contributed to invasive state surveillance and some of the worst human tragedies, most significantly in the diverse contexts of the Global South, which are marked by historical exploitation and injustices as well as persistent socioeconomic inequality.

Disinformation and vitriolic expressions on WhatsApp have received extensive media and public policy attention, alongside academic scholarship that has concurrently drawn attention to the broader and complex ecosystems of hate and disinformation in the digital age. Recent academic scholarship has taken note of the risks of internet communication for democratic systems, as regressive regimes around the world have weaponized online discourse for partisan gains during elections, to undercut domestic dissent or power up geopolitical contestations against "rival" nation states through targeted disinformation campaigns (Bayer and Bárd 2020; George 2016; Graan, Hodges, and Stalcup 2020; Krafft and Donovan 2020; Lee 2019). Acknowledging the role of digital networks in inspiring social movements to hold the power to account and question entrenched hierarchies, studies have simultaneously highlighted the worrying developments around vitriolic exchange, inauthentic content, and "antisocial commenting" that are breeding on affordable communication and the circumvention of legacy gatekeepers that digital media infrastructures have enabled (Shmargad et al. 2024, 220).

The extent of disinformation and extreme speech prevalent in online exchanges is notoriously slippery for quantification, especially due to a lack of access to social media company data, including comprehensive transparency reports, and various barriers that companies have raised for API-based data gathering and auditing (Freelon 2018). Facebook reported that "between January 2021 and March 2021, there was a 0.05 percent to 0.06 percent prevalence of hate speech, showing a slight decrease compared to their two previous reports" (Bright et al. 2021, 6), although processes of drawing such metrics remain inaccessible to researchers. It is indeed not common for social media companies to publish "prevalence metrics," and "evidence about the prevalence of hate speech on social media platforms remains incomplete, partly due to a lack of transparency and data access on the part of platforms" (2021, 6).

While no consensus in academic scholarship about quantitative indicators of "prevalence" has followed as a result, academic studies, policy reports, and regulatory measures have highlighted the significance of the issue both in terms of public perceptions and emerging patterns of public discourses. According to the World Economic Forum *Global Risks Report*, misinformation and disinformation "was perceived as a moderately severe risk" by its respondents (World Economic Forum 2023, 24), while the Ipsos and UNESCO report (2023, 21–23) identified social media as a "top source of information in every country" it surveyed, finding that "two thirds [of respondents] often encounter hate speech online." Users in the Global South regions themselves experience the impact of disinformation on their daily lives and political participation as a major problem. Research in several sub-Saharan African countries, for instance, has shown that perceived exposure to disinformation is high and is linked to low levels of trust in social and national media (Wasserman and Madrid-Morales 2019).

Importantly, studies have shown that disinformation, misinformation, and extreme speech feature prominently in political discourses during major developments or events such as elections, social movements, and referendums within diverse national, local, and translocal contexts as well as linked to global strategic interests of "foreign actors" (Bradshaw and Howard 2018). In the Anglophone crisis in Cameroon, Schumann (chapter 5, this volume) notes that online posts are made every day and some incidents (fifty-two at the time of publication) have been verified by reports that are archived in the Cameroon Database of Atrocities hosted by the University of Toronto (Borealis 2024). Nkululeko and Gagliardone (this volume) cite the *Disinformation in Africa 2024* report published by the Center for Strategic Studies, which has found 189 documented disinformation campaigns in Africa sponsored by foreign governments including China and Russia, noting that such campaigns have quadrupled since 2022. Incidents of misinformation, disinformation, and deep fakes, including the use of artificial intelligence, have been documented around the world (Kertysova 2018). Subsequently, the number of publications on hate speech and related phenomena has seen an exponential growth between 1992 and 2018 (Tontodimamma et al. 2021, 163) and it continues to rise (Walther and Rice 2024).

In the complex mix of factors that shape extreme speech ecosystems, encrypted instant messaging services such as WhatsApp, Telegram, and Signal constitute a unique constellation. Such messaging services lack some of the core features of networked communication that typify social media, most prominently the ability for "participants [to] have uniquely identifiable profiles that consist of user-supplied content, content provided by other users, and/or system-level data . . . [and to] . . . publicly articulate connections that can be viewed and traversed by others" (Ellison and boyd 2013, 533). Content shared in one WhatsApp group, for instance, cannot easily reach other groups, and users on the platform do not have the affordances to present profiles based on metadata such as likes and followers and algorithmically curated metrics; they also cannot publicly maneuver their lists of contacts. In the messaging app ecosystem, influencers are likely to be those who contribute the most or whose posts are shared and liked most frequently, but such metrics are not easily available to gauge nor are they appended to the profile. WhatsApp, like other instant messaging services, is also distinct for its chronological message display and the conspicuous lack of algorithmic feed that characterizes quasi-public social media platforms.

However, alongside the basic feature to "consume, produce and/or interact with streams of user-generated content provided by their connections" (Ellison and boyd 2013, 533), encrypted messaging is integrating social media-type functionalities such as group messaging, bulk forwards, user reactions, channels, and group lists, thereby transitioning from a strictly interpersonal form of communication to a social media-like platform. Importantly, in contrast to social media channels that promise publicity and visible, and even spectacular, disruptions of mainstream media and political discourses, encrypted messaging often flows below the ground and end-to-end, often slipping out of direct regulatory reach and academic scrutiny. The very encrypted nature that makes messaging services attractive as a secure communication infrastructure also complicates access for research and content moderation, thus raising serious methodological and regulatory challenges.

This volume takes this vastly popular and important form of internet-enabled communication for closer examination, to develop a global critical inquiry into entanglements between encryption and extreme speech. It approaches the problem with an empirical focus on WhatsApp—an end-to-end encrypted, cross-platform messaging service owned by Meta (which purchased the app in 2014)—which has emerged as a central communication tool for a large number of people, with more than two billion users and one hundred billion daily messaging the world over (Ceci 2022). While messaging services such as Signal, Telegram, and WhatsApp have comparable functionalities and pose similar regulatory challenges to "securely" screen encrypted messages (Guest 2023), the popularity, reach, and specific purposes to which such messaging apps are put to use depend on the broader media and political systems they are embedded within (Rogers 2020; Semenzin and Bainotti 2020).

This volume explores WhatsApp as a critical window to encrypted messaging as a globally prevalent form of communication, particularly influential as a communication channel and platform for social and political mobilizations in the Global South. The analytical and methodological lessons derived from studying this messaging app, which has the largest user volume among such apps globally, can provide useful perspectives to study other messaging services as well as digitally mediated disinformation and extreme speech more broadly. WhatsApp's popularity has been linked to low internet connectivity and high data costs in the Global South contexts, but its uptake in different parts of the world, including in the Global North regions, awaits systematic research on user practices, infrastructural conditions, and political deployments around encrypted messaging, and how such features have uniquely inflected disinformation and extreme speech environments.

With a set of cross-disciplinary studies on a range of national and transnational contexts, and focusing especially on the Global South, we address this gap and propose the concept of "lived encryptions." Turning to practices where WhatsApp intersects with vastly complex social and political fields and the lived worlds of users, "lived encryptions" stresses that encryption as a technological feature cannot be taken at its face value or as a central piece of the affordance as it is experienced; rather, it embeds different, often contradictory, social and political formations and interactions. This is evidenced, for instance, in the way the promised confidentiality of encrypted messaging is upturned completely when surveilling states seize the phones from suspected dissenters to download the data or how seemingly closed group communication is channelized to "broadcast" top-down political messages. In the Global South contexts, the emergence of "broadcast" WhatsApp groups testifies to novel ways of creating conditions of virality where unfamiliar senders invert the very logic of end-to-end encryption as a privacy- inducing feature and transform it into a subsidiary of community conversation that can render political messages as socially significant. As such, encryption does not unequivocally pave the way for feelings of safety and security; within conflict and authoritarian contexts, it triggers tense appraisals around safe and unsafe spaces. The conceptualization of "lived encryptions" foregrounds such tensions, accounting for irreducible user cultures and localized innovations for political propaganda in theorizations of digital communication, disinformation, and vitriol. It also highlights methodological difficulties of tracking them, and the value of ethnographic research in navigating intimate networks of messaging that are hard to access by other methodological means, as well as in contextualizing the varied contradictions of encryption.

In the rest of this introduction chapter, we will outline key observations around WhatsApp, disinformation, and vitriol in current scholarship, lay out the theoretical stakes of encryption and develop the framework of "lived encryptions." Throughout this discussion, we will closely converse with the contributions in this volume to delineate the contours of what we call "lived encryptions." The final section provides a description of the five sections featured in this volume—politics of divisive messaging, safe/unsafe spaces, infrastructure, method, and policy—and how they advance distinct yet interconnected lines of inquiry around WhatsApp.

Hate and Disinformation on WhatsApp

Scholars have documented the vast popularity of instant messaging services, describing WhatsApp, for instance, as a "technology of life" in the Global South (Cruz and Harindranath 2020). Studies on Africa have stated that "WhatsApp is the internet and vice versa" (Mare and Munoriyarwa, this volume), and scholars on South America have noted how WhatsApp, affectionately called by the name "ZapZap," is an entrenched everyday communication infrastructure in Brazilian favelas (Parreiras, this volume) and a primary communication tool in Chile (Santos, Ortiz Fuentes, and dos Santos, this volume). The vast popularity of WhatsApp has emerged from path dependency and low data usage, both tied to the political economy of digital media expansion in the South when companies like Meta (formerly Facebook) adopted a predatory path to establish new user bases with aggressive acquisition strategies and programs such as Free Basics (free internet in exchange for default provision of company's social media platforms and by limiting the access to other platforms). As a result, mobile operators in several countries in the South offer zero-rated data plans for WhatsApp and Facebook access, while the company has also introduced more features to offer commercial services to "business users" and digital payments in some markets (Cruz and Harindranath 2020).

African grassroots movements, civil society organizations, and concerned citizens have "leveraged the platform to raise awareness, advocate for issues, and mobilize support," as Olaniran points out in his contribution to this volume, citing social and political movements in Nigeria such as the #BringBackOurGirls campaign and the #EndSARS protest. The possibility to share news articles, videos, and personal accounts on WhatsApp, he points out, has enabled users to create a "network of engaged citizens who could drive change and influence public opinion." Wasserman and Madrid-Morales in this volume similarly discuss how WhatsApp has offered multiple avenues for communication among different communities in South Africa.

In contexts where other social media platforms are banned or restricted, for instance the Russian extremism law which does not allow Facebook, Instagram, and Twitter to operate in the country, users have moved to messaging services not only to hold private conversations but increasingly to access news and public information (Sauer 2022). Encrypted channels are also popular for "political talk" in Europe, especially among users who are "reluctant to talk about politics in public," as evidenced in parts of former East Germany (Valeriani and Vaccari 2017). The *Digital News Report* by the Reuters Institute for the Study of Journalism at the University of Oxford has provided evidence that more users are holding political discussions through WhatsApp, often considered as "private" discussions on politics in contrast to perceptions of public engagements (Newman et al. 2023).

Simultaneously, WhatsApp's role in electoral politics and ecosystems of spurious content has expanded (Garimella and Eckles 2020). As Cheeseman et al. (2020, 145) note, "In the space of just a year, countries as otherwise diverse as Brazil, India, and Nigeria were said—with varying degrees of accuracy—to have witnessed their first 'WhatsApp election,' with the dissemination of rumors, conjecture, and lies allegedly undermining the democratic process itself." Rossini et al. (2021, 2434) have found "clear evidence that WhatsApp has been successfully used to spread false and misleading information during elections in Brazil as well as in India and Indonesia." Social media and messaging platforms do not remain hermetically sealed, however. Cross-fertilization between platforms, such as amplification of WhatsApp messages on Twitter and Facebook, has been documented in contexts such as Nigeria and India. In addition, recent developments of "deplatforming" hate influencers from Facebook and Twitter have spurred a wave of platform migration, as far-right actors in Europe and North America have turned to encrypted messaging alongside smaller platforms to build "resilience" and sustain group mobilization (Rogers 2020).

Disinformation and hate speech scholarship has largely advanced inquiries around instant messaging services within a broader critical framework which posits that the rules, protocols, conventions, and default designs of social media platforms shape user interfaces to enable and constrain forms of communication (Manovich 2001). Aside from the central feature of end-to-end encryption, topical studies have foregrounded the significance of closed communication architecture in messaging services as opposed to a timeline-based news feed and the absence of algorithmic sorting of content common in quasi-public social media platforms such as Twitter or Facebook. In addition, instant messaging services offer ways to imbue messages with ephemerality with the functionality of "disappearing messages." In contrast to radical user anonymity and subcultural semiotics of niche small platforms such as 4Chan (Auerbach 2012; Knuttila 2011), sources of WhatsApp messaging are both known and unknown, as it spreads within closed communication groups and among members whose telephone numbers are visible, yet elusive when groups expand beyond direct contacts. In addition, the possibility to share texts, voice notes, still and moving images, and web-links creates a multimodal and flexible conversational environment. Empirical evidence on partisan and inflammatory content flowing through WhatsApp has occasioned the argument that platform features, including affordances of forwards, group chats, and group calls, have prominently contributed to the ease and amplification of disinformation, conspiracy theories, and vitriolic exchanges (Binder, Ueberwasser, and Stark 2020; Evangelista and Bruno 2019; Johns and Cheong 2021; Nizaruddin 2021; Recuero, Soares, and Vinhas 2021; Resende et al. 2019b; Rossini et al. 2021; Soares et al. 2021; Williams et al. 2022).

Qualifying platform-based analysis, anthropologists have adopted a media practice framework, asking what people do with media and how complex mediations of lived worlds cluster around, draw upon, and re-shape technological possibilities. Ethnographic and interdisciplinary studies on hateful speech and disinformation in the Global South have especially drawn attention to the political use and party deployments of WhatsApp (Pinheiro-Machado and Vargas-Maia 2023; Wasserman and Madrid-Morales 2022). As Olaniran discusses in his chapter in this volume, Nigerian politicians have exploited WhatsApp's affordance of connecting members of communities by creating WhatsApp groups that serve as hubs for their supporters and volunteers. He shows how, by fostering personal connections, politicians aim to cultivate loyalty, mobilize their base, and disseminate their political messages more effectively. Politicians also tap into existing social, religious, or community networks on the platform to amplify their messages.

Such political deployments of WhatsApp are also strikingly illustrated by the prominent role of Bharatiya Janata Party (BJP), the right-wing nationalist party in India, in engaging cross-media manipulation through organized circulation of "trend alerts" on WhatsApp groups (Jakesch et al. 2021), spurring grave incidents of mob lynching (Vasudeva and Barkdull 2020). Similarly, WhatsApp's deployment in right-wing conservative leader Jair Bolsonaro's campaigns in Brazil has been widely documented (Machado et al. 2019; Parreiras, this volume).

Lived Encryptions

Picking up insightful threads from this scholarship, we center the significance of encryption as a technological infrastructure, social condition, and regulatory target. Our emphasis on encryption stems from the core feature of WhatsApp, which distinguishes it not only from other "open" social media platforms but also the direct messaging function available on such platforms. Our point of departure is the framework of "extreme speech," which refers to speech acts (text, audio, video, multimodal) that stretch the boundaries of legitimate speech along the twin axes of truth/falsity and civility/incivility (Udupa 2018b; Udupa and Pohjonen 2019). Distinct from the universal conception of "hate speech" and the risks of its regulatory misuse, the extreme speech framework stresses on ethnographic sensibility to cultural variation, historical awareness, and ambiguity of vitriol in assessing the nature and implications of contentious content. Of primary concern is the ways in which users draw meanings and create networks of distribution of extreme narratives, and sociocultural and historical factors that coalesce to shape the trajectories and consequences of such narratives. Methodologically, it calls for multiorder analysis linking platform features, user practices, and historical and political contexts surrounding digital messaging.

Guided by the extreme speech framework, we center the analytical value of encryption and consider encryption not as a determining technology feature but one that is suffused with multiple articulations and ridden with contradictions, which we capture as "lived encryptions."

In its unfolding, lived encryptions embed intimacy in articulating with the closed messaging architecture of messaging services, thereby enabling a sense of community that emerges within chat communication. For sure, this sense of community is neither exclusive to nor bounded by chat architecture but the very boundaries that closed chat architecture afford can ease the way to get a sense of community. At a bare minimum, this community is a communicative formation; i.e., one belongs to the community because one reads, shares, likes, and posts messages with others and develops a degree of intimacy because of staying communicative within bounded loops of WhatsApp groups.

Closely tied to the conditions of intimacy is the possibility of trust and a shared feeling of "tight-knit networks" (Belinskaya and Rodriguez-Amat, this volume) and greater user control over contacts, if not always over content. Conditions of intimacy and trust are neither determined by nor reducible to technology design, but they are vitally linked to social narratives and shared perceptions. In Turkey, anthropologist Erkan Saka (this volume) says that "WhatsApp is assumed to be more social, more familial than other . . . services." In many cases, existing social groups duplicate as WhatsApp groups, transferring trust and intimacy along the way. In the favelas of São Paulo, WhatsApp users who spoke to anthropologist Carolina Parreiras insisted that Workers Party (PT) leader and former Brazilian president Dilma Vana Rousseff's character was questionable because they had received messages that portrayed her in a negative light from friends and acquaintances "she trusted" (this volume).

Conditions of intimacy and trust shaped by encryption-enabled closed architecture have encouraged political actors to enlist WhatsApp groups to create and disseminate extreme content in intrusive forms, enabling what is described as "deep extreme speech"—forms of discourse in which exclusionary content comes comingled with good morning greetings and pleasant messages, "simulating the lived rhythm of the social" (Udupa, this volume). Such socially sanctified messages become widespread when political parties link WhatsApp groups through volunteer and party mediators, drawing them into networks that connect the desired narrative across other social media channels as well as mass media, thereby paradoxically infusing virality into architectural features of end-to-end loops and encryption.

Technological and social conditioning notwithstanding, law and order objectives of the state can overrule encryption not only through traceability requirement clauses in internet regulations but also through brazen forms of surveillance, as evidenced in cases in the Global South when repressive governments normalize the practice of seizing mobile phones from dissenters and inspecting or downloading the data. At the same time, encryption has allowed dissenters to assert and safeguard

safe distance from state surveillance—a crucial communication infrastructure that has aided a large number of activist groups, from queer
activists in the Middle East and North Africa to Black Lives Matter
protestors in the United States and journalists critical of the regime in
Rwanda (Moon, this volume). Ambiguities around encryption are pronounced in actual practices, when physical phone searches, for instance,
spark a panoply of strategies to avoid or subvert such searches, puncturing
a sense of security that messaging apps might proffer (Schumann, this
volume). Indeed, as the chapters highlight, in several contexts, encryption is not an actively acknowledged technical feature or a centerpiece of
how the messaging service is experienced and appropriated.

While affording protection to political critics in some contexts and
disappearing as a feature of the media at the experiential level or invert-
ing the logics of closed communication in other contexts, also by utilizing the platform's "narrowcasting" feature to send out messages to up to
1,024 individual accounts (WhatsApp 2023), encryption has nonetheless raised greater barriers for antihate and fact-checking initiatives, as
access, storage, retrieval, and response to problematic content become
more challenging and resource intensive. In authoritarian East African
countries, security affordances such as end-to-end encryption serve as
"window dressing that reinforce the perception of the surveillance state
and make information verification more, and not, less complex" (Moon,
this volume). Encryption thus entails new hurdles for scrutiny and verification, breeding innovative practices of "informal" fact-checking and
gray interventions (Mare and Munoriyarwa, this volume). Even more,
social intimacy of lived encryptions not only eases circulation but impedes correction, since users avoid calling out misinformation to limit
social frictions, as evidenced by young users in South Africa not ventur-
ing into correcting the messages of elders in the group out of respect
(Wasserman and Madrid-Morales, this volume). This stands in contrast
to quasi-public forums where correction faces no such hurdles, although
the effects of fact-checking and corrections are not guaranteed and can
even backfire.

Finally, encryption in the diasporic contexts in the Global North
offers pathways to craft alternative channels of communication distinct
from majority dominated "mainstream" communication, affording

marginalized communities a way to articulate political matters and remain connected with families and publics in their homeland (Trauthig, this volume). Encryption here evinces, if only partially, the ideal of subversive speech within relatively well-guarded spaces.

We consider such multifarious unfoldings as "lived encryptions," holding vital significance, as the contributions in this volume illustrate, for how extreme speech spreads and entrenches in public discourse. As opposed to "encryption" as a descriptor of a distinct and bounded technological feature, "lived encryptions" foreground contradictions and multiple lived practices that surround the messaging application. By pluralizing the term, we emphasize that a contextualized understanding— of how closed architecture is broken open with political messaging and encryption disappears as a feature at the experiential level as well as the social sanctification of political content, the risk-reducing privacy feature, the collaborative potential of WhatsApp groups and so on—is critical to draw out the normative stakes of WhatsApp as a communicational condition that inspires intimacy at the starkly oscillating boundaries of bounded communication and networked action.

While "lived encryptions" highlight the multifarious dimensions and ambiguities surrounding encryption, it is important to note that the consequences of socially sanctified disinformation and rumors, riding on in-group socialites and cross-group message virality of WhatsApp, have been stark in the numerous cases we highlight here. For instance, mob lynching of Muslim minorities and oppressed caste groups has closely followed rumors, fake images, and misinformation circulating on WhatsApp in India while exclusionary narratives within xenophobic WhatsApp groups in South Africa have amplified hostilities toward people migrating from neighboring African countries seeking jobs and livelihoods (Sibiya and Gagliardone, this volume). Such impacts have contributed to what UNESCO has observed as a growing issue of online hate speech impacting the physical world, as evidenced by incidents of violence in Indonesia, Kenya, Bosnia, and other countries, where "cases of harmful yet lawful ('gray area') speech have often led to real-world violence" (Brant and UNESCO 2023, 45). The grave consequences of lived encryptions of WhatsApp in the form of physical attacks, state surveillance, and ideological fanaticism, while also offering spaces for everyday conversations for a vast variety of social activities, stress the need for a global conversation and multiple angles of inquiries around this vastly popular messaging service.

Structure of the Book

We have organized the contributions under five sections tracing "lived encryptions" across different national and transnational contexts and with distinct focal points, which we outline next.

Politics of Divisive Messaging

This section will explore contextual social and cultural conditions that amplify the cocreation, consumption, and spread of disinformation and extreme speech on WhatsApp. The chapters will examine divergent practices surrounding WhatsApp, and how they fold into extreme speech as habitual, deliberate, and lived forms of discourse and meaning. The key focus will be on divisive, xenophobic, and partisan politics that draw on WhatsApp cultures, and how political campaigns deploy this messaging service to trigger animosities, panic, and violence. This analysis will also highlight examples of problematic content and networked dissemination patterns on WhatsApp.

In the first chapter, Parreiras draws on her ethnographic study of WhatsApp use in the favelas of São Paulo; she delves into a thick social world of moral values and shared anxieties and how the right-wing "Bolsonarista" groups amply instrumentalized them with their conservative discourse around "God and family." The far-right discourses on WhatsApp are embedded within peripheralized areas of the city "marked by different forms of material precariousness, such as insufficient sanitation, unemployment or underemployment, precarious housing, food insecurity, and difficulty in accessing health care." In such a context, also marked by heavy presence of the military and the police, WhatsApp use and political propaganda propel "local chains for spreading fake news, misinformation . . . and moral panics."

The next chapter turns to the anti-immigrant group "Operation Dudula" in South Africa. As Sibiya and Gagliardone show, WhatsApp serves as an important channel for the "informational and operational objectives" of this group, mobilizing narratives against immigrants but also connecting with "broader civic activities." Users on such WhatsApp groups are more likely to distribute information incidents of crime involving immigrants from neighboring African countries with an objective to provide supposed fact-based information that could show the immigrants in a poor light. Xenophobic sentiment that shapes and binds such groups defies easy characterization of blind and misinformed ideologues. The authors reveal complex practices among majority Black South Africans who drive such xenophobic sentiments, showing how they embed their content within "fact-based" information as well as ironically express nostalgia for the apartheid and suspicion about pan-Africanism.

Highlighting right-wing nationalist messaging on WhatsApp groups in India, Udupa argues that WhatsApp's unique role in disinformation and vitriolic ecosystems in the Global South contexts of hegemonic politics lies not as much in the architectural features of encryption but around particular clusters of social relations it enters, entrenches, and reshapes. Describing this as "deep extreme speech," she suggests that it is "characterized by community-based distribution networks and a distinct context mix, which both build on the charisma of local celebrities, social trust, and everyday habits of exchange." This type of extreme speech, she argues, "belongs less in the problem space of truth or the moral space of hatred and unfolds rather at the confluence of affect and social obligation, variously inflected by invested campaigns."

In the final chapter in the section, Santos, Ortiz Fuentes, and dos Santos illuminate a highly tense political moment in Latin America: the writing of a new constitution in Chile and the final referendum that ended up rejecting the proposal. Based on quantitative and qualitative content analysis of discourses in a set of over three hundred WhatsApp political chat groups, their study reveals widespread circulation of conspiracy theories about election fraud. Although widespread, the circulation, they reveal, was "asymmetric." Right-wing groups were exposed to a more and larger diversity of misleading messages compared to left-leaning groups, and right-leaning WhatsApp groups were also "closely knitted" in that users were active in more than one group, thereby bridging the narratives across several groups. However, over time, hoax messages split over into left-leaning groups, suggesting how WhatsApp circulation can lead to society-wide disruptions in political debate and citizen participation.

(Un)safe Spaces

In this section, studies explore subversive speech practices and political ambiguities on WhatsApp amidst perceptions of safety, disengagement, and fears of surveillance that at once surround WhatsApp. Across all the chapters, the authors highlight the tension between perceived safety of encryption and its affordances to forge networks beyond majority-dominated communication on the one hand and the risks, on the other hand, of exposure, mistrust, and ambiguities that actors negotiate in contexts of political volatilities. Exploring different communities of users—from political dissenters in a conflict zone and young users in the Global South to diaspora members in Western democracies—the studies show the deep ambivalence of WhatsApp as sites of community networking and (dis)informational sources.

Turning attention to invasive social media policing of the Cameroonian state, Schumann informs that "government suspicion against Anglophones existed long before" state practices of policing their WhatsApp and social media channels became widespread. Set in this context of a long-drawn conflict, "arbitrary phone searches" are common, and so are different ways in which Anglophones attempt to subvert everyday surveillance by leaving behind their smartphones at home or deinstalling social media apps—tactics that more often raise suspicion among government authorities since they begin to question why an affluent-enough person would not carry a smartphone. As Anglophone WhatsApp users vacillate between a sense of security around encrypted messaging and vulnerability to state surveillance, they also find themselves in awkward and potentially dangerous situations when violent images of killing surface unexpectedly on their phones, even as they hold on to the messaging platform as "private spaces of exchange" to "discuss negative experiences with state forces."

Wasserman and Madrid-Morales show that contemporary forms of disinformation campaigns in South Africa are rooted in "older histories of colonialism and postcolonial authoritarianism" and longer tensions surrounding ethnic and social polarization. In this context, young

WhatsApp users find themselves torn between actively attempting to counter misinformation they receive on their phones through various corrective actions and purposeful detachment from messages of this nature. What is safe and unsafe is shaped by thick social norms that surround young users' actions, especially in relation to how challenging false information in close-knit WhatsApp groups could cause problems in family relations.

Belinskaya and Rodriguez-Amat explore the understudied aspects of encrypted messaging about fleeing Russian migrants in Europe, revealing how bans on prominent "public" social media platforms in Russia have driven many users toward WhatsApp and Telegram. Offering thematic analysis of sampled WhatsApp groups used by Russian immigrant communities in Austria and drawing a comparison with Telegram, they show that misinformation is common within these groups. They argue that communicative processes of rationalization and legitimization significantly shape the exchange of such misinformation.

In her contribution, Trauthig shifts the focus to diaspora communities in the United States, arguing that WhatsApp use among Cuban American, Indian American, and Mexican American communities has allowed for alternative avenues to discuss contested issues and create counternarratives "outside of . . . majority-dominated public discourse." In this analysis, she compares WhatsApp with the features of community-owned media that offered an "alternative environment for inclusion and representation," crediting the "inherent subversiveness of encrypted communication" for this potential. Disputing extant evaluations of WhatsApp as a hotbed of disinformation, this study considers its ability to create channels for diaspora members to engage in "political talk" in ways that merit protecting such messaging channels as "safe news spaces" that could foster the ideal of inclusive democracy.

Infrastructure

Approaching WhatsApp as a sociotechnical architecture, the chapters in this section explore its shaping in relation to journalistic reporting and fact-checking and as a site for regulatory intervention and corporate moderation. This section also highlights how technical features of the messaging app are often overshadowed by vast human networks deployed for political messaging in the Global South contexts, prompting the consideration of human networks as infrastructure in and of themselves.

The thrust of the section is on considering WhatsApp as physical networks that can enable the movement of information, ideas, and emotions in routinized, persistent, and standardized ways, akin to other physical infrastructures, and for this very reason, they are subject to state scrutiny. Building on key anthropological and communications scholarship on infrastructure, Moon examines the ways journalists in the authoritarian East African country of Rwanda use and are shaped by WhatsApp as an element of infrastructure: "a 'boring thing' that distributes justice and power in the background of visible work." WhatsApp use among journalists in Rwanda, she argues, is strongly impacted by far-reaching surveillance of the authoritarian state that operates by covert strategies of control over infrastructure, including messaging platforms.

Olaniran (this volume) expands the conceptual scope of "infrastructure" into what Larkin calls "people things" by showing how politicians in Nigeria "assemble a human infrastructure to create partisan environments and inflammatory messages to bolster their candidacy" (see also Nemer 2021). This very infrastructure of WhatsApp, however, has also been utilized by diverse groups of "bottom-up" fact-checkers in South Africa. Highlighting the "unofficial, uncoordinated, and unorganized process of verifying the factual accuracy of questionable content," Mare and Munoriyarwa show that informal fact checkers with no specific training in journalism or fact-checking rely on their "intuition, epistemic capital, social networks, media literacy skills, and investigative skills" to verify content circulated in closed WhatsApp groups. They use the conceptual device of "social correction" to explicate how WhatsApp infrastructure of closed chats partly necessitates "informal and provisional" fact-checking since it can venture into "gray spaces" with flexible correction tactics. Such tactics, they contend, can navigate the "unpredictability and uncertainty associated with circulation of mis/disinformation on WhatsApp groups" in contrast to organized fact-checking groups that encounter the stonewalling effects of encryption.

The difficulties of fact-checking WhatsApp are a part of the broader problem concerning the regulation and moderation of such instant messaging services. Content moderation on WhatsApp is hard to implement at the message level, and platform moderation therefore becomes limited to placing restrictions on group membership and size, content labeling rather than removal, and mechanisms to decelerate message spread by limiting the number of forwards. Taking a closer look at such regulatory and moderation measures, Sinha considers WhatsApp infrastructure as a site of state and corporate intervention, discussing how regulation of WhatsApp has developed in India.

Method

The chapters in the preceding sections reveal the diversity of methods employed in current scholarship on WhatsApp. Each chapter uses one or more research methods, and across the volume, methodological approaches are diverse and multidisciplinary, ranging from quantitative computational methods to ethnography to surveys. Taking up the methodological question as a central concern, this section turns the focus on a major hurdle in WhatsApp research, which relates to the methods to access and store "private chat" data. Most studies have highlighted the methodological difficulties of studying encrypted messaging services because of lack of public API access, closed source code, sequential chats that do not allow key word searches, and the "private" nature of groups that requires group moderators to approve researchers' request to join (Barbosa and Milan 2019). Although Telegram and WhatsApp can be accessed via web browsers, and limited metadata can therefore be obtained with scraping, such methods raise ethical challenges and issues of data privacy, in addition to challenges of seeking approvals by parent companies and navigating company stonewalling.

Lack of data access has resulted in empirical blind spots in terms of assessing the volume of users exposed to extreme speech and the proportion of problematic content vis-à-vis the total corpus of exchanges. In view of severe limitations to data access, methods have emerged especially within computer science to develop web interfaces to encourage users to donate data and to organize such content in a "privacy-preserving manner on a large scale," but the adoption of such promising models remains nascent because of "serious privacy, legal, ethical, and practical challenges" including meeting the standards of data minimization and anonymization principles (Melo et al. 2019). At the same time, epistemological implications of calibrating data donations with research goals are yet to be investigated.

Foregrounding vast ethical, practical, and legal challenges around accessing and analyzing conversations on WhatsApp and similar messaging applications, this section provides methodological pathways toward addressing them.

Inquiring into ethical dilemmas involved in "lurking" within private WhatsApp groups, Saka highlights the ethical risks of navigating access to such groups and specific challenges of carrying out ethnographic work. When researchers announce their presence on closed groups and seek to gain informed consent of participants, as he points out, they are likely to inadvertently change the "nature of conversations." Reflecting on interdisciplinary methods he adopted in his study on WhatsApp groups and anti-EU disinformation in Turkey, he suggests that methodological strategies for content and group analysis of WhatsApp should pay close attention to conditions set by national media environments and the platform's shifting policies.

Highlighting how WhatsApp data collection on a "large scale presents serious ethical and practical challenges," Chauchard and Garimalla take up the challenge of evolving methods that can gather, store, and analyze data in a privacy-preserving manner. On the one hand, data extraction is "technically easy" once a consenting participant extracts it for the research term. On the other hand, data gained through such means and subsequent analysis raise the risk of falling into the gray zones of the platform's community standards and prevalent legal protocols. The only way to address the hurdles, they contend, is by adopting a data donation approach, elaborating on different technical and practical steps involved in operationalizing such a methodological protocol.

Honing further into computational methods of gaining data access, Micallef, Ahamad, Memon, and Patil outline the challenges of "automating large-scale data collection from public WhatsApp groups." Identifying such challenges in the three activities of "discovering, joining, and maintaining membership" in public WhatsApp groups, they point out how they faced hurdles at every step of the process, for instance, when their code allowed their "phone number" to join but their inactive presence in the group raised suspicion and ended up being removed by the moderator. They emphasize that automating data collection for public

WhatsApp groups remains highly vulnerable to the evolving policies of the platform, which can shift quite suddenly and without notice for the researcher community. Cognizant of such challenges, they provide a detailed description of data gathering including setting up of the devices, extracting the WhatsApp SQLite database from the devices, manual identification of WhatsApp groups for research, semi-automated ways of joining public WhatsApp groups, and maintaining group membership.

In the final contribution, Bosch provides an overview of common qualitative methodological approaches in available scholarship on WhatsApp and political discourses, highlighting the significance of mapping this from the vantage point of the Global South. Noting that "the majority of research on WhatsApp and political activism originates from the Global South," her review finds virtual ethnography, interviews, and surveys of users to be the most common qualitative approaches. Approaching the internet as "place and text," the chapter reflects on the possibilities, challenges, and limitations of different methods that emphasize each or both aspects in relation to WhatsApp discourse. The chapter's central thrust is a decolonial approach to WhatsApp research, calling for historical awareness, self-reflexivity, and an ethics of care.

As the methodological contributions in the volume indicate, one of the key methodological tensions around WhatsApp research is what gets deleted from the dataset to preserve privacy, as the authors from different disciplinary backgrounds illustrate in this section and different chapters in the book point out with varying degrees of emphasis. While removing one-to-one threads for the sake of protecting privacy appears reasonable for large-scale analysis of WhatsApp content using computational methods, it raises the question of the profound ethnographic value of such conversations, as Schumann's study of Anglophone users in Cameroon and Udupa's study of deep extreme speech illustrate. Similarly, removing seemingly banal, insipid, and repetitive one-liners appears to be reasonable for computational methods for the "noise" they bring to the data, but ethnographers would consider them as valuable for examining the interactional dynamics and rhetorical devices that draw extreme speech into the everyday and the ordinary. While such methodological differences ultimately rest on the epistemological grounding of different projects, this volume highlights ways of bringing them in close conversation, to highlight them not merely as problems of data selection and data cleaning but a methodological way forward for interdisciplinary collaboration. WhatsApp's intimate contexts of conversation with multiple contradictions underscore the value of ethnography in no uncertain terms, but precisely because of the political consequences of cross-platform velocity and virality they have induced in various contexts, computational and survey methods are vital to track their multiple trajectories. Taken together, the chapters in this section therefore underscore the importance of interdisciplinary research that combine different methodological approaches and epistemological orientations to highlight the different dimensions of WhatsApp in the communication landscape.

Policy

The final section aims to address the looming question around what to do with the complexities of encrypted messaging and extreme speech. The key challenge is to design regulatory and policy frameworks that can account for the diverse use of WhatsApp globally, and the dilemma of encryption in relation to protections it can provide to minoritized communities facing threat under surveillance and authoritarian regimes on the one hand, and the growing evidence, on the other hand, that affective and instrumental engagements around encryption have seeded possibilities for hateful exchange and disinformation. Regulatory models that have emerged include outright bans on encryption, limitations on the permitted strength of encryption, weakening of encrypted technologies, requirements for traceability of users, and requirements for back doors to be built into services and products to enable government access to information, and mandates for proactive monitoring of encrypted content. These proposals have emerged across jurisdictions including Europe, UK, Turkey, India, and the United States. The Indian government, for instance, introduced the traceability clause in the new Internet Intermediary Rules (2021), mandating the platforms to divulge information about the source and identity of viral messages in response to law enforcement requests—a measure that WhatsApp challenged in court (Sinha, this volume). Hence, while legally mandating access to encrypted messages appears to be an easy solution, it comes with serious challenges including security and surveillance risks that are introduced by the existence of back doors, ways to balance the potential infringement on privacy, lack of clarity on processes to carry out regulatory mandates, and jurisdictional challenges that arise once a back door is introduced. In a long-drawn contestation with service providers, the UK's Online Safety Bill, for instance, shelved the proposal to inspect encrypted messages, admitting that the "technology to securely scan encrypted messages . . . does not exist" (Guest 2023). In addition, regulatory measures such as limiting message forwarding and even fact-checks have yielded mixed outcomes (Melo et al. 2020).

In this section, two fact checkers and one policy expert reflect on the challenges that WhatsApp has posed to civil society and regulatory efforts in curbing exclusionary extreme speech and disinformation. Through their daily navigations of WhatsApp discourse for fact-checking, Cayley Clifford from Africa Check in South Africa and Jency Jacob from BOOM Fact Check in India discuss how their meticulous and timely fact-checks face the danger of being drowned in the virality of polarizing and sensational content that spreads through community channels at a rapid pace. In India, for instance, WhatsApp messages morphed public awareness videos from Pakistan and bodies of little children shot in Syria to raise panic about alleged child kidnapping gangs and organ harvesting rackets. Published fact-checks could not mitigate the rapid spread of these messages, although fact checkers simultaneously alerted the platform to take action. Jacob points out that efforts to develop open channels for community collaboration are exhausting and resource intensive, as often they are flooded with spam messages, including cryptocurrency messages, while citizens who alert the organization are on edge to see immediate action. Such practical challenges are situated within a broader political climate of antiminority ideological politics in India, placing enormous pressure on independent fact checkers as they jostle between platform complicity and repressive politics.

Clifford similarly outlines the challenges as well as opportunities presented by WhatsApp for fact checkers, stating that it is not only difficult to access information within encrypted channels but also to counter disinformation since it is very likely that messages that flow through such channels are trusted. Highlighting their initiative *What's Crap on*

WhatsApp?, a podcast circulated as WhatsApp voice notes to debunk misinformation, she details how they established a tip line for users to directly alert Africa Check about suspicious content and prepare periodic podcasts to raise awareness. The organization vouches for constant interaction with subscribers for effective fact-checks that can also address backfire effects when fact-checks reinforce beliefs. Both Africa Check and BOOM run helplines for users to directly alert them on viral messages that appear dubious as well as collaborating with alert citizens who would contribute as "fact ambassadors" or "truth warriors."

Across political contexts, some of the key challenges that fact checkers face in relation to WhatsApp discourse have emerged from lax platform action and challenges posed by the platform's architectural design. In Brazil and India, for instance, Reis et al. (2020) discovered that even after popular fact-checking agencies had verified the information, misinformation continued to appear within public WhatsApp groups, as the platform lacked the capability to label previously fact-checked content. Consequently, fact-checking organizations, as Jacob informs, have demanded "in-platform mechanisms to fact-check high-volume forwarded messages without compromising encryption."

Outlining some of the major policy challenges concerning "coordinated harm" on WhatsApp, Scott Timcke stresses that platform governance alone cannot solve the problem since governments tend to divert attention away from social conditions and "their history of governing those conditions," thereby "co-opting platforms into projects which circulate narratives of hate." Furthermore, platform governance as a distinct policy measure is constrained in the case of WhatsApp because of the lack of public metrics to assess the flow and impacts of hateful narratives as well as the very structural conditions of digital capitalism that policies, including those of UNESCO, fail to address in terms of devising ways to decommodify global social media platforms. Pertinent to the discussion is what he defines as "a climate of stochastic violence," which refers to hazy boundaries between intentional actors, passive recipients, and unmindful onlookers that WhatsApp groups afford, raising the difficulties of striking a "balance around intent, primary audience, and harms." Considering these challenges, he suggests that policies turn to understanding broader sociological and sociotechnical processes to pin down the processes and consequences of harm.

The emphasis on ground realities that fact checkers and policy experts have articulated in this section returns to our opening argument about the need for multiorder analysis with field-based ethnographic approaches developing in close conversation with quantitative methods, and for contextualized understanding that considers users not as "targets" of analysis and policy but as historically situated actors who draw and imbue meanings within contradictory climates of encryption—of which some distinct forays have emerged in this volume. The global influence of WhatsApp as a communication platform, its increasing sphere of influence in the Global South, and its contradictory characteristics outlined in this volume call for further interdisciplinary and multimethods research, for which this volume has laid the foundations.

PART I

Politics of Polarization

1

Extreme Speech, Community Resonance, and Moralities

Ethnographic Notes on the Use of WhatsApp in Brazilian Favelas

CAROLINA PARREIRAS

The last few years have brought a plethora of publications, debates, and discussions about the role of technology in people's lives, especially due to the COVID-19 pandemic. This body of scholarship greatly emphasizes the everyday and mundane character of these technologies both in terms of theirs uses and when considering the technical infrastructures, devices, network architectures, and the many connections between humans and not humans. Terms such as "platformization" or "datafication of life" have become commonplace, pointing to the ubiquitous, pervasive, "embodied and embedded" (Hine 2015) character of technologies and the internet.

Faced with this situation, an increasingly intense use of one of the various digital platforms stands out: WhatsApp.[1] Though it is important to recognize that WhatsApp allows for a myriad of positive uses, it is also responsible for controversies and for the spread of extreme speech (Udupa 2023), which is a "conceptual framework" that covers hate and derogatory/exclusionary speech and misinformation.

With this in mind, my interest here is to think about the uses of WhatsApp in Brazil to spread hate discourses and fake news, specifically

---

1. I consider WhatsApp a platform based on the meaning given to this term by van Dijck, Poell, and de Waal (2018), who understand "platform society" as not only an economic or technological phenomenon, but a process in which platforms are "in the heart of societies" (2), intrinsically related to social structures. As the authors state, "Platforms do not reflect the social: they produce the social structures we live in" (2).

in peripheral, impoverished, and marginal areas of the country: the fave-las, places where my fieldwork research is situated. Although the contro-versies around WhatsApp are a problem for Brazil as a whole—with a series of political implications, such as Bolsonaro's election in 2018, I am proposing a localized view—or in terms of Wasserman and Madrid-Morales (2022, 210), a perspective based on "contextual knowledge and experience." I am also considering an intersectional approach, based mostly on social class, income, and place of origin/habitation.

The large popularity of WhatsApp in Brazil can be explained by two in-terwoven factors: the practice of zero rating by the mobile carriers and the multiple forms of communication allowed by the platform (text, audio, video, gifs, emojis, stickers—synchronous and asynchronous), which fa-cilitate use even in places where the connection is weak or when users lack digital literacy for other activities. As the ethnographic data presented will show, the functioning dynamics and the technical structure of the applica-tion are fundamental to understand the centrality that WhatsApp gained in Brazil, as well as the many controversies that surround its appropriation for the propagation of fake news, its uses in national political processes (elections and management of the pandemic, mainly), its transformation into an environment of disinformation, and even the legal battles around ways of regulating the platform in the country.

The daily and endless uses of digital media and WhatsApp create the false sensation that everybody is connected. What reality shows us, how-ever, is the opposite: not everyone is connected and, even among those who have some kind of connection, this does not happen under the same conditions, with the same possibilities and quality. This introductory note is important because the index of digital inequalities in Brazil—if we think only on the official data from PNAD[2] and ICT Households[3]—is still high. Just to give an example, during the period covered in this chapter (2016–2022), an average of 27 percent of households in the coun-try did not have any type of connection to the internet.[4] It is important

2. National Household Sample Survey.
3. Annual survey conducted by the CETIC.br, the branch of research of the CGI, the Brazilian Internet Steering Committee.
4. According to the ICT Households 2022, this accounts for about fifteen million households and thirty-six million people without access to the internet. https://cetic.br/en/pesquisa/domicilios/.

to note that there are important differences between regions and based on social classes within the country. Another important piece of data is that among the population connected to the internet, the mobile phone is the tool utilized for most of the connections. The pandemic also made the many digital inequalities more evident, as well as the countless barriers to overcome this inequality that do not simply concern having or not having technology but also other considerations such as domain, uses, and production of outcomes arising from this use. Finally, the ICT Households 2022 brought additional interesting data for the goals of this chapter, concerning digital skills: only 51 percent of users declared having used the internet to check the veracity of information. This points to something I encountered during fieldwork: the difficulty of checking news and especially accessing fact-checking agencies.

With this contextual background, my proposal is to think about the localized uses of WhatsApp situated in the daily life of favela residents that also help us to refine the ways we have thought about fake news, disinformation, and hate speech through the use of digital platforms. These behaviors, which gain scale due to its digital architecture and the many networks it allows, bring practical consequences to people's lives. It is exactly in this inflection between the use of technologies and the internet and the daily life of the subjects that I am interested in. Due to its widespread use among the Brazilian population, WhatsApp is a fundamental mechanism in this broader context, as today it is the most used platform in the country, configuring itself, as Cruz and Harindranath (2020) propose, as a "technology of life."

By looking at the "microhistories" (Das, 2020) narrated in my fieldwork research (carried out at the Complexo, a group of favelas in Rio de Janeiro), I intend to show some of the local dynamics of fake news circulation, with the generation of environments of disinformation and types of hate speech that are good indications of the moralities that circulate through this platform. Ultimately, my goal is to understand how WhatsApp is used daily in these favelas, based on its use by my research interlocutors, leading to issues such as digital inequalities, to community logics for the establishment and judgment of what is understood as truth, and resulting in a local chain for spreading fake news, misinformation, hate speech, and moral panics.

Method

As stated in the Introduction to this chapter, technologies and the internet appear to be increasingly mundane and even "banal" (Treré 2020). The term "mundane technology" was first used by Dourish et al. (2010) and appropriated by several other scholars to mark, as Nemer (2022) suggests, technologies that have become "commonplace." That is, they are widely incorporated into people's lives so that often their presence or use is not even noticed and is not even noteworthy. Hence my use of the adjective "banal." According to the *Oxford English Dictionary*, "banal" has "ordinary" as one of its synonyms. "Ordinary" is an important concept in the writings of Das (2007; 2020) and key for the association she makes between the everyday and the ordinary (which she calls "kindred terms" [2020, 6]). Although she is interested in themes such as violence and social suffering, I believe that her broader theoretical reflection also serves for other fields of study. Thus, something valuable brought by Das (2020, 2) and which I have sought to apply as a methodological stance in carrying out my ethnographic incursions is that understanding the ordinary requires "attention to detail." In this relationship between the everyday and the ordinary, Das (2020) suggests that the everyday is the place where the other becomes concrete and where her/his life (and ours) is engaged. Thus, it is an anthropological task to follow and trace concrete relationships and be attentive to everyday events. However, even if this also seems banal, Das alerts us to how much the ordinary nature of everyday life makes it difficult to "see what is before our eyes" (2020, 15).

Wouldn't that be the same challenge to understand the many technologies or even the various digital worlds that become mundane and seem almost invisible? WhatsApp appears here as a field of research, as a platform that allows for a series of relationships and is directly related, as a "technology of life," to basic operations in life: work, education, sociability, information, banking operations, creation of groups, organization of all kinds of activities, and business, among others.

When carrying out fieldwork research in the Complexo, it always caught my attention that even in conditions of material precariousness and low quality of internet access, there was a heavy use of smartphones and WhatsApp. So, even before I realized it, WhatsApp had become both an object of research and an essential tool for my contacts with interlocutors. WhatsApp was never a smaller or less important part of the research.

The idea of ethnographic sensibility has been one of my main concerns from the beginning since I understand it not only as a methodological premise but also as an ethical posture. In this matter, I call attention, for example, to the fact that WhatsApp creates the feeling of being always connected, making it difficult to separate research moments from those that do not fall within this scope. This is a situation I have experienced countless times in the field and that seems heightened by the fact that WhatsApp allows me to create an intimate relationship with my interlocutors, allowing different types of narratives and confidences. Relationships of intimacy and trust are fundamental for carrying out ethnographies, allowing us to go beyond pre-established scripts and facilitating the openness of our interlocutors to the discussion of sensitive topics. As I will show in the next item, the reflections carried out in this chapter were only possible due to the existence of trust and intimacy, since interactions via WhatsApp touched upon several common situations in the Complexo, as well as involved the sharing of print screens of various fake news and extreme speech contents received and sent by residents. I also need to say that I had the invaluable help of a privileged collaborator (Duda), with whom I was later able to reflect on the issue of fake news when we participated in a podcast together.[5] Without her help, it would have certainly been much more difficult for me to access some of the news and messages that circulated and were shared by residents of the Complexo. All my interactions with Duda also took place through WhatsApp.

Using the extreme speech approach (Udupa 2023), which is a methodology and an ethical principle, we can analyze how different actors are involved in the production and consumption of hateful content, showing

5. I have always adopted as a central criterion for ethical research to maintain the anonymity for my interlocutors. However, when invited by colleagues to participate in an anthropological podcast with the theme "Fake," they asked that a research collaborator also be present. I invited Duda and explained to her the implications of her participation, especially that she would henceforth lose the fictitious name I had used until then. She accepted the invitation, and many of the questions developed here came from this interaction through the podcast. Thus, to a certain extent, she also has authorship of the ideas presented in this chapter.

the importance of a contextual and ethnographic reflection. This framework helps us to go beyond fast and simple explanations, allowing for an understanding that considers "historical, cultural, and political variations" (Udupa 2023, 238). This perspective also allows for the researcher to consider the everyday practices of the subjects and the meanings they gave to them, based on the idea of ethnographic sensibility. In this sense, the extreme speech framework has an invaluable importance for my goals in this chapter, since it is essential to understanding how information and news disseminated by WhatsApp among my interlocutors reproduce moralities and create moral panics amongst them, or even help us to understand what they evaluate as true or fake. It is important to note that it also accounts for longstanding structures and power relations already present in these favelas. The technology and the use of digital devices and platforms serve as a mechanism to update these moralities, giving them scale and an unprecedent reach.

Of course, the interactions didn't come without problems or hiccups. I bring this issue up because it confronts us, firstly, with the versatility of using WhatsApp as a research tool, but also points to its limitations, especially when we consider social markers of difference such as class and schooling and issues of literacy and digital literacy. People are connected and find multiple ways on digital platforms to deal with facts of everyday life, but this is done in an extremely uneven way.

WhatsApp and Its Community Resonance: Extreme Speech and Moralities

Since 2016, I have been carrying out ethnographic incursions into favelas in the city of Rio de Janeiro, the second-largest city in Brazil. For ethical reasons and because they are sensitive territories, I chose to use the fictitious name Complexo to name these places. In Rio de Janeiro, the word "Complexo" refers to any group of favelas, being quite generic, which makes it difficult to identify my interlocutors and their place of residence. In general terms, favelas are peripheral urban formations—not necessarily geographical peripheries—marked by different material precariousness, such as insufficient sanitation, unemployment or underemployment, precarious housing, food insecurity, and difficulty in accessing health care, among others. Furthermore, as the researcher Farias (2020) suggests, I understand favelas as "militarized territories," marked by the ostensible presence of police and military forces, by the media, and by commonsense images of "urban warfare," which put into opposition good actors (police, state, army) and bad actors (favela residents, seen recurrently and entirely as bandits due to the presence of criminal groups linked to drug traffick- ing and cargo theft in this territory). However, as Das and Poole (2004) taught us, the relationship between the margins and the state is more com- plex than a pasteurized image of the margin would imply. Nevertheless, it is important to say that, in the Complexo, the police enter the neighbor- hood frequently, which often results in deaths, including of residents with no connection with the factions (the so-called stray bullets).

In these more than six years of research at the Complexo, I have tried to understand the daily experiences of women, especially with regard to different forms of violence, both private and intimate (marked by gen- der and sexuality), as well as public. More recently, I have also consid- ered their use of the internet and digital technologies. My main interest today is divided into two axes: digital inequalities and the centrality of WhatsApp in my interlocutors' lives.

In 2018, my interest in the issue of using WhatsApp to circulate ex- treme speech content arose, although this has never been my main point of research. This coincided with the election campaign of 2018, which opposed two different political projects: Fernando Haddad, from Work- ers Party (PT) and representing a progressive agenda, and Jair Bolson- aro, the conservative right-wing candidate. In a broader sense, the use of WhatsApp during the campaign was discussed and contested, especially due to the use of automated messages to disseminate extreme speech content in groups and among users.

It is important to mention that, in recent years, there has been an important stream of Brazilian academic scholarship showing WhatsApp as an environment for disinformation, spreading fake news, and also on its importance in political processes in the country, such as the election of Jair Bolsonaro (Cesarino 2019; 2022; Nemer 2020; 2022).

Likewise, attention is being paid to derogatory speech directed at mi- norities (poor people, LGBTQIA+, people of color, women, academics, and leftist parties as a whole) and especially against presidents Lula da

Silva and Dilma Rousseff, from the Worker's Party. In the case of Dilma,[6] since the public campaigns that culminated in the 2016 coup d'état, when she was deposed, numerous messages circulated on WhatsApp and Facebook that represented Dilma as ignorant, took her sentences out of context and, above all, utilized a highly misogynistic tone. The apex of such misogynist images was the circulation of a sticker, to be placed on the cap for the car's fuel tank, with an image of Dilma's face on a sex doll with opened legs. The region of the vagina coincided exactly with the hole where the supply hose is attached. This image went viral in Brazil and also in the Complexo.

One of my interlocutors confessed to having believed everything she saw about Dilma on WhatsApp groups, about her government's alleged theft of public funds, and shared this same image and many others, in addition to having defended the process of her impeachment. When I asked her what made her believe in this content, given that she had just mentioned that she didn't think her life was bad during the Dilma government, she replied that she shared it and believed it because she had received the content from people "whom she trusted," "acquaintances and friends." In this case, as in so many others, there was no doubt as to Dilma's questionable character or her guilt. This interlocutor ended by saying that she regretted having believed and, even more, having "passed on" the messages she received.

It was in 2018, on the eve of the second round of the presidential and state elections, that I came across the first mention of fake news and the automated sending of messages by WhatsApp in the Complexo. I had gone to the Complexo to carry out a "turn-the-vote" mobilization, with the intention of talking to the residents, explaining about candidate Fernando Haddad's government proposals, from the Workers' Party (PT), and warning of the dangers of electing a far-right candidate like Jair Bolsonaro. The decision to do this came from the realization that residents

---

6. Dilma Rousseff was the president that succeeded Lula da Silva. In her second term as president, she was impeached based on accusations of fiscal irresponsibility. This was the official explanation, but the real reason for her deposition was the confluence of opposing parties that had wanted to end the Worker's Party dominance in Brazilian politics since they entered office in 2002. It is important to note that the coup d'état was highly misogynistic with the use of Dilma's image to reiterate women as emotional beings, not able to govern a country. Her vice president, Michel Temer, became president, representing all the structural power relations of Brazil: male, White, and rich.

of the Complexo who in past elections had voted for progressive governments, especially due to the various social policies of which they were beneficiaries, particularly during the 2018 campaign—like a large part of the country, it must be said—presented a surprising turn to the right, demonstrating support for a candidate with no concern for social issues.

While talking to Isa, one of my oldest collaborators, she mentioned having received numerous messages on WhatsApp, sent from an unknown number with no apparent photo. These messages solicited votes for candidates Jair Bolsonaro (president) and Wilson Witzel (governor of Rio de Janeiro), in addition to making a series of negative statements about the Workers' Party. To my amazement, she unlocked the mobile phone, located the messages, and showed me all of them. As I read the content of the messages, I understood what she meant, insofar as the texts sent did indeed contain the campaign ads, but also a campaign of lies against Lula and his party. Words like "delinquent," "*mamata*"[7] and "Brazil becoming Venezuela" filled the messages, as well as the news of the supposed end of Bolsa Família Program. The latter is one of the most successful programs of the PT governments and based on transferring income to families in social vulnerability. This was of particular concern to my interlocutors as, in the Complexo, almost the majority of women are beneficiaries of this program.

Shortly after, when talking to Beca, another interlocutor, she pointed out a similar situation, but unlike Isa, she had received the messages in her church group, from an evangelical denomination. In her narrative, the same points appeared, especially the demonization of everything related to the left. The pastor, apparently very active in the group, requested votes for Bolsonaro and Witzel, considered by him "men of God" who valued the "family." As we will see in another example below, this appeal to an idea of an ideal heterosexual family is a constant in conservative discourse, and utilizes the tripod God, homeland, and family.

It caught my attention that I hadn't received messages of that kind, especially ones that were clearly the result of automated mass discharges. I also voted in Rio de Janeiro, I had a mobile phone with a Rio number, yet I was not the target of this type of content. Even if automated, the

---

7. Brazilian slang word to refer to "an easy way to get something, or at someone else's expense," really characteristics of politicians.

target of these mass messages was not random, but directed to certain populations, using artifices that made sense to the target audience (such as, for example, the supposed end of Bolsa Família Program). Somehow, there was a clear demarcation of class and income in the sending of these messages, which attests to the planned nature of the creation of an environment of disinformation, which, at the very least, generates doubt and confusion about the content of the messages.

All these questions did not appear again in my fieldwork research until 2020, with the onset of the COVID-19 pandemic. In 2018 Jair Bolsonaro was elected, and the country was immersed in the president's orders and excesses. The return of hunger and misery was on the horizon, as well high unemployment rates and recurrent cases of corruption (and without investigation). With the pandemic, all of this got worse, as there was no government management of the disease, much less of the prophylaxis to avoid further contamination.

In 2020, the first cases of fake news came from Mari, another interlocutor, and concerned the so-called emergency aid. In the first months of the pandemic, still without a vaccine, under pressure from leftist parties, emergency aid was approved to slightly lessen the effects of the pandemic. There was an application where the aid could be requested, but with the lack of information, a series of false links appeared, which led to malicious sites, probably for data theft. Mari had been deceived by one of these links, which she had received in a WhatsApp group, and asked me for help because during her registration the mobile phone "had crashed." I asked her to send me the link. She forwarded the message, and I immediately noticed that that URL did not make sense and did not have the credentials of the state bank responsible for the registration and payment of the aid. I explained to her that it was a fake link, that there were risks in clicking and filling out the supposed registration and that I would help her with the correct application. On that same day, I registered more than thirty residents of the Complexo who went through similar situations or just couldn't handle the app.

Days later, Mari and two other collaborators asked me for help again, forwarding a new link, this time from a supposed campaign to donate water made by Nestlé. Again, it was another fake link. This was followed by alleged donations of basic food baskets, water, and cleaning materials

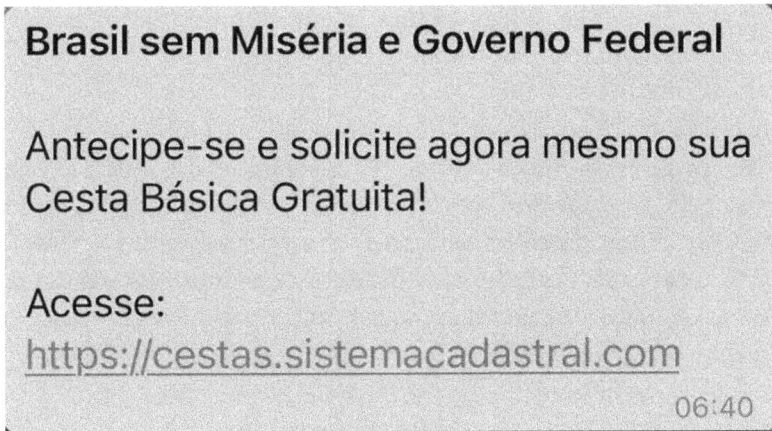

Figure 1.1: Fake ad from Brasil sem Miséria. Source: WhatsApp exchanges with a collaborator.

by Heineken, Carrefour, and even the federal government. One of these messages is reproduced above to illustrate how they are constructed.[8]

This message reads that the program Brasil sem Miséria[9] and the federal government would be donating "basic food baskets," food kits considered essential for the population's livelihood. By accessing the indicated website, you would have early access to this donation. When clicking on the fake link (which uses the word "basket" and, therefore, directly appeals to something commonplace and necessary to the recipients), a page covered with advertisements appears. Among these ads, some fields appeared to be filled in with the following data: full name, RG,[10] CPF,[11] telephone number, and address for the supposed shipment. This was probably also a page for data theft, as a document such as the CPF is the basis for virtually all operations in the life of a Brazilian citizen. If stolen, it can be used for a number of operations, some of them illegal.

8. All the images used in this chapter were received from my collaborators by WhatsApp, and they gave me permission to reproduce them here.
9. Program created during Dilma Rousseff's government (2011) to address the extreme poverty in the country.
10. Brazilian State Identity Number.
11. General registry number, similar to a Social Security number.

Upon noticing that it was another fake website and a fake assistance campaign, my interlocutor replied, "People play a lot with the needs of others," followed by angry face emojis. It is worth remembering that these links circulated in one of the worst moments of the pandemic and that the people who received them were experiencing different needs. They needed donations to feed themselves, to obtain cleaning and hygiene kits, and even protective masks. From what my interlocutors reported, these links were shared with all the residents who had contacts on WhatsApp, with a general urgency to carry out the registrations and obtain the resulting benefits.

In these specific cases, there was no doubt the disseminations of these messages were authentic, as not going hungry or guaranteeing the basic items for survival was, of course, more important. This reflection is useful to think about the very logic of dissemination of this type of content through WhatsApp. The sharing facilities of the app means that content can be disseminated very quickly and on a large scale. What starts in a small family group, for example, can reach the community as a whole. In addition, the messages are composed in such a way as not to raise doubts: the name of a large company or the federal government is mentioned, and they are all short and to the point, without a textual burden that people would not read.

Even though there are a number of fact-checking agencies in Brazil today and companies and the government have denied many of these messages, this does not prevent the spread of disinformation and fraud, not least because, in a context such as that of the Complexo, most of my interlocutors are not even aware of the existence of forms of checking the veracity of news, advertisements, and campaigns. This makes us think, without obviously failing to recognize the fundamental role that the fact-checking agencies play, that this is still a restricted type of access to knowledge, mostly targeted at people with higher education and from higher social classes.

In 2022, many of the issues that raised my interest in understanding how fake news and misinformation circulated on WhatsApp returned during yet another presidential election campaign. The two symbolic sides of the national polarization in recent years competed: Jair Bolsonaro, for reelection and in defense of the same conservative agenda, and Lula, the main leader of the left in Brazil, who served as president for two terms and represented the possibility of the return of progressive policies.

Hatred of Lula is old in Brazil. If in the case of Dilma, described earlier, there was certainly a gender focus, in Lula's case the main issues refer to his being from the northeast region of Brazil, his lack of higher education, and that he is a former metallurgist, coming from a less favored social class. Shortly after Dilma's coup, Lula was arrested based on arbitrary evidence of alleged corruption schemes. In 2019, he was released and later acquitted of the charges. All the attacks, which had already taken place since his two previous governments, returned during the 2022 campaign, with the highly organized Bolsonarista wing responsible for spreading countless extreme speech messages.

Cesarino (2019, 532), when analyzing the 2018 electoral process and the role played by WhatsApp in the election of Jair Bolsonaro, draws attention to the "populist mechanism" used by Bolsonaro's online campaign. In her research on pro-Bolsonaro groups on WhatsApp, she notes that some recurrences can be observed:

> Permanent mobilization through alarmist and conspiratorial content; inverted mirror of the enemy and return of accusations; and creation of a direct and exclusive channel of communication between the leadership and its public through the delegitimization of instances of production of authorized knowledge in public waiting (notably, the academy and the professional press) (Cesarino 2019, 533).

In my analyses, the characteristics she pointed out were exacerbated in 2022, especially since Brazil was still experiencing a pandemic and the "enemy" was Lula. Something remarkable in the Bolsonaro government were the corruption scandals that somehow involved the president. As he had privileged jurisdiction, no investigation was carried out. However, one of Bolsonaro's main agendas was precisely the fight against corruption, a situation that he imputed to PT governments, while presenting himself as a simple man of the people. At the same time, everything he was accused of, his digital extreme speech production machine, blamed Lula.

Going beyond Cesarino's excellent analysis, we can consider that there is in the production of this type of content what Taussig (2020, 2) calls a "smoke and mirrors game," in which the mirror works as the mimetic element and smoke is the element that creates mystery and makes the reflections of the mirror mix, in what he names "metamorphic sublimity" (2020, 2). For Taussig, Trump—whom he calls the shaman in chief—was one of those who bet on this "mimetic excess," amplified by connectivity (in Trump's case, his use of Twitter) and which generates an order close to a "dark surrealism." The same idea goes for Bolsonaro and the way he resorted to machinic instances added to his followers (therefore, sociotechnical[12]) to disseminate fake news, some that in fact flirts with a distorted and obscure surrealism and that, although fantastic or unbelievable, are read as true by a multitude of people who "read it on Zap" ("Zap" is a colloquial term for WhatsApp)—and if they read it on WhatsApp, coming from the president himself, this obviously becomes real and shareable.

To end this chapter, I would like to reflect on a last group of messages that circulated among residents of the Complexo in the 2022 election campaign: those that appealed to religious discourse and the issue of moralities, with central themes such as abortion, nudity, the "unisex bathroom," pedophilia, and the family.

Several colleagues (Facchini and Sívori 2017; Lowenkron 2015; Teixeira 2022) have analyzed this moral crusade promoted for years in Brazil, with greater reach since the so-called CPI on Pedophilia in 2008.[13] Throughout this period, the junction between churches (especially evangelical ones) and conservative moralist discourse gave rise to various controversies such as the so-called gay kit, to various moral panics, the most recent of them represented by the implementation of "unisex bathrooms" in all public places in the country. There is the constant accusation that the Brazilian left is promoting gender ideology, which is taught in schools and puts children and families at risk. All of this was resumed, from images, videos, and inflammatory speeches by leaders, such as Bolsonaro, throughout the electoral campaign. And it was

---

12. What Cesarino calls the "digital body of the king," meaning that Bolsonaro supporters do not need Bolsonaro to continue their hateful actions.

13. This was a parliamentary commission to discuss pedophilia and child abuse. It was proposed by Magno Malta, a right-wing senator, and precipitated by the heavy use of Orkut to disseminate sexual content with children involved. The idea of this commission was really great, but its results are doubtful.

this content that circulated the most on WhatsApp amongst the Complexo residents, seeking to raise votes for Bolsonaro, who was responsible for protecting Brazil from these threats.

The images below, shared with me by Duda, give an idea of the discourse used, as well as the appeal of the images themselves. Something interesting to think about is that most of these images circulated in individual conversations, in groups, but were also placed as "WhatsApp status" by Bolsonaro supporters. As WhatsApp started to allow the sharing of status as stories or Instagram feed publications or on Facebook,[14] this content also circulated outside of WhatsApp, which probably ensured greater reach and less restricted views. Another curious fact is that many videos originally came from TikTok, and it is possible to see the platform icon in almost all of them. This fact also helps us to think about the ways in which content circulates and is in flux between the various platforms.

In all these images, we can see a strong moral connotation, since, if Lula were elected, he would be the cause of family destruction (last figure), and he would promote the permission of abortion and a crusade against religions. All of this falls into two of the guiding threads of all Bolsonarista discourse: God and family. The idea of these images is undoubtedly to create panic, to make people feel threatened by an alleged violation of their religious freedom, life, and family nucleus (heterosexual, composed by a man, a woman, and their children). Although at no moment did Lula mention the authorization of abortion, being against the family, or the persecution of any religion, this was one of the most sensitive points throughout the campaign and a key focus in the fight against extreme speech content. In a context like that of the Complexo, with a strong presence of evangelical churches, these arguments were certainly quite convincing. Additionally, as my interlocutors mentioned a few times, if the message was received from known people, from local groups or from the church, the veracity was practically a given. In this sense, this point deserves reflection: What are the resonances of these WhatsApp messages in the territory, insofar as it is

14. Victor Hugo Silva, "WhatsApp Testa Opção de Compartilhar Stories no Facebook e Instagram," Technoblog, updated December 2021, https://tecnoblog.net/noticias/whatsapp-testa-compartilhar -stories-facebook-instagram/.

Figure 1.2: Lula da Silva smoking a marijuana joint. Source: WhatsApp exchanges with a collaborator.

trust in the community and in its members that make this kind of content gain the status of incontestable truths and linked to the agenda of customs and morals?

These last two images follow the same line as the previous ones, appealing to the moral argument. The first message reads: "When your daughter has to share a public bathroom with a man, then you do the L and everything is fine! Life goes on!" The second reads: "Out with

Figure 1.3: Ad against abortion. Source: WhatsApp exchanges with a collaborator

Bozo![15] Let's fight. We demand to abort, to use drugs, to abolish gender, vaccines, neutral language, LGBTQIA+ education." The peculiarity of these two images is to invert the ideas of the "left." In the first, they resort to a symbol popularized among Lula's supporters: make the L, symbolizing the gesture with the hands of making the letter L for Lula ("Make the L" became a WhatsApp sticker and has been used since January 2023 to

15. Word used to refer to Bolsonaro by his opponents.

Figure 1.4: A banned religious cross. Source: WhatsApp exchanges with a collaborator.

symbolize the changes of the Lula government). In the second, they even resort to the aesthetics of the "left," in shades of red with an art different from those used by the Bolsonarista campaign where the base was all in green and yellow, the colors representative of the homeland. In this case, there is also the distorted mimesis of progressive agendas, which in fact demands respect for gender identities and sexual orientations. Again, a mirrors and smoke game.

Conclusion

The scenes found in the field and presented here allow us to under-stand, at the community level, the ways in which extreme speech

Figure 1.5: The destruction of family. Source: WhatsApp exchanges with a collaborator.

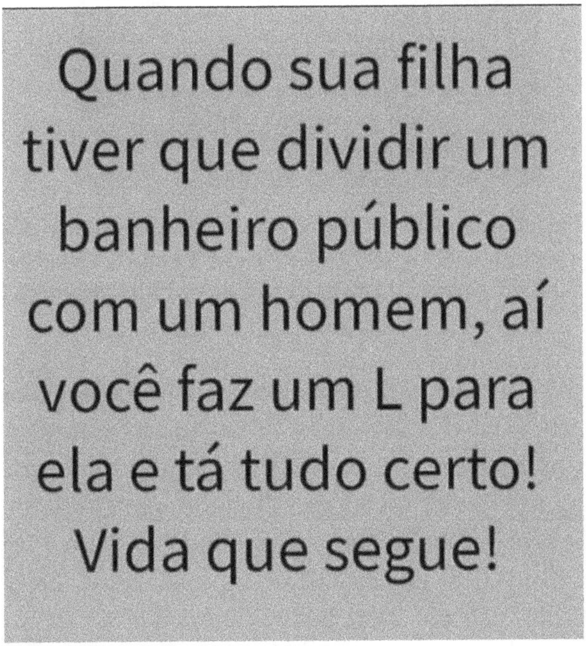

Figure 1.6: Text in the poster: "When your daughter has to share a public bathroom with a man, then you do the L and everything is fine! Life goes on!," evoking unisex bathroom polemic. Source: WhatsApp exchanges with a collaborator.

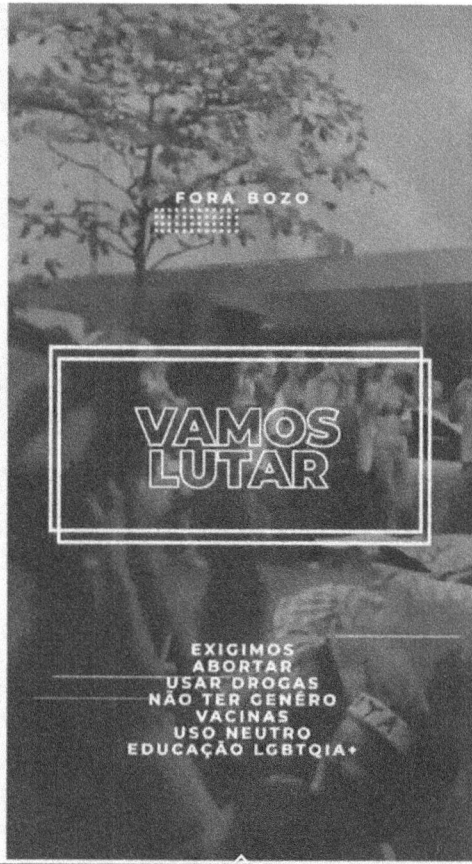

Figure 1.7: Image mimicking the left ads for Lula. Text: "Out with Bozo! Let's fight! We demand to abort, to use drugs, to abolish gender, vaccines, neutral language, LGBTQIA+ education." Source: WhatsApp exchanges with a collaborator.

(misinformation and hateful expressions) and moralities are intertwined and are directly related to the use of WhatsApp in these places.

The relation between the use of WhatsApp in impoverished places is marked by a series of social and digital inequalities. Since there is in Brazil the zero-rating policy concerning WhatsApp, the platform is the most used in these places and is responsible for the spreading of extreme speech. In Brazil, there is a widespread idea that "if I read it on Zap, it is true." Thus, there is usually no question about the veracity of the messages or even if it represents some kind of exclusionary posture against certain groups (the "left," LGBTQIA+, women who decided to abort, etc.). We can call this phenomenon the community resonance of the digital that creates a chain of practices and processes that reiterates the complementary nature of the relation between community territory and community digital groups.

The images and texts reproduced here also point to the moralities that circulate among residents of the Complexo. It is interesting to think how the defense of the family and of specific sexualities and gender conformations are the basis for the extreme speech contents and how this resonates at a community level. These are the same arguments used by people from privileged classes and mostly White.

Finally, the cases above help one think about the use of WhatsApp to conduct research, pointing to some possibilities, but also highlighting the possibilities in this use, especially when we consider social markers of difference such as class and schooling and issues of literacy and digital literacy. Since digital inequalities are present all over the world—and especially in the Global South—this suggests that we have to advance research about WhatsApp and its relation to social/structural inequalities.

Acknowledgment

The research that originated this chapter was funded by a generous grant from São Paulo Research Foundation: FAPESP (process n. 2021/06857–7).

2

Exclusionary Politics and Its Contradictions

Peddling Anti-Immigrant Sentiments through WhatsApp in South Africa

NKULULEKO SIBIYA AND IGINIO GAGLIARDONE

South Africa has some of the most progressive refugee protection laws and a constitution that defends the human rights of all those who live in it. However, migrant communities continue experiencing increasing violence and being targeted by antimigrant hate campaigns, some of which thrive in online spaces. Through an ethnographic content analysis of one of the oldest WhatsApp groups of Operation Dudula (OD), an antimigrant movement that emerged in South Africa in 2021 and has since gained significant traction in political conversations, this chapter highlights some of the paradoxes and contradictions of exclusionary politics in the Global South.

First it explores the coexistence of local and global forces influencing the cocreation, consumption, and spread of disinformation and extreme speech on WhatsApp. The South African context of extreme poverty, high levels of crime, corruption, a deep mistrust of the government, and a nostalgia for an apartheid past provides fertile ground for the proliferation of extreme speech against foreign nationals on OD's WhatsApp group. At the same time, despite OD's base in impoverished Black communities in South Africa, the movement has appropriated narratives and tropes adopted by the extreme right in Europe and North America with nostalgia for White rule and the former apartheid regime. This has gone hand in hand with exhortations to adopt civic behaviors, going from denouncing police corruption to demanding better services from the African National Congress (ANC)-led government, blamed for the corruption and the deterioration of the state.

The second part of our work further elaborates on how political campaigns and disinformation services engage and deploy on WhatsApp. We suggest that in the context of OD, WhatsApp is mainly used for informational and operational objectives. The platform allows members to mobilize and coordinate their offline activities. It also allows them to engage in debates about issues that plague their movement. This manifests through the continuous engagement of OD WhatsApp group members in a process of group and individual identity formation that is juxtaposed to the prevailing state or national identity.

Mis/Disinformation and Hate Speech in South Africa

OD is a movement that describes itself as "Patriots of South Africa engaged in a war against illegal immigration & crime in the Republic of South Africa" (Operation Dudula 2023). OD formally entered the South African mediascape on June 16, 2021, ushering in a new wave of antimigrant discourse that weaponizes militant extreme speech against foreign nationals. Despite OD's intention to disrupt South African politics— visibly embracing xenophobia as a rallying call, unafraid of public and political responses—the movement can be located in a long-term trajectory of mounting antimigrant sentiments, marked both by violent incidents against foreign nationals and by increasing mediated expressions of vitriol and extreme nationalism.

Xenophobia in South Africa can be traced back to as early as 1994 when armed youth gangs in Alexandra township, north of Johannesburg, destroyed the homes and property of suspected undocumented migrants and marched them to the local police station for deportation (South African History Online 2015). Between 1998 and 2000, there were at least nine recorded xenophobic killings in the country, and the years 2008 and 2009 saw some of the worst xenophobic attacks. More than fifty lives were lost and twenty thousand people were displaced, with numerous victims injured and robbed of their property (Human Rights Watch 2009; South African History Online 2015). From 2013 to 2022, various forms of attacks on migrants have been witnessed, such as the brutal killing of Zimbabwean national Elvis Nyathi in Diepsloot, a township notorious for such attacks (Gilili 2022).

Over the years, various explanations for the emergence and persistence of xenophobic sentiments have been advanced, ranging from the effects of a new nationalism to an uncritical media industry (Nyamnjoh 2010). Ariely (2017) has suggested that South Africans tend to have low levels of global identification, and thus do not perceive themselves as part of the global community, and this negatively affects how they feel about migrants. Gordon (2022) has posited that antimigrant sentiment and xenophobia among South Africans is a product of ignorance about the numbers of migrants in the country, which are overstated. The most noted of these reasons are socioeconomic issues fostered by a corrupt government and the propensity to criminalize foreign nationals (Choane, Shulika, and Mthombeni 2011; Solomon and Kosaka 2013).

The rise of social media has made xenophobic sentiments and incidents more visible and pervasive among South Africans, even before OD grabbed national attention. In sync with many other movements globally (Udupa, Gagliardone, and Hervik 2021), which started filling national conversations with unapologetic calls to drive the "other" out of their countries' borders, hashtags such as #PutSouthAfricansFirst had become nodes connecting political opportunists, fringe groups of political parties, as well as new parties, such as South Africa First Party, headed by Mario Khumalo (Findlay 2021). In a way, the form of exclusionary politics embraced and further promoted by OD emerged in continuity with these earlier expressions. Differently from them, however, OD sought from the start to combine the anger unleashed through online channels against foreign nationals, with forms of activism, online and offline, that belong to an earlier phase of online communication, demanding accountability from the South African government and seeking ways to coordinate locally to provide services to communities that national and local public authorities had been unable to offer for decades.

In recent years, the South African government has sought to find avenues to counteract the rise of online vitriol; for example, with the formulation of the Prevention and Combating of Hate Crime and the Hate Speech Bill, aiming to better define what constitutes hate speech and hate crimes in South Africa. More responsibility has been placed on the shoulders of group administrators, compelling them to monitor and delete any occurrence of hate speech. Under the new law, group administrators are held personally responsible for any message considered harmful (BusinessTech 2022).

At the international level, the United Nations believes perpetrators of xenophobic rhetoric and violence enjoy widespread impunity, resulting in a lack of accountability for severe human rights violations and the proliferation of racist and xenophobic political platforms. The United Nations further states that ongoing xenophobic mobilization in South Africa is broader and more pervasive than ever before, with some political parties using it as their central campaign strategy (United Nations 2022).

Social media platforms also bear responsibility for the proliferation of xenophobic messages. The Global Witness and the Legal Resources Centre conducted a joint investigation into the capacity of Facebook, TikTok, and YouTube to detect and remove actual instances of xenophobic hate speech directed at refugees and migrants in South Africa, and found that social media platforms are failing in their duty to safeguard vulnerable communities by not enforcing their own policies on hate speech and incitement to violence (Global Witness 2023).

As highlighted by fact-checking organizations, xenophobia also thrives on mis/disinformation, with foreign nationals often accused of crimes they either did not commit or that have been fabricated (Hiropoulos 2017). In South Africa, the dissemination of misleading and intentionally inaccurate information has become increasingly sophisticated, often crafted in various formats that include news stories that have been manipulated to skillfully blend truth and fiction in an attempt to incite outrage, mistrust, and sock puppet hate-mongering (Porteous 2023; Rousset, Maluleke, and Mendelsohn 2022). Prominent individuals and institutions like the South African National Editors' Forum have raised concerns about the destabilization of democratic institutions due to disinformation, a problem that became intensified on social media during the COVID-19 pandemic (South African National Editors' Forum 2021). A study conducted by Wasserman and Madrid-Morales (2019) surveyed 755 participants and found that audiences in Africa experience high levels of misinformation and often consciously participate in spreading it. South Africans believe that they regularly encounter inaccurate online stories about politics and government. Fifty-five percent of the participants assert that they often encounter such inaccurate stories, while 40 percent said they were exposed to these stories some-times. A report by the Africa Center for Strategic Studies (2024) suggests that disinformation and fake news campaigns in Africa are coordinated and paid for by foreign governments that include China and Russia, among others. The study posits that there are 189 documented disin-formation campaigns in Africa, and these have quadrupled since 2022.

Disinformation in South Africa is not just the purview of mischie-vous laymen. It is also a product of a news industry increasingly faced by the challenge of attracting and retaining attention, which may lead to publishing hyped or even false news for political or commercial gain, as exemplified by the globally viral story of the Tembisa decuplets. In this case, Independent Media, a leading South African media house, published a fake "exclusive story" claiming that a thirty-seven-year-old mother from the Gauteng Province had broken the Guinness World Record for the most infants delivered in a single birth (Isaacs 2022). Against this background, OD has developed a rather unique relation-ship with information manipulation. As our findings illustrate, exter-nally, OD members have disseminated false or misleading information to depict their opponents in a negative light, or to deepen resentment against specific communities or phenomena. At the same time, OD members have also promoted internal fact-checking of posts shared on the group, flagging incorrect, repetitive, and malicious information. The organic moderation of content by OD's WhatsApp group members is informed by the movement's desire to create and maintain a strong and credible brand image through the management and control of informa-tion, while pushing the boundaries of their extreme activities; a general need for accurate information; and high costs of data that force mem-bers to self-regulate, and discourage the posting and reposting of wrong information.

Methodology

I (Nkululeko) was introduced to OD by a community activist friend of mine when he invited me to my first OD meeting in May 2021. From this first encounter, I immediately realized this movement was going to be central to South African politics and its mediascape for some time. This was due to the topics being discussed on the day of the meeting—from outright demands for the removal of foreign nationals to everyday demands for basic services like electricity and the eradication of crime in the township—and the vigor with which those in attendance expressed themselves. I first saw Nhlanhla Lux, a member of the Soweto Parliament, and other OD leaders at this meeting. Lux, who was later referred to by local and international media as the leader of the movement, addressed the community, focusing on issues like the provision of electricity, illegal immigrants, drugs, and other civic issues. At the end of his speech, Lux and OD leaders encouraged the community to sign up to be added to a WhatsApp group to ensure they were part of the communication loop and to show up on June 16, 2021 to march to Eskom, the state-owned electricity provider, and to various parts of Soweto to help clean up the township and confront drug dealers and illegal immigrants.

When the meeting ended, one sentiment was clear as far as OD and those gathered at the meeting: the ANC government was perceived as failing the people, unable to address rampant poverty and unemployment, while illegal immigrants were accused of being major contributors to joblessness as well as the drug scourge plaguing the country. The WhatsApp group was formed days after the meeting, and I was invited into the group shortly after through a link sent by my said friend.

At first, I did not see the WhatsApp group as a site for doing research on OD or understanding migration issues in South Africa; however, as I got more exposure to the content being shared on the group, and experienced how OD captured mainstream media attention, I began to be interested in the group. Following conversations with Iginio about OD, my interest began to grow.

While I did not attend the march on June 16, the public noise and media coverage that followed the OD protest again solidified my resolve to study this movement. The press and other news media were littered with headlines like "Operation Dudula pushes ahead with hateful politics" (Bornman 2021) and "Enough is enough: Soweto residents commemorate June 16 differently" (Nonyane 2021). Based on my professional background, and the fact that I grew up in the area where the meeting occurred, I noticed there were differences in terms of the discourse in the WhatsApp group and what was being reported in the news media. Iginio and I were fascinated by the myriad of conversations taking place within the group because they touched on a number of civic issues, but at the same time contained and pushed vitriolic language and exclusionary politics against foreign nationals.

In order to address these phenomena and contradictions, we settled for an ethnographic content analysis (ECA) that involves the systematic analysis of textual, visual, or audio-visual data in an effort to understand cultural phenomena engulfing OD. ECA is useful for analyzing qualitative data and enables the acquisition of contextually rich insights into cultural practices, social interactions, and the meanings people attribute to various phenomena. It combines the principles of ethnography with those of content analysis, leading to analyzing and interpreting the content of communication or media in ways that are both systematic and contextually rich. It is distinguished by the highly reflexive and interactive nature of the research, concepts, data collection, and analysis (Altheide 2001).

OD has a number of WhatsApp groups. In our case, we decided to select and use the oldest and most active of the groups we had access to. By familiarizing ourselves with OD and its WhatsApp platforms, we ensured we had a clear understanding of the research context, the population under study, and the key concepts that were being explored (Bryman 2016). We then proceeded to select a limited number of posts in order to generate initial categories that guided the collection of data and the creation of a coding schedule. Given that our approach is thematic, the selection of these categories is interpretative; however, this interpretation was guided by the ongoing organic emergence of topics. In a reflexive back-and-forth process, some categories were dropped, while others solidified, as more and more themes and content aligned to each emerged (Bryman 2016). The data was collected over a period of two months from January 1 to March 2023 using dates as a guide to scroll through the content and screen grab the content chronologically. We identified several categories, and while these categories draw clear thematic lines, some topics overlapped and could not be fully boxed into one category. In such cases, subjective decisions were made on where to place certain pieces of data based on the initial theme or topic that started a thread we might be following. For the purposes of this chapter, the most relevant themes and topics are discussed in the findings.

Due to ethical considerations, the copying and pasting of posts from group participants required the cropping of images and the removal of any information that can lead to the identification of these individuals. The data shared on the platform comes in text, audio, and video format. These were downloaded, stored in password-protected folders marked by date, and assigned to specific categories. The process of analysis involved reading through the data, identifying themes and patterns, and developing a deeper understanding of the cultural phenomenon under study. This was operationalized through an exploration of narratives, discourses, and visual assessment of manifest content. We then proceeded to interpret the findings and draw conclusions based on our analysis, making connections between the collected data and our research objectives. This allowed us to develop a nuanced understanding of OD and some of the cultural phenomena that surround it. The most prominent challenge with conducting an ethnographic content analysis of this OD WhatsApp group was the effects of the disappearing messages function as well as the sheer scale of the content shared on the platform.

Results and Findings

Our work is aimed at exploring the social and cultural conditions that amplify the cocreation, consumption, and spread of disinformation and extreme speech on WhatsApp. The results confirm our hypothesis that the South African context of extreme poverty, high levels of crime, corruption, and a deep historical mistrust of the ANC-led government have provided fertile ground for the proliferation of extreme speech against foreign nationals. The categories that illuminated these social and cultural conditions the most are the "White supremacy and apartheid nostalgia" and the "SA [South African] crime and authorities." When looking at the data in the latter category, we found that there is a deep general mistrust of South African law enforcement agencies and the South African government.

No Trust for Law Enforcement Agencies

On the OD WhatsApp group there is a higher propensity to share posts about crime committed by foreigners than South Africans. However, this does not mean there is no posting of information about crimes committed specifically by South African authorities and members of the public in matters of national interest, or in specific communities. Content about the South African Police Services (SAPS) shows a lack of confidence in the police, which is often accused of being incompetent, prone to accepting bribes, and being corrupt. Members of the police force are further accused of collaborating with foreign nationals in illegal activities, while the South African National Defence Force (SANDF) is seen to be failing in its role of protecting the border. For example, January traditionally marks the start of the year and the return of holiday makers from various parts of the SADC region into South Africa. A shared post indicating that the SANDF border patrol was ready to deal with returning Zimbabweans attracted several comments that revealed the deep distrust and lack of confidence that OD members have of law enforcement agencies. In another video post related to the SAPS, a complaint was made that a Zimbabwean national was spotted driving a SAPS vehicle. This sparked outrage and replies that chastised the minister of police, Bheki Cele, and the government more broadly. The criticism of the government comes in the form of comparisons with other African states who are seen to be advancing the interests of their citizens by limiting access and the rights of foreigners. The South African government is seen to be incompetent and favoring foreign nationals over South African citizens. Those who identify as military veterans from exile argue that they were never allowed the same freedoms and movement allowed foreigners in South Africa during their stay in those countries.

Nostalgia for Apartheid Rule

It was quite surprising for us to find the consumption of, and identification with, far-right content and ideology. The awe may be triggered by the fact that numerous shared posts are racist and derogatory toward Black people, and the OD WhatsApp group is made up of majority Black South Africans; this made one wonder what may lead OD members to welcome and celebrate such content. It is in this content that we find evidence of nostalgia for the apartheid government and the return of Whites into power. The government and its institutions are compared to the National Party (NP) government, and the verdict is that the NP was better than the ANC and would have been better in power. In one

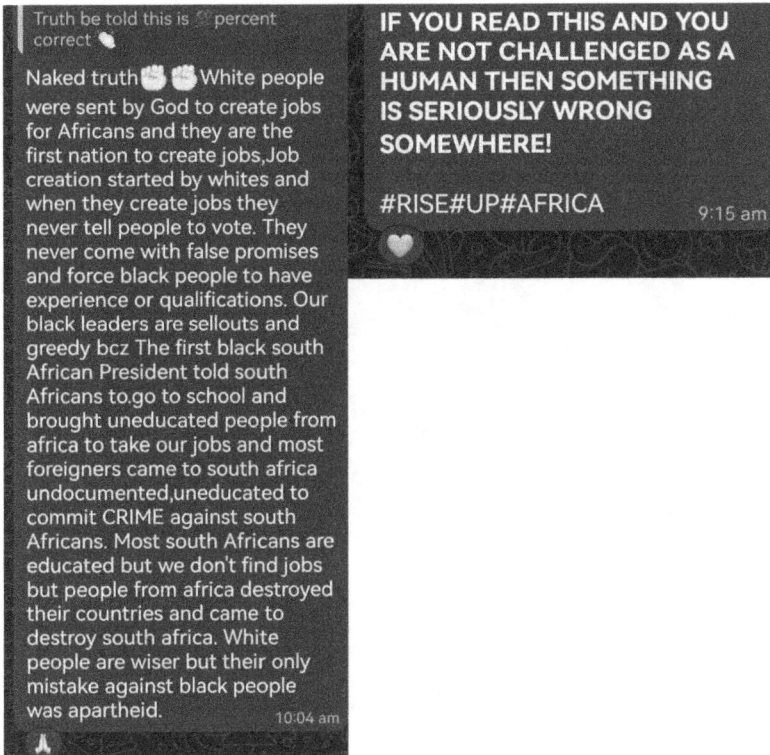

Figure 2.1: Posts celebrating White supremacy and displaying feelings of nostalgia for apartheid South Africa are frequently shared in OD's WhatsApp Group.

instance, a speech titled "Rise Up Africa," falsely attributed to former U.S. president Donald Trump, was shared on the group (Horner 2020). The central message of the speech is that a Black man is nothing but an alcoholic delinquent capable of nothing but corruption and disaster. The statement suggests that the Whites were chosen by God to rule over Blacks. This statement was celebrated by some in the group, with others even testifying to its truthfulness, despite it being disinformation as shown below.

In the group there are continuous, emotionally charged debates about how to solve the problems the country is facing, and one of the central solutions frequently offered is the removal of the ANC government from power in 2024. The responses to this question vary, with some

> I agree with you entirely,most African countries they put and protect their own citizens they tell you that their citizens are their priority but in SA i don't remember our leaders put South Africans first they don't even respect the people who fought for freedom,in other countries including the UK veterans are celebrated they are respected and well looked after but in our own country we just don't care it's so sad
>
> 11:03 pm

> Comrade it is so painful 😣
>
> 11:07 pm

Figure 2.2: Emotionally charged posts that suggest deep pain and a feeling of neglect by the South African government are frequently shared by OD'S WhatsApp group members. Source: OD WhatsApp group

suggesting that there is no alternative power that can replace the corrupt ANC, while some argue that such a response is fatalistic, and either propose that the current government be replaced with a political party sympathetic to OD or replace it with nothing at all. The ANC is accused of being the new oppressor and being led by tired old people, hence its failures. A viral video of ANC members walking at a conference that suggested that the members are too old to be in government was also shared on the platform.

OD identifies and sees itself as a movement that represents the poor, voiceless, and marginalized South Africans, and they argue that the plight of South Africans is rooted in the poverty, corruption, and crime that the ANC government has allowed to go unchallenged. The expression of this frustration with the government, crime, poverty, and unemployment is emotionally charged and palpable. Taking note of such frustration and anger can help in charting explanations of why and how extreme speech acts transform into violence. Above is a screenshot that shows an example of such emotionally charged posts.

Counternarratives of Identity Formation

The appeals to White supremacy, nostalgia for a return to White rule, and xenophobic pronouncements do not go unchallenged in the group, illustrating how group members are continuously engaged in a process of group and individual identity formation. For example, after the posting of a document titled "How to Destroy a Black Man," which had both the Confederate and the Ku Klux Klan flags on its cover, some celebrated the content without question, while others challenged the message, calling for the realization of the detrimental effects of White supremacy. It is noteworthy that such responses to extreme language and content do generate responses that seek to insult and remove those who disagree with the dominant views of the group. Another important thing to note is that there is a firm belief in some quarters that the Pan-Africanist movement is alive and well, and its central goal is to render Africa borderless through the annihilation of all the states that currently exist. OD members believe this movement is led by the ANC and the EFF. This suggests a complete disavowal of Pan-Africanism ideology.

Internal Content Moderation

We posit that the relationship between extreme speech peddlers and disinformation is a complex one and does not necessarily have a mutually reinforcing relationship. In the context of OD on WhatsApp, in certain circumstances, disinformation is regarded as detrimental to the operations and success of OD's antimigrant campaign. In some instances where OD stands to benefit, disinformation is ignored and encouraged. Evidence suggests that there is a need for OD to protect its brand and reputation, and thus content moderation is encouraged when the brand is in danger. There is also a general need for accurate information by those participating in the platform to safeguard resources. Lastly, the high costs of data force members to self-regulate, and discourage the posting

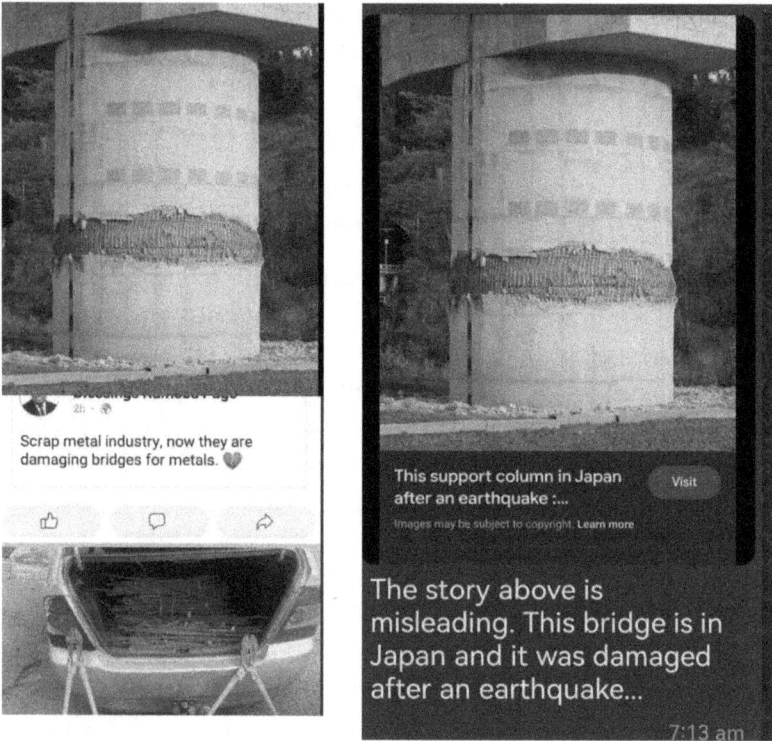

Scrap metal industry, now they are damaging bridges for metals.

This support column in Japan after an earthquake :...

Images may be subject to copyright. Learn more

The story above is misleading. This bridge is in Japan and it was damaged after an earthquake...

Figure 2.3: Fact-checking in action. OD members often engage in the verification of facts. Source: OD WhatsApp group

and reposting of content especially if that content is false. While there is a proliferation of content that seeks to blame foreigners for all crime in South Africa, there are some participants in the group who moderate content and flag inaccurate information. For example, a post showing a bridge that was seemingly blasted for the removal of reinforcement steel in one of its columns was shared on the group. The post suggested that it was the doing of Zimbabweans involved in the scrap metal industry. The story was fact-checked by a user who pointed out that the bridge was actually in Japan and that the story was false.

Evidence also shows that data costs are a critical matter to those participating on the platform. Repetitive old content is frowned upon in the group, and when posted, it is often met with responses of disapproval and complaints about data. While the moderation of certain pieces of content may be a positive sign, there are other forms of troublesome information that the members of the group are either unknowingly, or purposefully ignorant, about. There is also content that suggests that participants generate and/or distribute misleading information. For example, the "shutdown" campaign poster of March 20, 2023 of the EFF was taken and remixed to show signs that suggest that the EFF is mobilizing foreign nationals to take part in the shutdown. This is problematic in that it easily riles up members who take the content to be authentic. A comparison of the original poster and the remixed poster shows these salient differences; the remixed poster featured the Zimbabwe national flag in four different places on the poster. These were not present in the original. The campaign slogan, "#Nationalshutdown" in the original was reworded to "#Umzansi4Africa," which means South Africa for Africans. On the bottom copy of the remixed poster, EFF promotional copy is replaced with Zimbabwe, Mozambique, Somalia, Malawi, Congo, and South Africa to depict the EFF's affiliation with these countries.

The second part of our work elaborates on how OD uses WhatsApp for operational purposes related not only to mobilizing against immigrants, but also for coordinating other broader civic activities. We illustrate how OD uses WhatsApp to leverage the interest and energies activated by rallying calls to mobilize against immigrants and effecting change in their own communities. OD uses the platform to announce events, share documents, and report back to its members on issues discussed at offline meetings. Official announcements of the core business of OD are communicated through digital posters, which are standardized, and have OD colors, which are red, green, and white. Often these posters will have a beige background and have a call to action that asks its members to respond to a particular event. For example, one called for members to attend the court case of the Zimbabwe Exemption Permits (ZEP) for Zimbabwean nationals. These posters announce the date, time, and meeting place for all events. The language used has themes of patriotism, nationalism, militancy, and aspirations for progress. OD's slogan on these posters is "High discipline, high morale." The posters are generally accompanied by the South African flag. The OD logo with the black/brown hand with a clenched fist is always present on the posters.

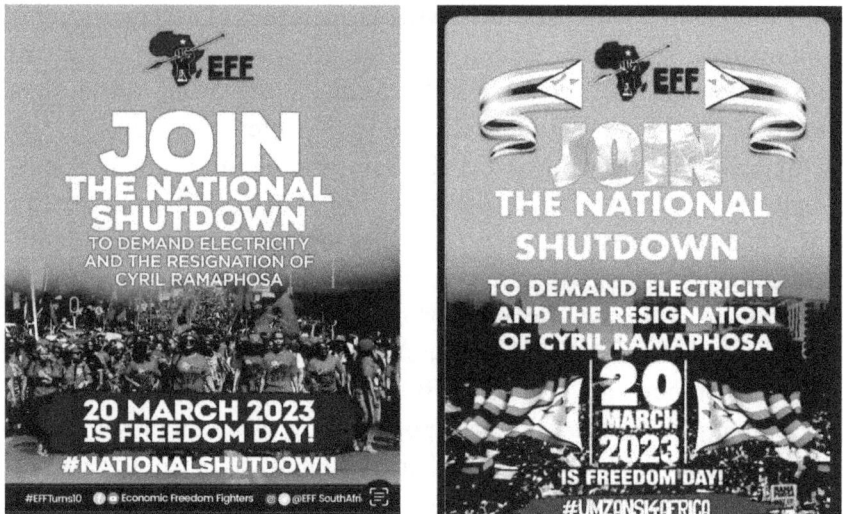

Figure 2.4: A comparison of the Economic Freedom Fighters' (EFF) original shutdown march poster of March 20, 2023 (left) and the fake or remixed poster with Zimbabwean flags created by OD members (right). Source: OD WhatsApp group

The slogan "Our economy our heritage" is also featured. There is a second logo that represents the women of OD. It looks like the dominant "male" or primary one, but the brown clenched fist looks feminine and the pose is slightly different. The logo has laurels forming a semicircle on both sides, with two spears at the bottom facing opposite directions. Between them is the word "*Izimbokodo*," which is a term that emanates from the resistance song "Wathint' Abafazi, Wathint' Imbokodo" (You strike the women, you strike the rock), which was sung by South African women during the Women's March against the restrictive pass laws in 1956 (South Africa History Archive 2002). Emojis of fire and of fists are often used on the posters. Other slogans that generally accompany the posters are: "South Africa for South Africans" and "Enough is enough." "*Phambili*," which means *forward*, is another term frequently used in the posters. They also feature scores of OD members marching.

Put South Africa First (PSA) sometimes features on OD posters, showing the collaboration that often happens between the two organizations. The posters also share contact information for those who need more information in relation to an event or issue. Analysis of the posters gives information about the campaigns that OD mobilized during the period in question.

OD has been massively invested in the ZEP saga, and it continuously mobilizes its members to attend court cases. The ZEP poster was also accompanied by a voice note from OD leader Dan Radebe calling on people to attend court. Court judgments on this matter were also shared in PDF format. A picket at the State of the Nation Address 2023 was communicated through a digital poster that called for the mass deportation of foreign nationals. A campaign to confront and remove foreign-owned spaza shops was also shared; however, no details of the date of the event were given. Public offline meetings are also announced through these digital posters; however, in certain instances some details are left out so it is the "in" group that is privy to certain pieces of information considered to be critical to operations.

OD uses multiple social media platforms, and often uses WhatsApp to amplify its message generated or communicated on other platforms first. It also uses the platform to notify its followers about interviews and appearances on mainstream media and live broadcasts on social platforms like Twitter Spaces. OD also activated a "back to school uniform campaign," which called on members of OD and the general public to contribute stationery and school uniforms nationwide. This was communicated via a digital poster and promoted on Twitter, launched by PSA. The platform is also used to share OD banking details whenever there is a need to contribute money.

Conclusion

Our work provides some of the building blocks to understanding the South African antimigrant group Operation Dudula. Through an ethnographic content analysis of one of OD's WhatsApp groups, the platform is examined as a critical component of OD's media arsenal. OD uses WhatsApp to mobilize its constituencies through emotive, arousing language on digital posters. A similar modus operandi has been seen in operation with right-wing extremist groups in the United States, who have used encrypted messaging platform Telegram to recruit new members and organize and plan their activities, including the January 6 attack on the White House (Grisham 2021).

The protection of individuals' identities and the encryption of messages by these messaging platforms has been a key element of appeal for groups like OD to use WhatsApp to organize their antimigrant campaigns. Our work shows that a deep mistrust and disenchantment with the ANC and law enforcement agencies have provided fertile ground for the proliferation of vitriolic language, xenophobia, and exclusionary politics. This has also generated nostalgia for the return of the apartheid past in the form of White rule, which is seen as better and less brutal than the rule of the ANC. This nostalgia is accompanied by the consumption of right-wing racist content. The disillusionment with conventional politics and consumption of ultra-right content on encrypted messaging apps has been experienced globally, fueled by different reasons, including immigration policies and failing economies. This disillusionment is seen as a global crisis by a number of researchers (Koutsokosta 2023), and OD's disillusionment with the government of the ANC and its consumption of White extremist content is not unique and specific to the South African context. Considering that WhatsApp has become a primary means of delivering political messages in many countries of the Global South since 2015, it is no surprise that OD has followed suit. WhatsApp's instrumentality in the political landscape of the Global South can be seen in the rise of former Brazilian president Bolsonaro (Avritzer and Rennó 2023).

OD's relationship with disinformation is a complex one, and our work shows that disinformation is discouraged and encouraged depending on the prevailing circumstances. OD has an interest in protecting its resources, brand, and image for its own political ends, and this includes protecting the brand from political impostors and criminal elements that use its brand to rob and take advantage of people.

In these circumstances, there is a desire to flag false information, and correct it. This is done by the leadership of OD as well as ordinary participants in the group. They moderate content and provide correct information not only for political ends, but also for the basic need of consuming accurate information. In certain instances where disinformation propels OD's ends, it is encouraged and manufactured internally; however, our evidence shows limited instances of extreme disinformation campaigns like those witnessed with the Bolsonaro campaigns in Brazil.

Further investigations into the differences in the nature of disinformation in different countries are required and are not part of our work. Ours provides one of many entry points of studying OD. Accessed through an ethnographic content analysis, this entry point provides a modest description of one of OD's oldest and active WhatsApp groups, its social context, and some of the cultural phenomena that surround it. The textual fields presented by the content on the WhatsApp group are vast and require more attention.

3

Deep Extreme Speech

Intimate Networks for Inflamed Rhetoric on WhatsApp

SAHANA UDUPA

"There are some people whom you cannot convince to go beat up some-body," observed Anusha, an Indian journalist. "For them, they [political party campaigners] use religion to sustain the political interest." In an ethnographic conversation in March 2023, the Bangalore-based jour-nalist was describing to us the vast number of WhatsApp groups that political parties have raised in India, and how such groups have lately transitioned from bundled-up masses of members to "specialized and nuanced" clusters. This fine-tuning has led to the parsing and slicing of voting populations based on age, region, gender, lifestyle, and assumed value systems, and offering content to align with such demarcated vec-tors within specially carved out or embedded WhatsApp groups.

Anusha went on to provide more examples. "Teenagers are primed to talk about violence," she said. "They are given false histories, for ex-ample, that Uri Gowda and Nange Gowda [warriors from the Kannada-speaking region in Southern India] killed Tippu Sultan [Indian Muslim ruler of the kingdom of Mysore in the eighteenth century][1] while those of us who have studied the history know it was the British who killed Tippu. Homemakers are treated with temple stories and questions of dharma [righteousness in the Hindu tradition]; for the Marathi-speaking community, content will be on Shivaji [Marathi warrior king who resisted British occupation], and for Kannada-speaking people,

---

1. The rewriting of history around Tippu Sultan's death sparked a controversy in Karnataka in March 2023, when the state was preparing for regional elections. "'Pure Fiction': Historians on Uri Gowda and Nanje Gowda Killing Tipu Sultan," News18, March 20, 2023, https://www.news18.com /india/pure-fiction-historians-on-uri-gowda-and-nanje-gowda-killing-tipu-sultan-7336759.html.

stories will be on Kittoor Rani Chennamma [Kannada warrior queen who fought British occupation]."

As research participants described to me in various ways, stories of historical and cultural significance are embellished with track beats to raise the level of attraction, especially among younger users who could be drawn to histories beyond the "boring" static formats of textbooks. The varied strands of targeted content are appended to the perception of correct political choices, i.e., the perception that by consuming and sharing certain types of content and by taking part in this sharing practice, one makes the right political choice. In this chapter, I delve into WhatsApp as the unique social infrastructure that makes this form of content sharing an important feature of extreme speech ecosystems, with particular valence and heightened significance in the Global South contexts.

A growing body of scholarship that has examined the entanglements among platform affordances, political propaganda, hateful speech, and disinformation has highlighted the role of encrypted messaging applications in terms of their disinhibiting effects upon users who share problematic posts as well as enabling impacts of in-group camaraderie and content-rich influence strategies (Bursztyn and Birnbaum 2019; Cheeseman et al. 2020; Evangelista and Bruno 2019; Garimella and Tyson 2018; Johns and Cheong 2021; Nizaruddin 2021; Scherman et al. 2022; Recuero, Soares, and Vinhas 2021; chapters in this volume). Qualifying platform-centered analysis, anthropologists and ethnographers have pinned their focus on what people do with media and how complex mediations of lived worlds cluster around, draw upon, and reshape technological possibilities of WhatsApp such as closed-chat architecture and end-to-end encryption (Cruz and Harindranath 2020; Williams et al. 2022). Ethnographic studies on hateful speech and disinformation in the Global South have especially drawn attention to the political use and electoral mobilizations on the messaging service. In Nigeria, studies have shown how WhatsApp differs from algorithmically shaped echo chambers on Facebook or Twitter, prompting politicians to "create partisan environments and inflammatory messages to bolster their candidacy" (Olaniran, this volume). In a study that explicates this point with ethnographic detail, Stalcup (2016) draws attention to styles of sharing on WhatsApp in Brazil that render the messages memorable, leading to a form of "aesthetic politics" that involves intentional deployment of platform aesthetics to political ends. Studies on WhatsApp in India have argued that WhatsApp groups help to "achieve homophily," preparing a fecund context for microtargeting "not at the individual level but at the group identity level" (Sinha, this volume).

Drawing upon this insightful scholarship, this chapter suggests that WhatsApp's unique role in disinformation and vitriolic ecosystems in the Global South contexts of divisive politics lies not as much in the architectural features of encryption but around particular clusters of social relations it enters, entrenches, and reshapes. In other words, rather than technical design features seen in isolation, WhatsApp might be better understood in terms of interactional and structural dynamics around social relationality, obligation, and kinship (Fortes 2005; Sahlins 2014; Tenhunen 2018), and how social relations across a range of contexts are reified and reproduced through WhatsApp communication, with significant ramifications for political discourse. WhatsApp as a social relational form represents a unique strand in the complex mix of factors that enable speech practices that stretch the boundaries of legitimate speech along the twin axes of truth/falsity and civility/incivility—practices defined here as "extreme speech" (Udupa and Pohjonen 2019).

The key argument is that messaging apps such as WhatsApp alongside domestic social media platforms in regional languages with group functionalities enable what might be described as "deep extreme speech." Deep extreme speech is characterized by community-based distribution networks and a distinct context mix, which both build on the charisma of local celebrities, social trust, and everyday habits of exchange. Deep extreme speech could be seen as the social corollary for technologized deep fakes deployed in political campaigns. This type of extreme speech belongs less in the problem space of truth or the moral space of hatred and unfolds rather at the confluence of affect and obligation, variously inflected by invested campaigns.

The forthcoming sections will develop the concept of "deep extreme speech" first by briefly outlining digital propaganda activities in the Indian context, followed by empirical sections on the two aspects of distribution and content, and a concluding note on policy directions to address deep extreme speech. The chapter builds on ethnographic fieldwork carried out in Bangalore, Mumbai, Delhi, and adjoining towns since 2013, especially the latest rounds of fieldwork in March and October 2023 among online users, journalists, political consultants, and activists in Bangalore and online interviews with political consultants and journalists, as well as content analysis of WhatsApp groups affiliated with three major political parties during the 2019 general elections.[2]

WhatsApp in Digital Campaigns in India

Digital social media has emerged as a critical election apparatus in India, following the ruling nationalist party's (Bharatiya Janata Party, BJP) pioneering use of social media channels for election propaganda and ideological (re)production in the last two decades. While the party enjoyed significant electoral success in recent years by being the first to capitalize on online networks, it now contends with the emergence of several other political parties vying for similar resources. In an atmosphere of a "permanent campaign" enabled by digital media (Neyazi 2020), political parties across the spectrum have ramped up their digital campaigns across diverse social media platforms. The Indian National Congress (INC), the chief pan-Indian opposition party, has research teams to prepare "counters" to the BJP's public statements and a social media dissemination team that distributes "research data" and composes content for voting publics. Regional political parties such as Samajvadi Party and Shiv Sena are increasingly enlisting the services of commercial political consultants and digital influencers to promote campaign content and leaders. Aside from social media channels attached to political party systems, individual political leaders of all major parties, including regional parties contesting the elections, are now actively recruiting social media teams for campaign work. While these activities indicate more clamor and flux in the online political sphere, such efforts are quite often overwhelmed by the BJP's heavily funded and rapidly adapting campaign structure and its large "volunteer" base. The vastly intricate digital influence operations of the party keep the momentum of party

2. My sincere thanks to Miriam Homer, Neelabh Gupta, Deeksha Rao, Amshuman Dasarathy, and Sudha Nair for their excellent research assistance. This research is supported by the European Research Council funding under the Horizon 2020 program (grant agreement number 714285 and 957442) and the Centre for Advanced Studies Research Group Funding (2023–2025) at the Ludwig-Maximilians-Universität München.

discourse and flows of audiences alive on social media, with war-like spikes in provocation but also routine everyday exchanges that repeat and reproduce the "party line." The party's dominance in the social media space is such a widely acknowledged reality that a major opposition political leader, during a closed-door meeting with academics in 2023 that I attended, remarked in frustration that "the narrative"—public apprehension of the political situation and collective imagination of the shared futures—is now completely, if not irrevocably, in the hands of the ruling party. While social media influence translated into electoral success for the ruling party in the previous elections of 2014 and 2019, the 2024 elections indicated that major factors weighing in on campaign efforts, including the economy, election financing, and welfare promises, rendered social media messaging an intense battleground among different parties competing to settle voter uncertainties in their favor.

The ruling party's digital presence, a dominant campaign structure, has multiple layers and dimensions spanning dissemination and content creation techniques across platforms in ways to dynamically adapt to polymedia environments and how users engage different platforms at different intervals and for different purposes, all while remixing them with newer articulations (Madianou and Miller 2013). Thus, the party has wide uses and campaign deployments across Facebook, X (formerly Twitter), Instagram, Pinterest, and Clubhouse as well as homegrown platforms such as ShareChat, Moj, and Kutumb. For instance, on platforms such as X (formerly Twitter), the party cultivates and draws from argumentative styles of online engagement that typify such platforms, paving the way for "volunteering" warriors and prepped-up tweeters who reproduce the key ideological tenets of the party by bickering with opponents and repeating simplified summaries of what they understand as the nationalist ideology (Udupa 2018a). The BJP's central Information Technology cell—the nodal office for digital influence operations—coordinates a range of top-down campaign activities attuned to different platform cultures. The official line of work is complemented by hired commercial consultants with promises of data-backed influence strategies and a breeding shadow industry that operates through gray practices of clickbait operators, hired influencers, and loosely knit networks of dispersed amplifiers drawn into precarious and informal labor arrangements crafted by ambitious mediators.

With more than 490 million users in 2021 (Degenhard 2023) and as the second-largest online platform for accessing news in India (Newman et al. 2023), WhatsApp has emerged as a major platform for party campaign activities. It figures in the digital campaign structure of the party—and increasingly other political parties—in two prominent ways: creating intrusive channels of distribution and tactical mix of content. Distribution and content aspects of WhatsApp not only feed on the messaging application's unique affordances to create and sustain groups on a subterranean level but also, more importantly, by replicating some of the core aspects of social relationalities.

Intrusive and Intimate Channels of Distribution

The ruling party's use of WhatsApp, developed over the last decade into a stable campaign structure, is a telling testimony for novel forms of distributing ideological and party-favoring content. The party's WhatsApp system consists of multistep distribution with horizontal networks connected to different vertically integrated nodes. The official network channel, headquartered in the capital city of New Delhi, is centrally controlled for disseminating national-level issues; the central IT team dictates the tenor and content of messages in the national space while regional and local units are allowed to compose messages around context-specific issues relevant to respective electoral constituencies.

However, this centricism is strategically positioned in relation to a more dispersed, flexible structure that can draw and retain volunteers. Together, these networks connect "official" workers with other official workers and connect official workers with "general sympathizers" or "well-wishers," allowing sufficient space for official workers to draw "general sympathizers" to become more committed, and general sympathizers to draw other sympathizers and fence sitters. Content flows from node to node. The flexible parts of such networks are not edges of a single core but constitute connected nodes of content building and influence enhancement. Giving an overview of the IT operations of the party, Amit Malviya, the national head of the Information Technology cell of the party, told me in New Delhi that the party has diversified the online communication channels, "in the sense that we have gone down to [regional] states and we are telling them, look, we have to have multiple layers of communication, we have to communicate at the central level, we have to communicate at the state level, and perhaps localize it even further." The party's strategies around WhatsApp fall squarely within this ambition for distributed networks of influence enhancement.

During a meeting in 2020 in Bangalore, a high-ranking party official in the Southern Indian state of Karnataka was in awe of what they had managed to raise. He exclaimed excitedly, "BJP had formed their own twenty-five thousand WhatsApp groups during the Karnataka [regional] elections. And there were fifty thousand WhatsApp groups during recent Lok Sabha elections in West Bengal [another subnational state]. Only West Bengal! Imagine, fifty thousand times two hundred, that is ten to fifteen lakhs [1 to 1.5 million people]. So, their message has been reaching ten to fifteen lakh people directly. And imagine the forwards which these posts received! This is huge!" He added that the INC, the chief opposition party, had also formed around five thousand WhatsApp groups during the Karnataka elections, and hence he sees "a lot of official WhatsApp groups being formed by party supporters of different parties." The INC runs a "Rahul Gandhi WhatsApp group" on the national level to facilitate direct interactions with the opposition leader (LiveMint 2024).

In 2023, more WhatsApp groups were added and fine-tuned to achieve greater correspondences between group membership and assumed characteristics of members. In the words of one of the interlocutors, "WhatsApp worlds are more streamlined," as several types of strategists, including former journalists, are consulted to "devise plans to reach out to people." During ethnographic conversations in October 2023, journalists turned political consultants added that WhatsApp groups draw upon the sphere of influence that local actors wield within particular communities, building on the social ties of "nano-influencers" (Joseff, Goodwin, and Wooley 2020). Within large residential apartment complexes in urban areas, for instance, influential "uncles"—kin-based authority figures among residents—are encouraged to steer WhatsApp groups for party-favoring narratives, including top-down messaging such as the "Letter from the Prime Minster," which was sent out during the 2024 elections for cascading circulations on the messaging platform (LiveMint 2024).

A vital feature of WhatsApp campaign work is thus the way it helps create intrusive channels for inflamed rhetoric of different kinds. The key motivation for political parties in this regard is to combine top-down "broadcasts" with "organic bottom-up messaging." They have sought to accomplish this by installing "party men" within WhatsApp groups of family members, friends, colleagues, neighbors, and other trusted communities. "WhatsApp penetration"—defined as the extent to which party people "organically" embed themselves within trusted WhatsApp groups—is seen as a benchmark for a political party's community reach. Local musicians, poets, cinema stars, and other "community influencers" have been recruited to develop and expand such "organic" social media networks for party propaganda. Typically, a party moderator would find his way into a WhatsApp group through local connections or by leveraging community work such as local brokerage to help people to access state benefits and so on, and once admitted, he would relay party content in unobtrusive ways. "WhatsApp groups are intentionally generated," observed Tabassum, a journalist based in Bangalore, and party workers and activists added that right-wing groups meticulously go by the electoral list and start WhatsApp groups for every "sheet [in the electoral list with voter names]."

Such distribution patterns on WhatsApp reveal that while online extreme speech circulation is driven in part by technological features of virality and putative encryption, a significant part of this circulation operates by tapping social trust and cultural capital at community levels, often making deep inroads into the "intimate sphere" of families, kin networks, neighbors, caste-based groups, ethnic groups, and other longstanding social allegiances. Such types of vitriol rely on and rework localized community trust as the key lubricant for the networked pipeline of extreme speech and hate-based disinformation.

To be sure, trust and mistrust are socially negotiated, and not all content that comes on WhatsApp groups is likely to be trusted at once. Kishor, an activist in Bangalore, recounted his experience of interacting with school children in Shimoga, a semiurban area in Southern India. "Schoolchildren who were studying in the Kannada [regional language] medium schools said they were aware of algorithm-based tweaking and that social media narratives have a right-wing bias. 'WhatsApp University' as a derogatory term to signal deliberate creation of distorted knowledge is also common among users. Yet, we are not able to break the cycle of networks sharing such content." As philosopher Hartmann (2011) suggests, trust and mistrust are not mutually exclusive categories;

they evolve dynamically. For instance, anthropologist Beek's (2018) ethnography on "romance scammers" in Ghana illustrates that people on online dating sites engage in romantic relationships with their contacts despite mistrusting them. In other words, social practice exceeds subjective assessments of trustworthiness.

WhatsApp content flows are also not without friction. Sample this conversation between a political consultant and Deeksha, our research assistant in Bangalore, during one of the interviews:

> GM: I am speaking as an individual now, not as a political consultant. I used to receive a lot of forwards [on WhatsApp] from my own family members. So, I used to have debates, discussions, everything with them.
> DR: Do you finally do what I did? I left the group for my own sanity.
> GM: So, in the beginning I used to do that. But later, I thought let us give it back to them.
> DR: So you were a fight person. I was a flight person. So, it's good to know someone is fighting battles.
> GM: Yes, yes, we have to fight.

The resolve to fight or the urge for flight notwithstanding, the conversation attests to the deep inroads that partisan content has managed to make in and through messaging applications like WhatsApp. "There are often intense political battles within families," said Tabassum about her own experiences. "My uncle fumes at me for toeing the line of the ruling party, and I hold ground despite pressure."

Simulating the Social

If community trust embedded in WhatsApp eases the flow of extreme speech, the manner of crafting content for electoral and political influence is no less significant. To examine content patterns on partisan WhatsApp groups, we carried out content and thematic analysis of a corpus of messages sourced from eight WhatsApp groups. The groups were affiliated with or explicitly supported three major political groups: nationalist BJP (groups identified here as B4I, JSR, and ABP); the opposition Indian National Congress (KLI, Inspire, Align) and Aam Aadmi Party, an urban political party and the ruling party of Delhi (AAP). Our research assistants joined the groups either using the phone number publicly announced by the parties with an invitation to join the groups or by obtaining oral consent of the moderators. Data was gathered during the 2019 general elections, which were held between April 1 and May 19, 2019. The length of the data-gathering period varied for different groups and ranged between September 2018 (oldest) to January 1, 2020 (latest) for the sampled election period. The messages, which contain text and media (still images, GIFs) in English, Hindi, and mixed registers of Hindi and English, were qualitatively analyzed by identifying thematic categories and tones through bottom-up coding, and with a binary classification for extreme speech and lists of types and targets of extreme speech (Tables 3.1–3.4). Weblinks in the messages were not included in the coding. A total of 30,887 messages were coded (*KLI* 9,550 text; *ABP* 4,701 text, 495 media; *AAP* 4,216 text, 281 media; *B4I* 3,583 text, 394 media; Inspire 1,893 text; *Align* 924 text, 67 media; *JSR* 658 text, 280 media).

Table 3.1: Overview categories

Category Key	Category Key Name
1	Nationalism/patriotism
2A	Religion (Hindu)
2B	Religion (Islam)
2C	Religion (Christianity)
3AA	Politicians (local, regional) NATIONALIST
3AB	Politicians (local, regional) INC
3AC	Politicians (local, regional) Other
3BA	Politicians (national) NATIONALIST
3BB	Politicians (national) INC
3BC	Politicians (national) Other
3C	Politicians Modi
3D	Politicians Rahul Gandhi
4	Development
5	Personal wellness/greeting
6	Historical
7	Party symbols
8	Any other
9	Inhuman/violence

Table 3.2: Overview of tones

Tone Key	Tone Key Name
1A	Sarcasm (with humor)
1B	Sarcasm (without humor)
2	Informational
3	Greeting/personal
4	Confrontational
5	Graphic
6	Warning
7	Allegation
8	Soothsaying
9	Eulogizing
10	Any other

Table 3.3: Extreme speech types

Extreme speech Key	Extreme speech Key Name
1	Offensive to community
2	Call for violence
3	Violence through reference
4	Call for exclusion from nation/community
5	Sexist
6	Casteist

Table 3.4: Extreme speech target groups

Target of Extreme speech Key	Target of Extreme speech Key Name
1	Muslims
2	Women
3	Dalits
4	"Pseudoliberals" (progressive-liberals)
5	General public/unmarked audience
6	Right-wing groups

A quantitative summary analysis provides an overview of the distribution of thematic categories and extreme speech types across three political groups, which were calculated as a percentage of the total number of messages per group to highlight the significance of different content types within the group. In terms of message types, AAP and nationalist WhatsApp groups relied more on images in comparison to the Congress WhatsApp groups.

Across all the groups, an interesting finding is that a very small number of users sent the greatest number of messages, confirming other studies that have documented the prominence of "super users" in online political networks, who "account for the vast majority of posts and of extremist language" (Kleinberg, van der Vegt, and Gill 2021). Figure 1 shows the twenty most active users based on the number of total messages (including text messages, media, and web links) they shared. The users were ranked from 1 to 20, with 1 referring to the most active user. Many groups had a single user who was significantly more active than their fellow top users, which is especially evident in the case of *ABP* (the number of messages sent by the first and second most active user differs by more than half: sender 1: 1212 messages; sender 2: 507 messages). This pattern is visible in other groups as well, although less starkly in *AAP* (sender 1: 789; sender 2: 570), *KLI* (sender 1: 773; sender 2: 630), *Align*

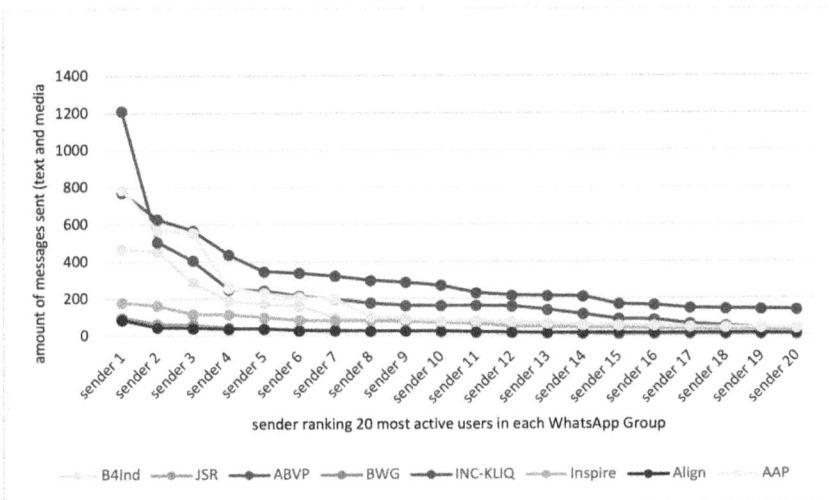

Figure 3.1: The twenty most active users in WhatsApp groups. Source: Author

(sender 1: 89; sender 2: 47), *Inspire* (sender 1: 182; sender 2: 166), and *JSR* (sender 1: 102; sender 2: 66). This illustrates the "long tail" phenomenon in online political groups, especially within extreme speech scenarios, highlighting how discussion communities are sustained largely by the overactivity of a handful of ringleaders with a long tail of relatively underactive followers and signaling the possibility of sponsored or highly motivated actors.

Themes

Text messages and media were analyzed based on eighteen thematic categories, and up to three categories were included in a single message if multiple strong thematic indicators were found. The residual "any other" category included empty messages, advertisements for consumer products and services, notifications of members joining or leaving the group, and others. Thematic categories were mapped across three ideological/political groups (nationalist, Congress, and AAP).

Discussions around politicians were a core thematic strand across all the groups. Prime Minister Narendra Modi was the most widely cited politician (nationalist: 760 times; Congress: 862 times; AAP 415 times). In contrast, mentions of Rahul Gandhi were fewer (nationalist: 242; Congress: 383; AAP 28). Regional politicians also featured across the groups, revealing the locally specific concerns of the groups.

Nationalist WhatsApp groups stood out for their higher prevalence of discussions around religion. More than 10 percent of the messages were about positive portrayals of Hinduism (720), followed by references to Muslims (314 messages). Personal wellness and pleasant greetings were more common in these groups (950 messages) while the Congress groups had 659 messages in this type (a smaller percentage of the total messages within the group compared to nationalist groups). As a percentage of the total number of messages in the group, development topics featured most frequently in *AAP* (153), followed by nationalist (269) and Congress (84) groups. Historical themes were used within nationalist groups (194) and Congress (166) groups, but only sparsely in AAP (38). Party symbols were more common in nationalist groups (197), and so was the common symbolism of "Jai Shree Ram" (chant for the Hindu god). Patriotic themes were prominent in the nationalist group (556), far more than Congress (180) and AAP groups (123). Violent topics were more common in nationalist groups (329), followed by Congress groups (138) and AAP (75).

Extreme Speech

A binary classification for extreme speech that depicted offense, exclusion, or violence revealed that only less than 3 percent of the total messages belonged to this category. The highest occurrences were in the nationalist groups (2–3 percent), while they were less than 1 percent of the total message instances in the other two groups. These messages were further coded for six types of extreme speech (Tables 3.3 and 3.4). In the smaller volume of extreme speech instances in AAP and Congress groups, the messages largely comprised offenses to communities (Congress: 18; AAP: 5). Align, in comparison to other Congress groups, had one instance of extreme speech that was casteist and four instances of calls for an exclusion from the nation or community. AAP also showed one instance of extreme speech that called for violence.

Among nationalist groups, more than 20 percent of extreme speech instances (53) had calls for violence and about 10 percent (27) used acts of violence as a justification for violence. They also had instances of casteist (1) and sexist (3) speech. Aside from offenses to communities (57), calls for exclusion from the nation or communities (72) were common in these groups.

Six different targets of extreme speech were considered for analysis (Tables 3.4). In the case of AAP, extreme speech was largely targeted at right-wing groups (4). Others were toward a general, unmarked audience (2). Similarly, extreme speech within KLI (8) and Inspire (4) groups were directed toward right-wing groups. Extreme speech in the Align group was toward general, unmarked publics (5), with one case of extreme speech toward "pseudoliberals" (a derogatory term for progressive liberals). The most common target group for extreme speech among nationalist groups was Muslims (65 percent), and the remaining messages were targeted at general, unmarked publics (53), "pseudoliberals" (18), and women (2).

The mix of content observed in the groups prompted an inquiry into the temporal flow of different types of content. When content types were plotted sequentially, the graph revealed an interesting pattern of shifting themes and tones, which oscillated between provocative and pleasant content (see Figure 2 to see this pattern in an exemplary nationalist WhatsApp group). Combined with ethnographic interviews, this message flow pattern indicates how party workers, once admitted into WhatsApp groups, would relay party messages in unobtrusive ways, often embellished with jokes, good morning greetings, religious hymns, microlocal municipal issues such as water or electricity supply, and other kinds of socially vetted and existentially relevant content. The temporal flow of such messages—amplified and articulated by ordinary users—is characterized by the sudden appearance of explicitly hateful messages against Muslims in the midst of an otherwise benign sequence of pleasant or "caring" messages. The flow of content thus simulates the lived rhythm of the social.

While one interpretation of the "simulation of the social" points to deliberate attempts at camouflaging the context of extreme speech dissemination, another interpretative frame would be to recognize the normalizing effects it can have upon everyday WhatsApp conversations. There is an attempt to "dilute the context of delivering extreme messages," observed an interlocutor. "That is how you normalize." Considering that just about 3 percent of the total messages on WhatsApp groups in this study were found to be extreme, WhatsApp content mix and temporal flow reveal unobtrusive ways of embedding divisive content by re-creating the familiar worlds of everyday social exchange.

Deep Extreme Speech

Intimate networks of distribution and socially vetted content flows discussed in the preceding sections highlight the specificity of WhatsApp groups as a communication form in shaping the ecosystems of extreme speech and partisan political messaging more broadly. The embedding of this type of content especially within nationalist WhatsApp groups shows how social trust and context camouflaging have emerged as key aspects of disseminating and normalizing contentious content, leading to what I describe as "deep extreme speech."

Deep extreme speech centers community allegiances in distribution logics, framing extreme content as part of the everyday. With distinctive

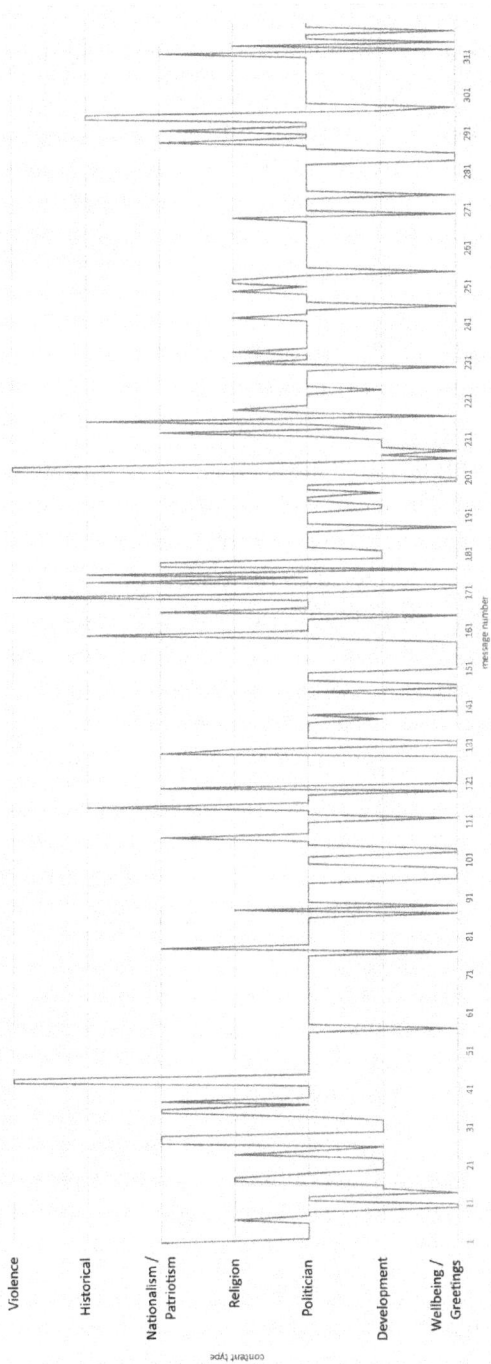

Figure 3.2: Chronology of content in a nationalist WhatsApp group (3,583 messages). Source: Author

distribution patterns, content mix, and temporal flow, such forms have placed extreme speech at the confluence of affect and obligation, thereby delinking it from the impersonal constructions of truthfulness or moral constructions of hatred. In other words, when messages are embedded within personalized, trust-based networks, what ensues is not as much a problem of truth (whether it is true or false) or a problem of morality (whether it is good or bad) but an emotional or obligatory urge to share them and be in (if not with) the flow.

The affective dimension of extreme speech, for instance, is starkly evidenced by the fun cultures of online exchange when people who peddle exclusionary discourse take pleasure and celebrate their collective aggression (Udupa 2019). In the context of deep extreme speech, fun cultures are amplified by social trust and the familiar language of in-group exchange. The aspect of obligation is pronounced in the case of deep extreme speech, as users within kin or kin-like networks feel the need to share and respond to the messages they receive. Any form of inaction on received content conflicts with the sense of obligatory ties and reciprocity that define socially thick networks of deep extreme speech. In such circulatory milieus, responsible action is itself conceived as circulation—the sense that by forwarding the messages one has done one's duty. These social relational dynamics reconfigure trust in extreme speech contexts, as intimate networks become imbued with the intensities of political discussions.

Deep extreme speech that works its way through intimate channels of kin and kin-like relations amid a tactical mix of content might be seen as the social corollary for technologized deep fakes and commercial digital influence services. Analytically distinct but intermingled in practice, these forms are reconfiguring, rather than dismantling, structures of trust in political discourse.

While offering a vital infrastructure for deep extreme speech, WhatsApp communication has ramifications for political discourses beyond the seemingly self-contained worlds of in-group conversation. "There appears to be some sort of coordination across WhatsApp groups of extreme actors since politically relevant contentious content jumps between them in quick time," observed two activists in Bangalore, offering a perspective from their daily navigations of online discussions and interactions with user communities affected by hateful speech.

Moreover, WhatsApp sustains cycles of circulation around specific incidents that would otherwise have disappeared from public memory as a singular incident. "Take the case of a hijab-wearing student who was yelled at by a college lecturer in Karnataka,"[3] said one of my interlocutors. "Someone filmed this chiding, and the video started circulating on WhatsApp. A one-off exposure to politically charged extreme conduct then becomes a prolonged experience for the victim."

Studies have also gathered evidence for the ways in which the ruling political party manipulated social media narratives by using thousands of WhatsApp groups with dispersed volunteers and "loosely affiliated online supporters" to engage in "trending" campaign-friendly hashtags on Twitter (Jakesch et al. 2021). Through such "cross-platform media manipulation" tactics, "hundreds of trends were fabricated" during the 2019 elections (2021). These trends were later picked up by other media outlets, contributing to coordinated amplification of the ruling party's campaign line. By linking content across platforms, therefore, digital influence strategies have sought to evade the limits of encryption and boundaries of closed communities, embedding socially sanctified content derived from WhatsApp communication within multiple streams of digital discourse.

Patterns of deep extreme speech are noticeable in other parts of the Global South, as studies in Bangladesh, Indonesia, Kenya, the Philippines, Brazil, and Turkey, attest (Ong and Cabañes 2018; Saka 2018). In these contexts, large networks of actual, real-world people peddle and amplify animosities online, raising a vast "human infrastructure" for disinformation (Olaniran, this volume). Actors who moderate and lead WhatsApp groups are not highly visible Twitter celebrities but represent smaller nodes that are aplenty and dispersed. During interactions in the AI4Dignity project, a collaborative coding project for AI-assisted content moderation I steered, Gilberto Scofield, a leading fact checker in Brazil, described such community level partisan digital actors as "hyperlocal influencers." They are tasked with "maintaining" about two thousand people on different WhatsApp groups and send out "very useful and local messages," similar to Hindu nationalist moderators

3. Divya Arya, "Karnataka Hijab Controversy Is Polarising Its Classrooms," BBC, February 15, 2022, https://www.bbc.com/news/world-asia-india-60384681.

conveying helpful information on the local water supply and road development. Importantly, such hyperlocal influencers live in the same areas
as targeted communities and have a "good understanding of organized
crime militia and political parties." These "independent guys" also from
the favelas are seen as the "voice of people." The content they provide is
seen as "quality information."

Conclusion

WhatsApp's role as a key conduit for deep extreme speech in diverse
contexts underscores the need for directing regulatory and policy efforts at the manifold impacts of the vastly popular messaging service.
While content moderation is without doubt a significant regulatory
measure but hard to implement for encrypted messaging, the analysis
presented here stresses the need for a renewed focus on networks of
actors and practices as they develop on-ground in closed loops but can
scale with active digital influence strategies. Aside from platform governance measures around design features, moderation, transparency, and
accountability, which also need to be monitored for potential regulatory excess by state agencies as well as regulations in campaign financing, actor-focused measures would involve regulating the practices of
precarious information warriors and data brokers and their role as dispersed amplifiers, curators, and creators of content within WhatsApp
networks.

In a recent concept note on online hate speech, the European Commission (2017) observed that while cooperation between governments
and civil society organizations is necessary, "cooperation of IT companies with civil society also plays a fundamental role in counternarra-
tive campaigns." According to the note, there is evidence that deploying
social media advertising tools to "target audiences" improves awareness
and engagement and leads to a substantial increase of NGO's social
media presence. To this policy direction, it might be added that civil
society and industry should actively involve local WhatsApp groups already constituted by political parties by crafting organic interventions
that can *repurpose* existing groups. Such measures will also be crucial to track evolving tactics of digital influencers, such as the instru-
mentalization of "status update" and other newly introduced features

of WhatsApp, including ways to circumvent the barriers the company has raised in terms of capping the number of forwards and labeling forwarded content by spawning multiple human networks.

Convening self-styled political trolls, local politicians, and commercial digital influencers for awareness-raising activities and sensitizing them about global human rights standards and the dangers of digital campaign manipulations is a necessary step, and so are efforts to strengthen grassroots anti-hate communities to report online extreme speech to social media companies and monitor progress once complaints are raised. Strengthening local communities to petition lawmakers is another important measure, while concurrently supporting local organizations and groups in establishing hate-monitoring dashboards constitutes an equally significant endeavor. In the Global South context, a multimedia strategy is crucial in ways that community trust as the key lubricant for extreme speech is recalibrated through television, radio, and other popular media, and the epistemes of the "WhatsApp University" can remain substantively contested rather than serving as the new normal.

4

Misinformation behind the Scenes

Political Misinformation in WhatsApp Public Groups ahead of the 2022 Constitutional Referendum in Chile

MARCELO SANTOS, JORGE ORTIZ FUENTES, AND
JOÃO GUILHERME BASTOS DOS SANTOS

Studies on misinformation have extensively focused on public figures, fact-checking or "open" social media, but in the Global South, the role and effects of instant messaging apps on political processes and social cohesion is a growingly pressing issue, as this book depicts. This chapter sets out to study the circulation of misinformation on instant messaging during a relevant political event in Chile.

On September 4, 2022, Chileans voted on a referendum to approve or reject the new constitution written by a group of representatives elected for that purpose. In the one-year period between July 2021 and July 2022, the elected legislators worked on the proposal for a new constitution for the country. The final document was finally rejected by over two-thirds of the population. Amidst numerous accusations of disinformation campaigns during the referendum, we ask to what extent messaging apps, in particular WhatsApp, were also a source of misinformation.

Findings point to an unprecedented asymmetry regarding misinformation, as right-leaning WhatsApp groups displayed a disproportionately larger amount and diversity of misinformation. Alongside the magnitude of misinformation, there were morphing dynamics in the most circulated falsehoods, which ultimately led them to spill over to the left-leaning groups. By examining these major episodes, this chapter will discuss the implications of WhatsApp growth for civic and political life and will conclude by outlining the limitations of the study.

The Problem of "Uncheckable Content" in Misinformation Research

It has been established how difficult it could be to find and discern quality sources of information on politics as it gets distributed and shared (but also forged and altered) in digital media and social networks (Lewandowsky, Ecker, and Cook 2017). Newman et al. (2022) estimate that 32% of the Chilean population use WhatsApp as a news source. Considering this high rate of use, it is worth looking at what is being discussed about politics in WhatsApp groups and how misinformation coexists with interpersonal communication on those platforms.

WhatsApp was originally designed as a messaging application, intended to be used for interpersonal communication as an alternative to SMS, but over time it was equipped with new functionalities and sociotechnical connotations. Since the platform is more opaque than other social networking services, the task of detecting misinformation becomes more complex compared to services such as Twitter or Facebook, and the circulation of uncheckable content may go unnoticed by traditional information watchdogs such as journalists and fact checkers (see Mare and Munoriyarwa, this volume). In fact, it is in instant messaging applications where the least is known about the phenomenon of misinformation (Kligler-Vilenchik 2021).

In Chile, mobile penetration was 137 percent in 2021 (Data.ai 2022), while WhatsApp's penetration surpassed 90 percent (Pascual 2021). The messaging app kept growing as it was the most downloaded app in the country in 2021 (We Are Social 2021, 30), and the mobile app with the greatest number of active users during the same year (Data.ai 2022). Commercial agreements that allow people to use certain apps like WhatsApp without incurring data charges—a practice commonly referred to as "zero rating"—help explain its widespread use and traversal penetration in the country and the Latin American region more broadly. Yet, literature on politics and WhatsApp is extremely scarce in Chile and the existing studies rely on self-reports via surveys to inquire on the orbit of political behaviors and effects that its usage may entail (Scherman et al. 2022; Valenzuela, Bachmann, and Bargsted 2021).

Though there are many obstacles to researching encrypted messaging apps based on their content due to methodological and ethical difficulties (Barbosa and Milan 2019), one technique to observe what happens on this platform is monitoring public groups that discuss politics (Garimella and Eckles 2020; Reis et al. 2020). This chapter advances this approach by analyzing two different moments in a set of over three hundred public WhatsApp political chat groups in the two distinct periods prior to the historic referendum in Chile. We set out to explore what was going on behind the scenes—i.e., conversation and information dynamics—in these messaging groups, addressing one critical node in the misinformation phenomenon.

Misinformation in opaque environments cannot be treated in the same way as other digital social networks. In many cases, fact-checking is not possible since forms of discourse such as future predictions, conspiracy theories, and similar assertions based on unrevealed sources go beyond the parameters of "fact-checkability." Uncheckable content is a serious problem in a post-truth environment amid hybrid media platforms (Suiter 2016), where "alternative" facts are wrongfully balanced with "factual" facts. This is a more acute problem in opaque environments such as WhatsApp group conversations because professional fact checkers encounter difficulties in authentication and validation, and the resources accessible for general users to scrutinize and verify are even more limited.

The discussion on unverifiable type of content is largely absent in available literature, partly as a result of how misinformation is usually operationalized in extant inquiries (Guess and Lyons 2020). It is common for surveys, for instance, to display a set of false claims and ask respondents whether they were exposed to such content, whether they have shared it or believed it (e.g., Valenzuela, Muñiz, and Santos 2022). This procedure has minor variations such as including "placebo" true content to test response bias (Allcott and Gentzkow 2017). In addition, scholarly work to identify confirmation bias shows that people might tend to believe in false content that reinforces their previous beliefs on controversial issues, for example regarding climate change (Zhou and Shen 2022). Uncheckable content has received less attention, and we suggest it should receive greater scholarly focus. In the context of messaging apps where uncheckable messages may also carry misleading information, we will use the term "misinformation" to refer to the information that is verifiably false, incorrect, or dubious, as validated by the

International Fact Checkers Network (IFCN)–accredited fact-checking agency Fast Check CL in Chile; but we also include types of content that were coded as uncheckable by fact-checkers.

Chilean Constitutional Referendum

In the current study, we focus on public groups that discussed the Chilean referendum for a new constitution, which posed the question, "Do you accept the newly redacted constitution?" and a binary response option: yes/no. The referendum took place on September 4, 2022, when Chileans rejected the new *carta magna* with almost two-thirds of the votes.[1] The new proposal contained strong support for the enhancement of social rights, but widespread concerns around provisions such as higher environmental protection, interpretation of private property rights, and plurinationalism, among other factors, ultimately lead to its rejection at the ballot. The referendum was the end of a process that began on October 19, 2019 with the most important social uprising in the recent history of Chile, the *estallido social* (Salazar 2019). After months of clashes with the police and amid hundreds of accusations of human rights violations, politicians from almost all parties in Chile signed the National Peace and New Constitution Agreement on November 15, 2019 to kick off the new constitutional process (Senado.cl 2019).

In the binary ballot, accusations of information manipulation were abundant (Molina 2022). The diffusion of misinformation related either to the constitutional process, to its representatives, or the content of the new *carta magna* happened almost exclusively via social media (Santos and Plataforma Telar 2022). It is reasonable to infer that less scrutinized communication environments such as instant messaging would also be instrumentalized as channels to seed misinformation. Such scenarios led us to the following research questions:

RQ1: To what extent were users of Chilean WhatsApp public political groups exposed to misinformation about the 2022 constitutional referendum?

1. Equipo Emol (2022). "Resultados Plebiscito Constitucional 2022—Una cobertura especial de Emol.com." Emol.com. https://www.emol.com/especiales/2022/nacional/plebiscito-salida/resultados.asp#19001

RQ2: What kind of misinformation circulated via WhatsApp in Chile regarding the 2022 referendum?

RQ3: How did misinformation spread from one cluster to another on the network of WhatsApp groups?

Method

This study employs a sequential mixed method approach to analyze data gathered from public WhatsApp groups.[2] The systematic aggregation of groups commenced during the 2021 presidential campaign in Chile. This involved collating publicly available groups from social media platforms such as Twitter and Facebook, as well as from WhatsApp directories. The method further built on snowball sampling: as more groups were identified within the original dataset, they were subsequently incorporated into the collection.

In the first step, we coded the data using content analysis, with the main purpose of identifying misinformation and other types of political messages. The messages coded by the researchers as misinformation were then submitted to IFCN-accredited fact-checking agency Fast Check CL for analysis.

The enriched database was scrutinized with a mix of computational methods including visualization with complex network analysis, an approach that uses algorithms based on network structure to analyze viral dynamics of relevant messages (such as the most repeated misinformation messages) and how its topology favors specific actors and information flows. The results show the dynamics of the flow of the most viral misinformation messages, contrasting both networks: one that constitutes all the groups for the *apruebo* option (approve the new constitution) and those for the *rechazo* option (reject the new constitution).

Datasets

This work builds on two datasets extracted from the collection of groups. The first dataset is from May 2022 (*May dataset*), three months

2. As defined by Eckles and Garimella (2020): "Any group on WhatsApp which can be joined using a publicly available link is considered a public group" (p. 7).

before the referendum. For the *May dataset*, we collected messages from 464 public WhatsApp groups, usually themed around the referendum, either for (*apruebo*) or against (*rechazo*) the new constitution. These groups were the organic evolution of the right-wing presidential candidate's groups into pro-*rechazo* (reject the constitution proposal) and, conversely, the left-wing candidate's groups turned into pro-*apruebo* (approve the constitution proposal). Such a dataset has two purposes: we used it to develop the code for content analysis and to carry out the intercoder reliability tests (ICR), but also to understand if there were any traces of persistent misinformation on the data that would be present in the weeks prior to the referendum.

The main dataset (*referendum dataset*) was built on the two weeks before the referendum (between August 22 and September 5, inclusive), considering that during this period conversations would be likely to revolve around the upcoming major political event. The referendum dataset resulted in 619 messages with more than four repetitions across the groups.

Content Analysis

During a preliminary analysis, we reviewed the most repeated messages during May 2022 as a way to test if there was misinformation in the data, and if they went viral to some degree. To achieve this, we applied two filtering criteria to our dataset. Firstly, we excluded all messages that contained less than thirteen words to avoid simply emotional or phatic messages, as it was highly unlikely that these short messages contained relevant falsehoods. Secondly, we grouped messages according to their Levenshtein distance, using a threshold of 98 or higher. Levenshtein distance is a metric used to calculate the difference between two strings of characters based on the minimum number of insertions, deletions, or substitutions required to transform one string into the other (Levenshtein 1966). By applying these filtering criteria, we aimed to exclude messages that were less likely to be repeated and at the same time identify similar or identical messages, grouping them based on their semantic similarity rather than their exact wording, which may vary due to misspellings, typos, linguistic variations, or insertion of short commentaries and emojis.

Table 4.1: Types of political information

Label	Description
Creative	Content that builds on creativity to express political points (memes, jokes, satire, poetry, and so on)
Incivility	Content characterized by rudeness, name calling, cursing, and other forms of toxic speech directed toward either a person, an organization, or a group of people
Information	Informational content about politics with no falsehoods, such as data, links, explanations, and so on
Misinformation	Content that uses factual content that is not completely truthful. Includes conspiracy theories, fake, imprecise, or uncheckable content
Mobilization	Content with an explicit call to action
Opinion	Content that expresses the user's opinion or point of view, with no facts to support it

The result was a corpus of 178 unique messages, repeated across groups from four (minimum) to forty-three (maximum) times. We then coded the messages: first between "political" (presence of either political figures, themes, or processes) and "nonpolitical" and subsequently coded on a second level only the subset of political messages (eighty-seven, or 49 percent of the messages). Though the main dataset is temporally the one closest to the referendum, we also offer an analysis of the *May dataset* to help understand continuities and discontinuities from one period to the next.

To perform the ICR, two of the authors coded sixty-two messages from the May dataset (10 percent of the referendum dataset) on two stages: first, discerning political from nonpolitical messages, then coding for referendum-related (yes/no) and type of political messages according to a set of categories that emerged bottom-up as we got familiar with the data (see Table 4.1). The overall agreement calculated with ReCal (Freelon 2010) was 85.5 percent and the average Krippendorf's alpha for both sets of variables was 0.78.

By mid-August, there were 427 groups still active from the previous sample. Due to the low levels of political messages in the first round of analysis (circa 7 percent without the filter on the group level described in what follows), we decided to run a previous filter at the group level, realizing many of the groups had been transformed into nonpolitical groups (such as religious, sales, regional, or thematic news and others).

We randomly selected the last one hundred messages from each group from the day of the referendum backward and filtered all groups that did not display any message with political connotations within the sample, resulting in 315 groups (74 percent of the initial sample). The outcome was much more efficient, for the proportion of political messages went from less than 7 percent on the *May dataset* to 88 percent on the *referendum dataset* ($N=545$ out of 618 total messages). Of those, 90 percent were about the referendum ($N=491$).

Social Network Analysis (SNA)

Social network analysis (SNA) involves a structural analysis based mainly on the topology of networks and its relationship with processes happening on the local scale. Its application in this study reveals where users stand in the network,[3] who they are connected to, and how the groups are or are not interconnected. With the content and network analysis over time, a clearer picture of how the misinformation spreads and contaminates other groups emerges.

To perform the SNA, users and groups were adopted as knots and the messages as edges in a bipartite network, where some actors connect one group to the other because they participate in both (Dos Santos et al. 2019). Additionally, each message connects a user to a group and thus creates the networks. SNA is used here to calculate and visualize the network structure as to which user or group is more relevant for the dissemination of misinformation and/or to understand dynamic patterns of dissemination over time.

Results and Discussion

May Dataset

The prevalent form of political messages was neutral, journalistic, or journalistic-like information: 34 messages, 39 percent of the sample of political messages (see Figure 4.1 for details). They referred to events,

---

3. Individuals on a network may or may not hold positions of centrality, belong together to a certain group (cluster), bridge different groups, and so on. This is analyzed mainly with graph interpretation and network metrics of both.

Types of Political Information
(May dataset)

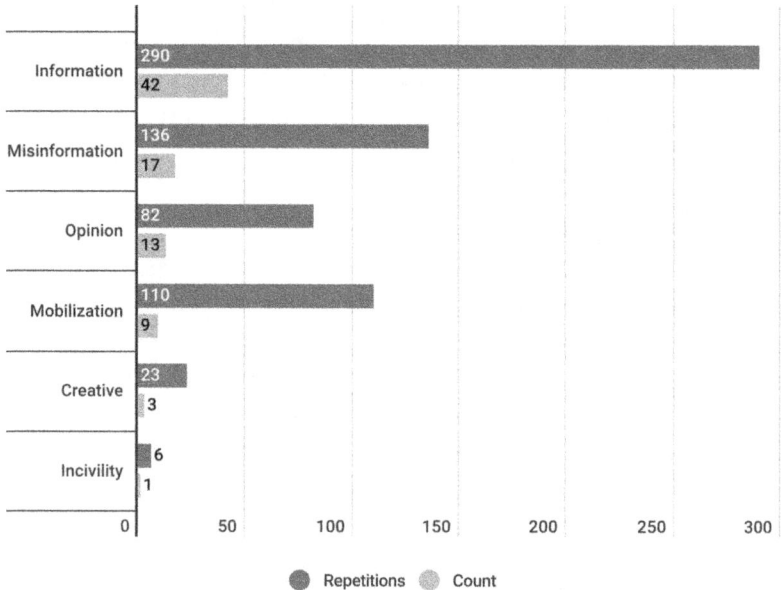

Figure 4.1: Characterization and magnitude of the political messages (total repetitions across groups and unique different messages) within the collected messages of public WhatsApp groups during May 2022. Source: Authors

news, or other informative messages deemed relevant to the group by the user. Next were messages coded as mobilization ($N = 24$; 28 percent), opinion ($N = 99$; 18 percent), and misinformation ($N = 10$; 11 percent). Two messages were creative manifestations of political views such as jokes and memes, and one was an uncivil message.

Political Misinformation Asymmetry

Though the collection dates from over three months before the referendum, the dataset displayed high prevalence of misinformation. Additionally, 100 percent of the misinformation detected was circulating in groups supportive of rejecting the new constitution in a clear display of political asymmetry. The most popular misinformation was a fraud accusation, probably modeled after Trump's refusal to accept the result of the ballot in 2020 in the United States, which was also copied by former president Jair Bolsonaro's supporters in Brazil (Rocha 2022). While in the United States, the incitements led to the Capitol riots (Leatherby et al. 2021), in Brazil things got worse, as thousands of followers of the losing party, far-right former president Jair Bolsonaro, invaded the capital's official buildings during the summer recess, in what has been labeled symbolically as terrorist attacks (BBC News 2023) with the accusation of coup intent (Kirchgaessner et al. 2023). This pattern held a mirror to misinformation strategies used by the extreme right in the Americas, as similar rumors have circulated in the last ballots in Argentina, Colombia, Mexico, and Perú (Fast Check CL 2020), all countries where left-leaning candidates won the last ballot. In the Chilean case, it could be argued that this persistent disinformation campaign, i.e., misleading content created and/or disseminated intentionally, was aimed at convincing the voters that in the eventuality of a defeat, they would suspect the electoral system and the role of some variation of a fraud for the outcome. Such a scenario did not occur since the proposal was rejected.

The Viral One

From the political messages coded as misinformation, nearly half were about the referendum or the constitutional process, all of which were *against* the new constitution. This points to what could be perhaps an unprecedented misinformation asymmetry in a binary electoral process. Though for this dataset we do not have the political orientation of the group that created such content or started such conversation threads, the relevant fact is that the political inclination of the misinformation messages is explicit, directed toward mobilizing fears in the *rechazo* supporters. As we show in the next section, during the weeks prior to the election, the same fraud hoax took a more general form—more verisimilar—thus flooding the other side of the political spectrum.

By far, the most viral message was a conspiracy theory, claiming there would be an electoral fraud by the supporters of the *apruebo*. With nearly fifty repetitions among different or repeated groups, the six hundred-word message represents half of the diversity of misinformation and had an accurate description, with minor variations, of what

URGENTE

Aunque el rechazo llegue a más del 90%, IGUAL VA A GANAR LA APROBACIÓN, no sean ingenuos.

HAY QUE TRABAJAR ESTO AHORA con temas publicidad $$$,
campañas
VIENE ....

Hay un fraude monumental preparado.
Estos tipos no han improvisado.
Todo el show que muestran y que parecen payasadas, son solo distractores.
Esto viene preparándose hace muchísimo tiempo y no está en sus planes perder.
Los que nos hemos visto siempre anticipados y sorprendidos somos nosotros.

Analiza la votación del triunfo de Borić.
"Aparecieron" de la nada un millón y medio de votos "nuevos".

Eso no ocurre espontáneamente.
Son "milicias" muy disciplinadas.
Listas para actuar y entrar en acción cuando se les da la orden.
El fraude es imposible de parar, a menos de que se intervengan el Registro Civil y el SERVEL.

Figure 4.2: Reproduction of part of the most circulated political message in the sample. Source: Authors with data extracted from public WhatsApp groups

would be an electoral fraud to come (Figure 4.2). This is equivalent to half the volume of total misinformation posts detected around the referendum. An excerpt from the message reads as follows:

URGENT

Even if *rechazo*[4] reaches more than 90% of the vote, *APRUEBO* WILL WIN, don't be naive.

(...)

There is a monumental fraud in course

These guys have not improvised.

All the spectacle that looks like clowning, are just a distraction.

This has been preparing for a long time and it is not in their plans to lose.

---

4. *Rechazo* stands for rejection of the new constitution. The opposite is *apruebo*, which stands for "I approve."

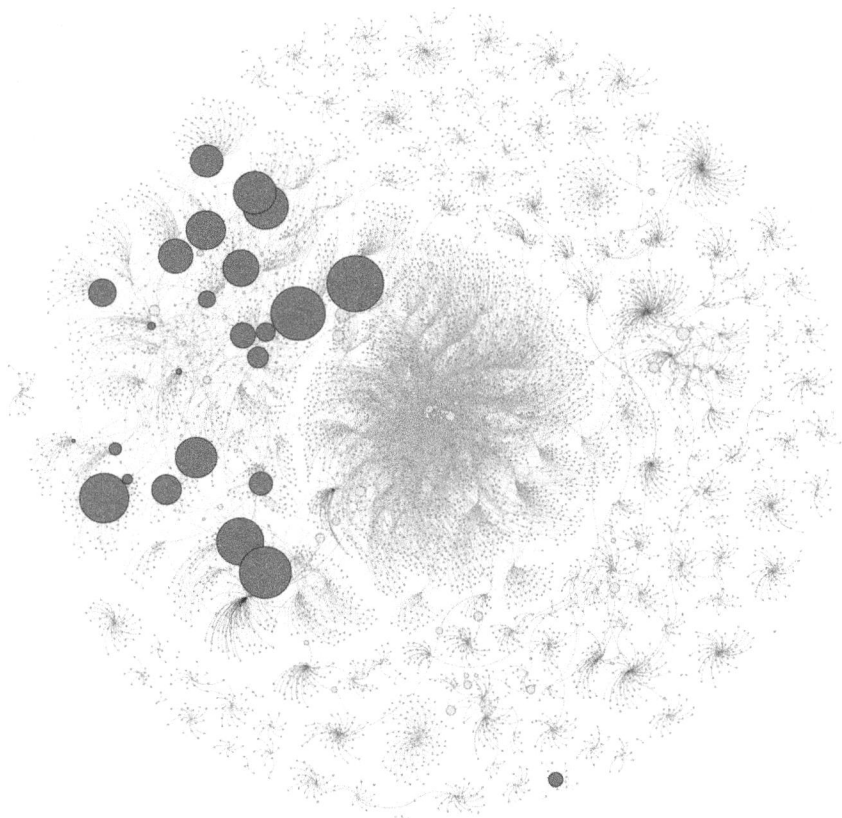

Figure 4.3: Contagion of the network with the misinformation about electoral fraud. On the left side, the cluster of *rechazo* groups, while to the left of the central cluster the *apruebo* groups are not contaminated. In the central cluster, political groups that ceased to discuss political matters. For a version that distinguishes the clusters by color, see https://bit.ly/NYUP_Fig03. Source: Authors

CHILEANS, LET'S ACT NOW, TOMORROW WHEN WE ARE UNDER CHILEZUELA,[5] WE WILL ONLY CRY FOR DOING NOTHING AND GIVING OUR COUNTRY AWAY

A portrait of the diffusion of the message can be visualized in Figure 4.3, where some observations about the network structure can also

---

5. Neologism that unveils a supposed threat to turn Chile into Venezuela, a common way to scare anti-communists, anti-Chavists, and anti-socialists.

be inferred. The graph was created with ARS software Gephi and Force Atlas 2 algorithm, a force-directed layout where "nodes repulse each other like charged particles, while edges attract their nodes, like springs" (Jacomy et al. 2014, 2), until it reaches some stability. In the resulting graph, the *rechazo* network (concentrated on the left side of the graph) is closely knitted, as the different micro-networks (WhatsApp groups) are close and more connected to each other, i.e., users take part in more than one group, acting as bridges between groups. On the other hand, the *apruebo* network is more distributed. The central network is groups that were created during the 2021 presidential elections but ceased to be political and began to have other functions such as religion, sales, and others, showing that groups not directly dedicated to politics can acquire or maintain a central position inside political debates. The large dots represent the groups contaminated with the fraud hoax. As previously stated, the image reflects the asymmetry: all the contaminated nodes belong to the *rechazo* network, on the left side of the graph. The inverted size of the node represents the order they were contaminated; that is, the larger the node, the earlier the network was infected.

Referendum Dataset

Since this dataset is adjacent to the referendum, there are more messages around this political event. The dataset was coded following the same categories, and the analysis is based only on the political messages related to the referendum ($N = 491$). Expectedly, 44 percent of the messages related to the referendum were mobilization content, mainly related to the previous week's campaigning and recruiting of poll observers[6] ($N = 216$, see Figure 4.4 for complete descriptive data). This is not unrelated to the misinformation issue because the importance assigned to the recruitment of poll observers by both sides of the referendum is related to the perceived threats of electoral fraud, which constituted the most popular hoax in the data. The following message illustrates the relationship:

---

6. Supporters of one party or in this case, option, that act as observers during the voting process to ensure the legitimacy of the process.

Types of Political Information

(Referendum dataset)

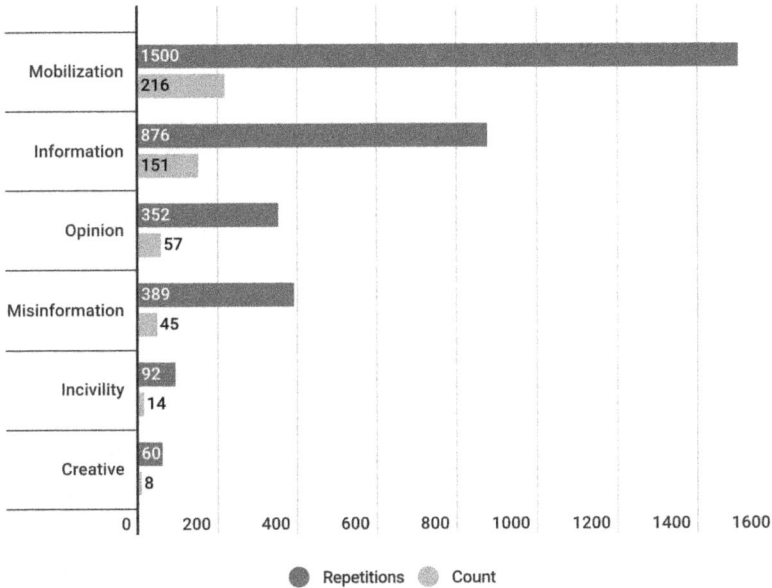

Figure 4.4: Characterization and magnitude of the political messages within the collected messages of public WhatsApp groups two weeks prior to the September 4 referendum. Source: Authors

If there are no poll observers . . . we will allow communism to carry out the fraud intended without them.

Seriously, YOU, w ur arms crossed, watching, seeing how freedom escapes from our hands when the communist yoke is imposed, due to the lack of observers .. YOU WILL STAY HOME AND BE A PASSIVE AGENT OF COMMUNISM THIS SEPTEMBER 4TH. WILL YOU GIVE YOUR FREEDOM AND THOSE OF YOUR CHILDREN TO MARXISM?????

(our translation)

Such messages sought to instrumentalize the established fraud hoax to mobilize poll observers. One important difference with the fraud theory in this dataset, though, is that this time it reached both sides of the political spectrum, *rechazo* and *apruebo*. Another recurring mobilizing message was to organize a gigantic human flag by the supporters of the *rechazo*, which indeed was successful.

Next, there was informative content with 31 percent ($N=151$), followed by opinion ($N=57$, 12 percent) and then misinformation ($N=45$, 9 percent). Some messages ($N=14$, 3 percent) were incivil, just venting out emotions or negative emotion against individuals or organizations on the other side of the political spectrum. Finally, 2 percent were "creatives" using images and artful depictions in political messaging ($N=8$).

Political Misinformation Asymmetry

As explained earlier, in the *referendum dataset* we did a more thorough analysis, previously coding the groups according to their political orientation (*apruebo* or *rechazo*) or other characteristics (news, sales, etc.). This process allowed us to more accurately identify a possible asymmetry between groups in terms of exposure to misinformation. Though there are twice as many *apruebo* groups than *rechazo* ones in the sample (193 against 99), the exposure to misinformation (number of groups exposed to misinformation) and diversity of misinformation circulated (unique content coded as misinformation to which each cluster of sympathizers were exposed to) in the *rechazo* groups is strikingly higher. Adding each exposition of *rechazo* groups to misinformation, the sum is 278 while the same operation adds up only to 124 among *apruebo* groups. This means that if one belonged to a *rechazo* group, they are about four times more likely to be exposed to misinformation.

Furthermore, there was a diversity of misinformation content that circulated within each respective network. While *apruebo* groups were exposed to eleven distinct instances of misinformation, *rechazo* groups were exposed to a total of forty unique instances of misinformation. Of those, eight were common to both political clusters, meaning there were only three instances that "belonged" exclusively to the left, while thirty-two circulated exclusively on right-leaning groups.

The Viral One(s)

The fraud accusation is by far the most circulated misinformation in the data. Nine different variations were detected in a span of little more than twenty-four hours, less than two days prior to the referendum, contaminating seventy-eight different groups. While in May there was a blunt accusation against the left and the *apruebo* supporters, this version, with all its variations, was not directed toward the left, and was worded in more refined and indirect expressions—a tactic that helps explain the spillover.

> Keep an eye on this . . .
> VOTE FRAUD:
> I will try to explain in simple terms the fraud that is done with the ballot, after voting. After voting and when you come to deliver the vote, the procedure is:
> 1) the vote is handed out to one of the assistants
> 2) the assistant cuts out the serial number and delivers it to another assistant who writes it down in their book, while the one with the vote replaces it by one previously marked (THE CHANGE), but you don't realize because they have you engulfed with the serial number annotation.
> 3) When you come back for "your" vote, it has been replaced.
> How to avoid that . . . ???
> Tear off yourselves the serial number and give it to the assistant, instead of giving the whole ballot. Then you make sure the vote that goes into the ballot box is yours.
> You have to pay a lot of attention in this procedure

This hoax, the most viral content in the dataset, also reinforces the perception of asymmetry observed in the whole dataset and in the May dataset. The message was repeated 146 times, but 100 of them were in the *rechazo* groups, accounting for more than two-thirds of the repetitions. Considering again that the sample has twice as many *apruebo* groups, this is yet another display of the political asymmetry in the case and the datasets studied.

Figure 4.5 displays the contagion dynamics: the more visible knots are those contaminated, and the larger ones are those who first shared the fake message. The preponderance of the *rechazo* groups, clustered in the center of the graph, is evident, especially during the initial spread (larger knots), as shown in the graph. The twenty-first spread—i.e., the twenty-first time the content is published in the WhatsApp groups over

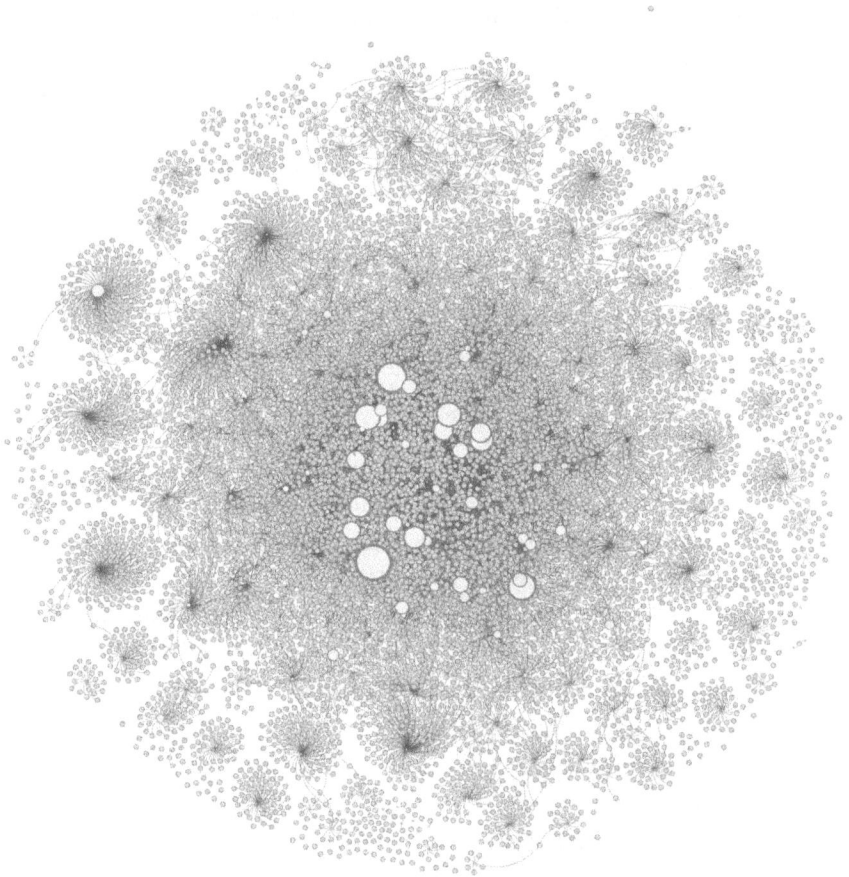

Figure 4.5: Contagion of the referendum network with the misinformation about electoral fraud. The *rechazo* groups are concentrated in the center, while the *apruebo* groups are spread out to the margins of the network. For a version that distinguishes the clusters by color, see https://bit.ly/NYUP_Fig05. Source: Authors

time—is when it reaches one of the *apruebo* groups spread around the edges of the graph, as in the upper left corner. It should be noted that there is also an early message on the *apruebo* network (second message that mentions the hoax) that alerts users that this is a strategy from the *rechazo* campaign to "delegitimize the elections," an attempt to control the spread of the hoax by supporters of the *apruebo* option. It is also relevant to note that messages coded as misinformation were reproduced, on average, 8.6 times. In other words, a single misinformation message was repeated in the same or in other groups almost nine times (repetitions divided by counts). Following in the ranking are creative messages with 7.5 repetitions per message. In the *May dataset*, this relationship was inverted, as creative messages had a repetition ratio of 8.1, while misinformation was placed second with 7.9.

Conclusion

There is no doubt that WhatsApp is a means for political participation and the data from this study provides further empirical support for this observation. As the date of the Chilean referendum to vote for the new constitution approached, there was more overall activity and many more political messages were found in the second dataset, both from a relative and an absolute point of view. What is concerning, however, is the quality of information that circulates in these networks: while most of the messages in the May dataset were to inform and in the referendum dataset they were aimed at mobilizing, in both the cases, the high volume of misinformation warrants critical scrutiny.

Though there was misinformation on both the networks and some are shared on both political clusters, there are two important aspects that differentiate them: first, the relative volume of misinformation, even in the case when it contaminates both sides of the political spectrum, is disproportionately asymmetric; second, the directionality is also asymmetrical, for there is evidence to imply that one network (*rechazo*) contaminated the other (*apruebo*). We have detected two cases where misinformation starts on the right-wing groups (*rechazo*) and spills over to the left-wing ones (*apruebo*).

The first is the accusation of electoral fraud, which was detected as the main instance of misinformation circulating in *rechazo* groups during May (Figure 4.2). Nevertheless, the message grew in sophistication and in September it appeared on both groups (though predominantly on the *rechazo* ones), as the message topped having an embedded political orientation (e.g., "The *apruebo* will cheat!") and became more general (e.g., "There will be fraud!"). It suggests that belief in conspiracy theories could be, to some extent, contagious even between opposing ideologies, even though we found one message on the *apruebo* groups that calls it a "trap made up by the right-wing." Overall, though, the spillover effect speaks poorly of the trust in the institutions that control the electoral processes—though there has been no evidence of important failures since the restoration of democracy in 1990—and such lack of trust could be fatal for democracy, as we have witnessed recently with the Capitol riots in the United States on January 6, 2021, and the terrorist attacks on Brazilian main governmental institutions two years later on January 8, 2023. Both the attacks were from right-wing extremists, and it would be important to monitor diverse kinds of extremism to understand how they engage misinformation that circulates in their networks, in particular those that conspire against democratic institutions or democracy as a whole. Either way, the symptom is evident: the crisis of democratic institutions is a matter of urgency.

A brief perusal of Twitter posts in the same period observed in this study shows a series of similar conspiracy theories around the idea of fraud, again inclined to right-wing supporters of *rechazo*, but not exclusive to the group. This indicates the opportunity and need to extend this inquiry to incorporate cross-platform research, although the question around the genesis of this kind of rumor remains a challenging area of inquiry. Cross-platform research aided by computational methods to identify similar texts on a timeline perspective could help further shed light on this complex problem and advance analysis, including ways to trace the origins of conspiracy theories and other misinformation variations within closed (e.g., WhatsApp) or open (e.g., Twitter) networks.

Another pressing issue that this study triggers is the urgency to measure the effects of persistent exposure to misleading information: it is not the same to propagate a hoax for a couple of weeks as it is to persist over a longer period on the same issue, as the present example illustrates. This persistence was not only evident in the specific fraud accusation case we examined but was also observed during the whole constitutional process, as accusations of laziness of the constituent representatives and political parties and different defamatory messages aimed to strengthen the perception that the constitutional process was not trustworthy. More than an organic or social phenomenon, available reports from other countries suggest a deliberate tactic that is becoming more frequent (Fast Check CL 2020). Since the constitution was rejected, the magnitude of the effects on voter decision resulting from persistent misinformation campaigns against the constitution, such as those detected in this study, is a topic for open discussion and merits further inquiry.

The most viral messages in terms of the messages that spread, on average, to more groups, were those coded as humor or as misinformation. This prompts serious evaluation of the implications of WhatsApp as a source of political information. It is noteworthy that the content related to the fraud hoax exhibited a greater degree of variation and repetition compared to other messages. This suggests that users were not only politically triggered but also influenced by an anxiety over the potential victory of the opposing faction within the context of an extremely polarized electoral event.

This work takes one more step into the literature on misinformation and social networks, specifically via the use of messaging apps in the Global South in a relevant political process such as the 2022 Chilean referendum. These results, however, do not imply that all WhatsApp political groups are plagued with misinformation. Nor do we aim to imply that these findings are generalizable in other ways. Yet it serves as a serious caution for the upkeep of democracy. After all, political participation is as legitimate as the sources of information it stands on.

Funding

The second author received support from Chile's National Research and Development Agency (ANID) with a grant from the Millennium Science Initiative Program (grant ICN17_002).

Acknowledgments

The fact-checking was performed by the professional IFCN-accredited organization Fast Check CL, and the authors thank specially Fabián Padilla for his valuable contribution.

PART II

(Un)safe Spaces

5

Delete This Message

Media Practices of Anglophone Cameroonian WhatsApp Users in the Face of Counterterrorism

KIM SCHUMANN

Members of Cameroon's Anglophone minority are aware that, in the context of the so-called Anglophone Crisis, they have been constructed as a risk to national security. To protect themselves from suspicious security forces and accusations of terrorism, many social media users have adapted their practices and withdrawn into private WhatsApp chats, but even here their conversations are not completely safe. This chapter shows the considerations and coping strategies of Anglophone civilians in dealing with invasive policing and discusses how needing to hide and erase digital content leads to nontransparent flows of information and further alienation of a suspect community from the state.

In May 2022, I was sitting on a balcony in Yaoundé, the capital of Cameroon, with a colleague's relative who was visiting from the English-speaking North-West Region. For over five years, her home region, as well as the likewise Anglophone South-West, have been the scene of an armed conflict between the Cameroonian state and armed groups affiliated with a separatist movement that seeks to attain independence for the territory of former British Southern Cameroons. While we were talking about cultural differences between Anglophone and Francophone Cameroonians, she received WhatsApp messages from her cousin, seemingly containing videos of arson attacks that had occurred in the northwestern city of Bamenda, perpetrated, according to her, by the Cameroonian military. She turned her phone around to show and explain the videos to me before rewatching parts of them herself, shaking her head, and deleting all related messages from the chat. She said she did not "want to see this anymore" but based on everything I already knew about the context, she also did not want anyone else to see this on her phone and infer any separatist political leanings.

In the following sections, I will describe the perspectives, experiences, and coping strategies of Anglophone Cameroonians regarding invasions of their private correspondences which, although illegal, are common under the state's counterterrorism approach to managing the so-called Anglophone Crisis. Under the threat of state surveillance and accusations of separatist terrorism, many users withdraw from more public forms of digital communication into encrypted WhatsApp chats. But manual searches that circumvent encryption require these users to go much further in their practices to safeguard secrecy. Users avoid politically charged WhatsApp groups, are highly selective about who they share information and opinions with, and delete or hide sensitive content. I argue that the need for self-censorship and feeling categorized as suspects by default stokes preexisting frustrations among Anglophones and contributes to their alienation from the Cameroonian state. At the same time, the restricted communication patterns have an impact on the public discourse about the Anglophone Crisis, which might reduce visible government-opposition and mass mobilization but does not contribute to resolving the grievances fueling the conflict.

This chapter is an offshoot of my PhD project on the Southern Cameroons independence movement, for which I conducted hybrid ethnographic research in Cameroon, Germany, and various online spaces from 2021 to 2023. Most referenced encounters happened in Cameroon in December 2022 and January 2023, but the issues discussed came up as recurring topics during an earlier field stay in 2022.

The Anglophone Crisis

As briefly mentioned in the introduction, the Anglophone Crisis is a conflict between the Cameroonian state and armed separatist groups. These groups are the militant flank of a broader separatist movement that seeks to attain independence for the territory of former British Southern Cameroons. The territory became part of the otherwise Francophone Republique du Cameroun through a plebiscite in 1961 which, among Anglophones, is controversial to this day. Within eleven years after reunification, the then-president Ahmadou Ahidjo turned the promised decentralized federation into a centralized one-party state (Konings and Nyamnjoh 1997). For decades, Anglophones have accused the Francophone-dominated state of marginalization and attempts at assimilation, stoking several protest waves with reformist and separatist factions that, despite violent repression from the government, achieved some concessions. These include a return to multiparty politics, the foundation of the first Anglophone university, and the appointment of more Anglophone ministers. Still, all these measures seemed like symbolic appeasement to many government-critical Anglophones. Although the 2000s and early 2010s were comparatively quiet, grievances remained, and separatist ideas kept being discussed in movement organizations like the Southern Cameroons National Council (Konings and Nyamnjoh 2019).

In 2016, several trade and student union protests were violently shut down by state security forces (Mougoué 2017; Pommerolle and Heungoup 2017). Descriptions and videos of state forces brutalizing protestors spread through social media, igniting outrage and solidarity, which led the government to impose a three-month-long shutdown of internet access in the Anglophone regions (Gwagwa 2018). This did not stop the escalation of violence between state forces and groups of Anglophone youths who were arming themselves, first with rubber guns and hunting rifles, later with modern weapons stolen from the military or smuggled in through the hinterland.

Although the initial protests did not center separatist demands, many protestors, including some prominent actors, were separatists. Following the state's aggressive reaction, their stance attracted broad sympathy, opening the way for separatist organizations to lay claims to governance of Southern Cameroons and frame the developing guerrilla conflict in terms of a struggle for liberation (Pelican 2022). From the government's perspective, on the other hand, the separatist movement is terrorism in nature.

Social Media in Cameroon

To set the policing of social media into perspective, I will provide an overview of social media use in Cameroon. Cameroon's internet penetration rate has been booming in recent years but, at 34 percent, it is still among the lowest in the world (Datareportal 2021). Internet users are concentrated in the urban middle class, where having a smartphone and access to mobile data is the norm, whereas "nonsmart" mobile phones remain common in rural areas and among less affluent users.

While Facebook has around four million Cameroonian users (equating to 23.7 percent of the population), other social media apps like Instagram or Twitter only have user counts in the hundreds of thousands (Datareportal 2021). The Cameroonian NGO #DefyHateNow (2020) assumes WhatsApp to be the second most popular app after Facebook but, based on my observations in the Anglophone regions, WhatsApp seems to be the main communication app whereas Facebook is only used to scroll through entertainment and news. Either way, the boundary between platforms is permeable as Facebook posts and screenshots are often shared to WhatsApp chats and statuses. Thus, when using the term "social media" in this chapter, I am referring to WhatsApp and Facebook being used in combination.

Despite its moderate adoption, as Awondo (in Miller et al. 2021, 152–153) observes, social media has opened a new space of public discourse in Cameroon in which users feel it is their civic duty to have an opinion on current matters. The Cameroonian government has long been uncomfortable with this situation and subsequently tried various strategies of guiding or suppressing political online discourse. In 2009, the then-government spokesperson Issa Tchiroma made a public statement calling diaspora members in France "cyberterrorists" for announcing a protest critical of the Cameroonian government (Tande 2011a). In 2011, the government made a short-lived attempt at banning MTN's Twitter via SMS service for allegedly posing a risk to national security (Tande 2011b). In 2016, three teenage boys were sentenced to ten years in prison by a military court for "nondenunciation of terrorism-related information" because they had forwarded a text message joking about youth unemployment and entry requirements for Boko Haram (Amnesty International 2017). These examples are not comprehensive, but they show a pattern of limiting and criminalizing freedom of expression in the digital realm.

In the context of the Anglophone Crisis, the government has enforced a temporary internet shutdown, as mentioned above, as well as a throttling of the bandwidth around the 2018 presidential election, which separatists had urged Anglophones to boycott (Gwagwa 2018). This has been defended as an attempt to reduce the spread of lies and misinformation (Ngange and Mokondo 2019, 60). Indeed, the information shared in posts about the Anglophone Crisis every day is hard to verify and likely contains errors and some intentional misrepresentations. While some incidents of atrocities have been verified by contributors to the Cameroon Database of Atrocities hosted by the University of Toronto (Borealis 2024) or organizations like Human Rights Watch (e.g., 2022b; 2022c), these analyses take months, sometimes years, to conclude and only cover a tiny fraction of incidents. Most smaller instances of violence as well as political developments or rumors about persons involved go uninvestigated. However, cutting access to all online communication hardly improved access to reliable information.

My interlocutors furthermore told me that they suspect government surveillance of public Anglophone Crisis-related online spaces like Facebook groups, leading most people to follow these silently, if at all, for fear of terrorism accusation if they post publicly. This has pushed the bulk of personal communication about the Anglophone Crisis into the private sphere: into face-to-face conversations, phone calls, or WhatsApp chats. But even encrypted chats are not necessarily private.

Even though this constitutes a breach of the Criminal Procedure Code, members of the security forces routinely conduct manual searches of civilians' smartphones without first acquiring a search warrant. It is not known whether these searches are ordered or merely tolerated, but they have reportedly led to arrests and, given the power imbalance between civilians and armed representatives of the state, victims of illegal searches have little option but to comply, regardless of the rights they have on paper.

Before going into more depth about phone searches and their effects on digital communication among Anglophones, I will locate the relationship between security forces and Anglophones within the framework of socially constructed target populations.

Theorizing Phone Searches

Policies and their implementation affect their respective target populations unequally. According to Schneider and Ingram (1993), this is due to a social construction process in which policymakers are influenced by the perceived power and positive or negative reputation of target populations and, as a result, attempt to benefit, appease, protect, or control them in different ways.

The way the Cameroonian government conceptualizes and interacts with Anglophone movements that agitate for change is double-layered. At times, the government has been applying means that Schneider and Ingram (1993) observe as patterns used in the management of "contenders": target populations that policymakers perceive as powerful but with a negative connotation in the public imagination, such as minorities, unions, or the rich. Examples of this are the appointment of Anglophone ministers in positions with limited leverage or, in the context of the Anglophone Crisis, closed-door meetings with Anglophone trade unions and the holding of a Grand National Dialogue, all of which my interlocutors described as symbolic gestures and minimal concessions intended to appease critics and thus reduce further protest and preserve the status quo.

Simultaneously, the Cameroonian government liberally categorizes its perceived opponents as "deviants," more specifically as suspected terrorists. According to Schneider and Ingram (1993), policies affecting deviants tend to bind them, with little regard for their civil liberties or dignity, to supposedly protect other segments of the population. In the context of the Anglophone Crisis, the harsh course against separatist terrorism has led to all Anglophones being constructed as a suspect community (Hillyard 1993), members of which are assumed to likely have ties to or information about separatist activities, making them vulnerable to invasive policing and accusations of being terrorists themselves.

The drastic categorization of political opponents as terrorists is baked into the Cameroonian counterterrorism legislation. Law 2014/028 has been widely criticized for its vague definition of terrorism that includes nonviolent disruptions of public order, like unauthorized protest (e.g., Johnson 2014; Kindzeka 2017). It calls for the trial of suspects in military court and prescribes harsh penalties including life in prison or death. While death sentences are rarely executed, the OMCT World Organisation Against Torture (2021) warns that their use against three alleged separatist fighters in 2018 signals a willingness of the government to further escalate the conflict. Amnesty International (2015) and Ashukem

(2021) have raised additional concerns about the Cameroonian government abusing the broad legal brush of Law 2014/028 to dispose of political opponents.

Even though phone searches without a search warrant are not sanctioned by Law 2014/028, they need to be viewed in the context of the Cameroonian government's counterterrorism approach to the Anglophone Crisis and the resulting construction of Anglophones as a security threat. It is precisely this construction that allows members of the security forces to circumvent encryption and invade private correspondences through simple manual searches without considering the legality of these procedures. Security forces going through private WhatsApp chats is thus a sociopolitical problem, not a legal or technological one.

Dealing with Phone Searches in Practice

The prevalence of phone searches is difficult to estimate. My interlocutors are under the impression that searches were more common from 2018 to 2019, a phase often described as the peak of the conflict, even though recent reports seem to indicate that things have hardly been cooling down (Human Rights Watch 2022b, 2022c; Schumann et al. 2023). In those peak years, arbitrary phone searches were frequently conducted at checkpoints along intercity roads in the two Anglophone regions. Checkpoints exist throughout Cameroon and, if stopped, car and bus passengers are expected to present their national ID card. According to George, a young man from Buea who has experienced two phone searches, it is only after asking for the ID card and holding onto it that members of the security forces demand to see phones, presumably to stop suspects from fleeing the procedure (Name changed. Conversation, Buea, 01/13/2023).

Since 2019, searches are said to have become less common, but they certainly still occur, especially in areas near armed group encampments or the scenes of recent battles. Additionally, they seem to have expanded into university campuses and student quarters (Cameroon News Agency 2023)—a trend that suggests motives besides counterterrorism. George's second phone search took place in the student quarter Molyko. He says: "They went through messages, saw my communications, looked at pictures. I have some pictures with Whites. They asked me, 'What are you using these pictures for? Are you a scammer?'"

Cameroonian colleagues who provided feedback on drafts of this chapter reminded me to mention that members of the Cameroonian police and armed forces are well-known to solicit bribes by pointing out real or made-up violations of regulations. Phone searches are merely a current and particularly intense iteration of this pattern that was enabled by the Anglophone Crisis. Under the guise of protecting national security, it can also facilitate bribe solicitation for other reasons, such as alleged scamming, as in George's case above. According to a staff member of the Centre for Human Rights and Democracy in Africa (CHRDA), phone searches have also been used to target homosexual people (Interview, Buea, 05/27/2022). But, while anything the police officer, gendarme, or soldier finds can be used against the "suspect" in question, evidence of military atrocities is still "the worst thing they can find on your phone," according to George. Due to the high stakes, bribes will either be exorbitant or no longer offered as a way out.

What happens in such a case is exemplified by the arrest of the Anglophone lawyer Nicodemus Nde Ntso Amungwa in May 2021, whose phone was seized and searched while he was attending to a client at a gendarmerie station in Yaoundé. The gendarmes claim to have found videos of military atrocities in the Anglophone regions and, rather than making a complaint about the acts in the videos, arrested Amungwa and kept him in pretrial detention for ten days until his court date at the military tribunal, where his case was luckily returned to investigation and eventually dismissed (Bureau of Democracy, Human Rights, and Labor 2021, 15). According to Gilbert Mbaku, a former regional coordinator for the Ayah Foundation, a Cameroonian NGO supporting victims of arbitrary arrest, it is common for pretrial detention to last much longer, sometimes several years, without a court date or a formally filed case (WhatsApp voice message, remote interview, 01/10/202).

Avoiding Searches

As already hinted at above, it is practically impossible for a searched person to provide a nonsuspicious phone. Considering that phone searches are illegal, the most obvious strategy to avoid the invasion of one's privacy would be to avoid the search. The illegality of phone searches without a warrant was confirmed to me in a legal commentary I requested from barrister Acho Wilson Yuh. He wrote that, according to the Criminal Procedure Code, searches of digital devices always require a search warrant from the respective authority. Cameroonian law recognizes the right to privacy and, depending on the exact scenario, accessing a person's phone without consent or warrant could constitute tampering with correspondence. Nonetheless, he closed by recommending complying with the search procedure to avoid additional harm.

In contrast, the Cameroon Human Rights Council called on the population to refuse handing over their phone and instead report the incident to help end the practice (Adams 2023). I consider this advice problematic, given that just a few days after the campaign in January 2023, a peace activist from Buea told me about how she was called to the police station to bail out her son, who had been arrested for refusing a phone search (Remote interview via WhatsApp call, 01/27/2023). More drastically, it is well known that security forces regularly shoot supposed warning shots at civilians for noncompliance (having stayed opposite a checkpoint in Bamenda in 2018, I can confirm this to be a daily occurrence) and have killed several people, including children, in this manner (e.g., Kouagheu 2021; N. 2021).

During the peak of phone searches along bus routes, some Anglophones would leave their smartphone behind altogether or swap it out for a mobile phone to avoid being searched without having to explicitly refuse. Unfortunately, this would not protect them from accusations either, according to Mbaku:

> At some control posts, they will profile some people and tell them: "You have the means of having a smartphone, why are you not having one?" If you're not having one, you're a prime suspect. And you're being arrested. They're claiming that you've left your smartphone behind because you didn't want them to have access to read (WhatsApp voice message, remote interview, 01/10/2023).

The same goes for de-installing social media apps or doing a full factory reset: not having WhatsApp or Facebook and no personal data on a phone is unusual and thus suspicious. "And then, you can imagine what people lose whenever they're doing a factory reset of their devices. You can imagine what it entails to lose all your data in your phone," Mbaku adds. I will get back to the issue of losing personally or juristically relevant data below. For now, the relevant observation is that digital safety requires Anglophone Cameroonians to view their own phone through the eyes of a security officer. They must provide a phone that seems authentically used but harmless.

Withdrawal into WhatsApp chats

While encryption does not protect against phone searches, the preference for WhatsApp, over more public apps, might have been amplified by fear of surveillance. This is not to say that the app was not immensely popular before the Anglophone Conflict. Rather, posting publicly has become unfeasible in the risk calculation of many social media users. The countless Anglophone Crisis-related Facebook pages and groups still exist and post regularly, but groups/pages with thousands of members/followers—seemingly real accounts—barely get likes and comments in the double digits. Yet, this does not mean people are not seeing their posts. They might merely take their discussions to WhatsApp.

After sending me a Facebook post about the concerning condition of the political activist Abdul Karim, who was at the time in detention without any known charges, a friend commented:

> Right now, I would have really loved to post this, but the bullshit government, through the yes-men that [they] have here and there who constantly report to them about what others think or say about this crisis and other political issues might likely report to them. You don't just know everything about everyone within your circle (WhatsApp message, 01/18/2023).

Karim's detention has been condemned by Amnesty International (2022) and Human Rights Watch (2022a) but because he has expressed separatist positions, speaking up for him is risky for Anglophone Cameroonians. This friend thus restricted himself to sharing the post in one-on-one chats with close friends, even though he was concerned about Karim's seemingly deteriorating health and thought that public outcry might increase his chances of being released.

Similarly, in an interview about her experience as a political prisoner, Beatrice, a Francophone supporter of the opposition party Cameroon Renaissance Movement, shared her view on the Anglophone Conflict and how it was changed by video evidence shared through a small WhatsApp group:

> I never see something [on the internet] and then I take it seriously. Because I thought that it's a lot of propaganda. They can take something from Uganda, from Kenya, they come and say that it's here. But there are things that a friend that was in Bambili sent to us and: What I saw the army doing—[. . .] She sent it to us in a group, she just said "Look at what they're doing." That video never went on the internet. Because if we sent it on the internet, they would have traced it back to her; maybe she would have been in prison today. Because there are things that they don't even want people to know. She was in a very responsible group: It never leaked. Never. Never leaked. We stayed with it (Name changed. Interview, Yaoundé, 12/28/2021).

On the one hand, Beatrice highlights the importance of keeping certain information off the public parts of the internet to protect the safety of everyone involved in the chat. At the same time, the quote shows a downside to this privacy: it leads to a distortion of public discourse in which the conflict seems less relevant than it is in private discussions; this illusion is convenient for the government, which presents itself as in control of the situation.

Avoidance of WhatsApp Groups

Despite the informative value, many Anglophones avoid being in WhatsApp groups that could contain conflict-related messages for three main reasons. Firstly, as explained by Beatrice, political content should be shared in small circles of trusted contacts, and a group can easily exceed that scope. One careless or duplicitous person could put all others at risk. For this reason, my interlocutors could not add me to any private groups they were members of, which was frustrating for me at first but ultimately illustrates the sociopolitical climate.

Secondly, simply being in certain groups could lead to terrorism accusations, as Mbaku explained:

> We had a WhatsApp group, a WhatsApp forum, with like 250 members in it from all over the world. [. . .] these were sympathizers of the Ayah foundation, donors, and we noticed after some time that everyone who was based in Cameroon had to exit the groups because they didn't want to be tagged as terrorists or having affiliations with the Ayah foundation, which is being tagged as a terrorist organization already. So, everyone that was in Cameroon became very, very scared of the fact that at any point in time they could find anything concerning the Ayah foundation in their phones and they could get arrested (WhatsApp voice message, remote interview, 01/10/2023).

Thirdly, conflict-related messages pop up unexpectedly, making them difficult to prepare for, emotionally and in terms of data management. While Cameroonian news reporting sometimes includes photo or video evidence of crimes and accidents that might be considered graphic, a lot of content about the Anglophone Conflict lies outside this accepted explicitness and, at least partially, fulfils what Pohjonen and Udupa (2017) have termed "extreme speech." For one, the images depict violence not commonly seen in Cameroon before 2017, such as explosions and resulting bodies or injuries, close-ups of facial shot wounds, and decapitations. Additionally, many of them are produced by the perpetrators (both state forces and separatist fighters) with the intent to shock, brag, dehumanize the victims, and to encourage additional violence. Others are posted by witnesses to call attention to acts they find outrageous. While some users want to see and hold onto such messages, many avoid groups where they could encounter them to protect their well-being.

Personal boundaries aside, this type of content can turn into an acute danger for the recipients when found on their phone during an arbitrary search. Even though it is commonly said that sending conflict-related messages is more dangerous than receiving them, receiving is risky enough, as Mbaku assured me: "Believe me: I have met so many people who were arrested because they just received a message. Because they simply *received* a message they didn't respond to; they hadn't even read the messages in their phone yet" (WhatsApp voice message, remote interview, 01/10/2023).

Viewing and Deleting

The potentially high rate of traffic in WhatsApp groups makes it difficult for users to see and delete all messages that could turn a phone search from an expensive situation into a freedom- and life-threatening one. But even without groups, plenty of conflict-related messages find their way into the phones of Anglophone WhatsApp users, ideally from trusted sources they know personally—and many of them are immediately deleted. Not only is deleting messages necessary for the recipient's safety; recalling Beatrice's comments about trust in WhatsApp groups, it also protects the sender, who could be traced if their message is found on someone else's phone.

However, as mentioned above, this means losing relevant information. Organizations like the Buea-based CHRDA have used smartphone recordings to verify incidents of human rights violations by separatists and state forces alike and have published reports that undermine government attempts at downplaying or misrepresenting the conflict (e.g., CHRDA 2022b, 2022a). For these, they depend on people recording and sharing evidence, as well as their own right to own copies of and handle the files.

Journalists likewise explained to me that their listeners and readers send a lot of sensitive material through WhatsApp for them to cover. They need to keep these lines of communication open but, because they cannot rely on security forces respecting the freedom of the press, they need to manage the data they receive, for example by deleting the WhatsApp message but keeping the media file and moving it to a password-protected folder.

Of course, passwords do not protect against phone searches. If they did, searches would not occur since most smartphone users lock their screen with a password or pattern. But phone searches are a stochastic risk, and a hidden folder that might be overlooked further reduces the chance of being found out. Additionally, they provide a sense of agency to the users, who can feel slightly more prepared for a search than they otherwise would. Still, secure folders and alternatives like moving sensitive files to encrypted USB drives are a technological solution to a sociopolitical problem. Much like WhatsApp encryption, they provide no actual safety because the risk is not hacking; the risk is being forced to divulge private information by an armed representative of a state that cannot be trusted to abide by its laws. Because of this, deletion remains the safer option.

However, deletion becomes difficult when sensitive data is at the same time personal, sentimental data. This became clear to me during an interview with a student at the University of Buea. I asked about her attitude toward graphic content and whether it should be shared since some people might find it disturbing. She responded by admitting she had some and pulling out her phone. She opened the gallery and swiped to photos of two corpses laying on the ground somewhere outside. Both had obviously died violent deaths. She apologized in case the photos scared or disgusted me and added, "I know I should delete them. But these were my direct neighbors. I knew them. So, I like to keep them" (Interview, Buea, 04/06/2022).

Visually, an extreme image kept to grieve for, remember, or process an event is no different from one circulated for separatist propaganda. The Anglophone Conflict is part of people's day-to-day lives. The vast majority of Anglophones have received messages about arson attacks in their hometowns, men from their village having joined armed groups, or loved ones hiding from a gun battle outside. Needing to delete or hide these messages conveys to them the impression that what is criminalized is not terrorism but their lived experience and identity as Anglophones.

Conclusion

In a conflict context in which publicly posting about political subjects has become unfeasible, WhatsApp offers private spaces for exchange in which Anglophones feel safe enough to voice their opinions, discuss negative experiences with state forces, and share photos and videos of occurrences in their home regions. Due to illegal phone searches, this privacy has its limitations. To avoid accusations of terrorism, Anglophone civilians who think they might be searched need to leave suspicious groups and selectively delete Anglophone Crisis-related messages and media. To avoid being exposed to unwanted content or having their sensitive messages leaked, Anglophones rely on interpersonal trust, avoid groups, and restrict who they have political chats with. Technological solutions like secure folders might provide a feeling of safety for Anglophone users but, ultimately, no technological coping strategy fixes the underlying sociopolitical problem.

Chat encryption does not matter when state forces are breaking the law to physically search phones. Diligently deleting messages does not protect a user when encountering an officer who is willing to use a suspicious lack of evidence against them. What puts Anglophones at risk is thus not primarily their social media use but the fact that the Cameroonian government's counterterrorism approach to the Anglophone Crisis preemptively treats them as potential threats, rather than citizens worthy of protection. This categorization is not lost on Anglophones who must hide and delete an aspect of their lives and identity to protect themselves from the state.

Even though few Anglophones are willing to engage in public government criticism, let alone separatist activism, the widespread negative experiences with the Cameroonian state, of which phone searches are just one, have led to deep-seated anger, frustration, and distrust that will continue to impact Cameroonian politics, regardless of how the Anglophone Crisis is resolved.

6

Engaging and Disengaging with Political Disinformation on WhatsApp

A Study of Young Adults in South Africa

HERMAN WASSERMAN AND DANI MADRID-MORALES

Concerns about the potential consequences of the spread of false and misleading information, particularly on social media and messaging platforms like WhatsApp, have become a recurring fixture of contemporary political life (Carlson 2020). While in many African nations, including South Africa, political processes have long been marred by the circulation of harmful information such as political propaganda, hate speech, and conspiracy theories (Dwyer and Molony 2019), the advent of digital media platforms has exacerbated the rapid spread of these damaging forms of communication, posing a threat to elections and the legitimacy of democratic processes (Boyd-Barrett 2019). These threats are exacerbated by other longstanding sociopolitical setbacks, such as different forms of democratic backsliding and challenges to press freedom (Okoro and Emmanuel 2019). Furthermore, even though political disinformation is not new to South Africa, since 2020 the country has faced new types of coordinated efforts, driven by both external forces and domestic actors, aimed at influencing elections and shaping political agendas (Gagliardone et al. 2021; Wasserman 2020).

Disinformation in South Africa spreads through a wide range of media platforms, from some radio stations and newspapers to digital spaces such as X and Facebook (Rodny-Gumede 2018). However, it is most often on WhatsApp that falsehoods become viral (Newman et al. 2021). WhatsApp has contributed to widening the reach and impact of campaigns of deception by enabling the sharing of disinformation across networks and at scale. A common and well-evidenced pattern of diffusion of false information looks like this: Disinformation gets amplified when a central source, either a domestic political actor or an agent participating in a foreign influence operation, supplies the content to a select group of influential individuals who then push the content or seek to disparage well-known activists and journalists (Wasserman 2020). The motivation for such amplification of disinformation may be to gain political influence, or financial profit (e.g., through clickbait aimed at generating advertising revenue), or a combination of these, when bloggers or influencers are paid to post disinformation by an actor seeking to extend their political influence (Ong and Cabañes 2018).

The enabling power of digital technology combined with the recurrent political narrative of a ubiquitous "information disorder" (Wardle and Derakhshan 2017) might explain the fact that, when asked about the prevalence of political disinformation, South Africans report higher levels of exposure to information that they believe is made up than citizens of other countries, including Nigeria and the United States (Wasserman and Madrid-Morales 2019). Exposure, however, should not be unequivocally equated to engagement. In fact, in the context of disinformation studies, some have argued that, in South Africa, only a small minority willingly engages with political disinformation online (Ahmed, Madrid-Morales, and Tully 2023; Tully et al. 2021).

This chapter seeks to contribute to research in this area by focusing on the ways young adults in South Africa engage and disengage with political disinformation on WhatsApp. Drawing on the analysis of focus group discussions with young adults across four South African provinces, we show how certain contemporary practices of political disinformation on WhatsApp map onto longer social, political, and cultural histories, and how young adults navigate the information environment, either by actively engaging (i.e., correcting, sharing, confronting, blocking, reporting) with political falsehoods on the platform, or disengaging with this type of content. We contextualize the empirical findings of this chapter against the backdrop of South Africa's fragile political settings, where electoral disinformation presents an especially dangerous and vexing challenge.

Political Disinformation in South Africa

While disinformation is a global problem, it can best be understood within particular contexts (Wasserman and Madrid-Morales 2022). We can only fully understand why disinformation appeals to audiences, how they consume and spread it, and what the best responses to disinformation are when we understand the particular social, political, economic, and historical context within which it appears. This is also the case when considering political disinformation. Although agents seeking to influence elections may operate across different countries or regions—for instance the team of Israeli contractors, so-called Team Jorge, who are alleged to have manipulated more than thirty elections around the world (Kirchgaessner et al. 2023)—the success of this meddling often depends on being able to exploit local social, cultural, and/or political dynamics. An example is Bell Pottinger, a British public relations firm hired in 2016 by political allies of former South African president Jacob Zuma to deflect allegations of corruption (sometimes referred to as "state capture") against him (Rensburg 2019). Their campaign focused on using the persistent economic inequalities in the country to foment racial tensions through an online campaign utilizing terms such as "White monopoly capital," which they coined (Cave 2017). This exploitation of racial tensions for the purpose of getting a foothold for political disinformation has also been a tactic employed by foreign actors. Numerous covert activities by countries such as Russia and China in which cultural and political cleavages are exploited to sow discord have been uncovered in countries across sub-Saharan Africa (but not in South Africa) in recent years (Graphika and Stanford Internet Observatory 2022).

The threat of disinformation, particularly on social media and encrypted messaging platforms, derailing South African elections was already identified during the municipal elections in 2021, when the local fact-checking organization Media Monitoring Africa teamed up with the Electoral Commission and secured the cooperation of Google, Facebook, Twitter, and TikTok to identify and eliminate disinformation on their platforms (Independent Electoral Commission 2021). This multistakeholder collaboration recognized that while big tech platforms have a responsibility to regulate disinformation, such regulation must be informed by contextual knowledge, through local stakeholders, for it to be successful.

In South Africa, as elsewhere in the Global South, new forms of disinformation have been emerging in recent years, partly because of the rapid uptake and diffusion of digital and mobile media technologies. However, when disinformation is considered as a form of social and cultural practice, it becomes clear that, in its contemporary manifestations, disinformation in South Africa has its roots in much older histories of colonialism and postcolonial authoritarianism. As noted above, current disinformation campaigns often exploit the ethnic and social polarizations inherited from earlier eras. The practices involved with spreading such information, however, also have historical roots (e.g., Mhlambi 2019). The tendency to imbue rumors, gossip, and jokes with trust, and to share such information freely, may also be traced back to communication practices during colonial and postcolonial authoritarian regimes when official channels of communication or state-owned media were not trusted, as well as the strong oral culture in African societies (e.g., Ellis 1989).

The rise in disinformation has coincided with a crisis for established news media. Around the world, fewer people consume news media, with some avoiding it completely (Villi et al. 2022). Trust in the news media is also on the decline generally, while young people tend to access news media mostly via social media platforms, where news coexists with other, less accurate, unverified, and sometimes outright false information. Unlike many countries in the Global South, where authoritarian governments restrict freedom of the media and make it difficult for journalists to counter false narratives with accurate information, South Africa has strong constitutional protection for media freedom. While trust in South African news media is also relatively high, it is social media platforms such as Facebook (52 percent) and WhatsApp (43 percent) that are the most popular sources of news in the country (Roper 2022). WhatsApp is also being used by a major newspaper, the *Mail & Guardian*, to distribute a PDF version of a current affairs magazine about Africa, titled *The Continent*, and readers have served as informal fact checkers by responding with feedback via the WhatsApp line (Allison 2022, 190). Disinformation on the platform among South African users is, however, rife—the largest share of complaints received by the fact-checking service Media Monitoring Africa during the COVID-19 pandemic was related to WhatsApp (Smith and Bird 2020).

Audiences' (Dis)engagement with Disinformation

The general belief is that, in South Africa, like in many countries in the Global South, there are two types of particularly vulnerable populations to disinformation: older adults and people living outside urban centers (e.g., Chakrabarti, Rooney, and Kewon 2018). What these two populations are said to have in common is that they are not technologically savvy and might, therefore, have lower levels of digital media and information literacy. The evidence to support these claims is, however, patchy. While there are studies in the United States that indicate, for example, that older adults did engage in disinformation behaviors more prominently than younger adults (Guess, Nagler, and Tucker 2019), the evidence in other parts of the world is more anecdotal (Duffy, Tandoc, and Ling 2020).

Previous research in South Africa and five other sub-Saharan African countries has shown that young media users often blame older family members for the spread of disinformation, while older adults tend to place blame on younger generations. Building on these findings, Madrid-Morales et al. (2021) remind us that, when thinking about "vulnerable populations" when it comes to disinformation, it is important to differentiate between types of disinformation (e.g., those vulnerable to political disinformation might be different than those vulnerable to financial scams), and to avoid making assumptions about the connection between certain types of digital inequities, such as age or place of dwelling, and disinformation behaviors. In other words, while there is evidence that shows that there are indeed gaps in digital access between some of the groups most thought to be vulnerable to disinformation (for example, older adults), there isn't robust evidence to support the claim that these gaps translate into disinformation vulnerability. Therefore, it is important to establish the diversity of experiences of media users empirically within specific contexts. These experiences also link to the expectations audiences may have of the media, the levels of trust in media and other forms of information, and the ways in which audiences engage—or disengage—with certain information platforms and content.

In the remainder of this chapter, we describe one such demographic, namely young adult media consumers in South Africa. This demographic group could potentially yield valuable information about the social and political dynamics of disinformation in South Africa, given its complexity and contradictions: these users are particularly active on social media (Bosch 2013), and are therefore often viewed as vulnerable to online disinformation (Makananise 2022); yet, at the same time, it can be assumed that they are also digitally savvy and should be able to critically navigate the digital sphere as it pertains to messaging platforms such as WhatsApp (Dlamini and Daniels 2023). However, younger adults have also been noted to disengage from media and to find less resonance in mainstream news media, which may render them vulnerable to disinformation on messaging platforms when they are unlikely to corroborate such information with reference to established news platforms (Boulianne and Theocharis 2020).

Against this backdrop, in the next pages we discuss the circumstances under which young adults in South Africa decide to engage or disengage with political disinformation on WhatsApp, and we examine the extent to which different groups of young adults in South Africa experience political disinformation on WhatsApp differently. Our discussion draws on data from focus group discussions convened in August 2022 at four higher education institutions in South Africa: University of the Free State (UFS), Cape Peninsula University of Technology (CPUT), University of Forth Hare (UFH), and University of KwaZulu-Natal (UKZN). Our sample includes a range of historically disadvantaged universities and well-established urban institutions in five provinces. Thirty-eight students participated in the discussions, with group sizes ranging from six to twelve participants. The focus groups, which lasted between sixty and seventy-five minutes, were audio recorded, transcribed verbatim, and analyzed thematically.

Most participants in our focus group discussions have been using mobile phones for a long time, and therefore saw themselves as avid and apt consumers of information, even if political information was not top of their agenda. The majority seemed to have extensive exposure to multiple types of false information, from scams and fake job ads to lots of information around and during COVID-19, which was a time where social media use was high. In this context, and in their experiences engaging with information, much of which was through mediated forms of communication, participants shared a perception that rumors, disinformation, and biased content appear to be pervasive. In some cases, students referred to the orality of rumors and disinformation. Speaking about disinformation related to COVID-19 and rumors that vaccines were not safe, one participant at CPUT noted:

Nothing happened, but, still, it [my town] is a small community so . . . and they keep gossiping, so obviously, they are old, and they won't believe me. They are old. When they test, or when they get vaccinated, they are going to . . . till today they haven't tested so I guess they still believe in that.

Aside from discussing the orality of disinformation, participants seemed to converge on the idea that the virality of disinformation needs to be attributed to social media. As another student from CPUT explained:

I had a gap year last year. I was in Pretoria with my family. And there was this thing going around in the news that there were like one million graves dug up by the government in Pretoria and most of the people in Pretoria, like, were not vaccinating because they were asking "what are those graves for, they're obviously planning to kill us with the vaccination" (. . .) It was on the news also. It was trending on Facebook, on WhatsApp on like almost every social media platform. Because even I know about that. Yes, in the communities that's how things go around. So, someone sees it on Facebook, and they send it on WhatsApp. "Oh, did you see this?" Yeah, then that's how it spreads.

When faced with content on WhatsApp and other platforms that they felt was inaccurate, participants had a range of responses. The majority would be in favor of ignoring the content, a form of disengaging, and moving on, but some acknowledged that they do sometimes confront the person who posted the inaccurate piece of information.

Ignore it, [because it] is a lot of . . . as you can see, or you can always see when it's forwarded and you know. You can see when it's been screenshotted a lot of times. And also sometimes if you're talking to, like, people you know, or like older family members and you try and tell them that it's fake, you know, it's fake and then they argue with you and they're like you're too young, you don't know anything, I'm right, you're wrong and then you just ignore it afterwards because you don't want to get into a fight. (CPUT)

If I have evidence, I will just tell them, "It's wrong information. Don't send it." Like if I can do my own research, and . . . I do my own research rather before sending it to other people. Maybe watch the news. Maybe if it's a COVID-related story, I'll watch the news. Maybe go on social media and check and just see how people are reacting to the news. (UKZN)

In some of the focus groups, there was some discussion on whether it might be difficult to confront some family members or elders, but there seemed to be different attitudes on this, with some students at UFS having no problem calling elders out:

INTERVIEWER: What happens when somebody, you're in a WhatsApp group, and somebody said something that is a scam?
STUDENT: If it's a lie, it's a lie. They should also know like that's a lie. And they're going to continue spreading . . . spreading it so you just trying to warn them. Shut it down.

Others at UKZN resort to blocking relatives to avoid the confrontation, which could be seen as the ultimate form of disengagement:

STUDENT 1: They [parents and family members] were the problem mostly because, you know, they don't even check. They're just they don't get scared and panic even more. So, yeah, I've had to block them and just call them if I need anything.
STUDENT 2: To me, I feel like it's not a problem anymore. Like I do not receive . . . those messages anymore. I do not know why. Maybe I'm like my classmate I blocked everyone, but nowadays, I just don't receive those messages anymore.

As the examples above illustrate, engagement with (dis)information on WhatsApp and other platforms tends to be linked to family connections. In many of the examples that were discussed during the focus groups, the main link between participants' exposure to political disinformation (or what they believe to be political disinformation, as we discuss in the next section) is family. It is through family members (not so much through friends) that they appear to be exposed to this type of content. This is best expressed in the following quote from an undergraduate film production student at CPUT:

> It [the false story] was all the way from the UK, I think. My aunt sent it me. It's someone talking about the Bible. Jesus is going to come back to everyone. [It] was like eight minutes long. There was a story going around about how in the Cadbury chocolate there are worms, and stuff and how there is poison in . . . I think it was bananas or something, and now people have to stay away from it. And it just didn't make sense. My mommy gets it, and I just look at it. I didn't share it. I never finished it. It was like eight minutes long. Scaring people off, like he's going to come . . . he's going to be holding a knife or whatever. They gonna come back and this is gonna happen on this day so be aware and, then you are like what? [Another participant: Then this day comes, and nothing happens . . .] Exactly!

Domestic and Foreign Dimensions of Political Disinformation on WhatsApp

WhatsApp use is widespread, but it's not equally spread across the country and, more importantly, different other social media platforms are more common entry points for students to get in touch with the news. For example, when asked about the uses of WhatsApp, students at UFS would say that they mostly use it for "school purposes" where they are in groups that exchange notes. They also use TikTok and, still to a significant extent, Facebook. The process of devaluing Facebook as being associated with older generations hasn't occurred fully in the life of most of our informants. Media consumption is gendered, age-related, and education-attainment related, as our data shows.

The relevance of WhatsApp as a source of "everyday information" was apparent in multiple discussions. For instance, a participant from UKZN noted that during the exam period, they deleted all the apps from their phone except for WhatsApp because so much important information, particularly about school, comes through it. Also, the uses of WhatsApp, or the type of information that participants associate with WhatsApp, tends to be less "news-like," and more "practical or everyday type of information." The following examples from some of the participants in the CPUT discussion are a good illustration of this:

> STUDENT 1: Varsity information like we have a game this weekend? And but when it's said we have class, in person class, this week? What day, what time . . .
>
> STUDENT 2: And also sometimes you just go to other people's statuses . . . let's say you missed something on social media, probably you were busy. And then they post something. "Oh, this is happening. You go to another social media platform to see if it's happening."
>
> STUDENT 3: Like when there was load shedding, load shedding times, the schedules. When is that? Okay, let me go check the actual website.

Some of these types of information are perceived by participants as examples of "political" disinformation. In other words, political disinformation, for most of the young adults we talked to, was mostly connected to everyday disinformation. Other examples of political disinformation were connected to how politics affects them, such as what they are to gain from getting involved in politics:

> CPUT STUDENT: So maybe like the DA, for instance, just an example that they are going to do certain things for the community. So yeah, just for the community members to vote for them, especially during the voting period. They would be like we are going to supply water and sanitation for everyone. We are going to build homes for everyone. The ANC gave us microwaves.
>
> INTERVIEWER: For real, or was it like a fake story? [Participant: For real.] Or real? And so who do you think are the sources of that? Like, who sent those messages?
>
> CPUT STUDENT: I think definitely people who want . . . I'm pretty sure they have a media department for marketing purposes. [It's] definitely false, because they never really keep those promises. They say these things, and they don't really commit.
>
> INTERVIEWER: Do you ever share this kind of stuff on WhatsApp or on Facebook?

CPUT STUDENT: I used to, though. Like the COVID-19 thing. Yeah, I used to share that because I feared dying.

I think it was on Monday. My mother sent me . . . it was a Capitec [bank] thingy for internships and all that. So she sent it to me and I'm like, the email address is not . . . it's not it (Capitec). So, I went to Capitec. There was like no internships and nothing. So I sent it to her and I'm like "Ma, this is not true. This is a scam." (UFS)

I'm not sure if you guys noticed, but there's that fake Takealot survey that constantly pops up. And there is obviously an incentive that's offered on it. And so, people think it's true. So, they shared and shared and shared because it says you have to share it to about ten people. And in terms of those kinds of incentives, they don't understand what on the site looks wrong. Or the comments on the section . . . on the page. You can see that it's obviously trying to make it look good, but it's fake in terms of those surveys. I think we've also, we've seen it, we've been there, done that. So, we can confidently say it's not true. (UKZN)

Very few students followed the most recent election, which at the time of data collection was municipal elections, which tend to have lower levels of involvement. Discussions around political (dis)information were not very rich, because most students saw themselves as uninterested in politics. One student at UFS said that "politics make one angry." When asked about domestic politics and recent elections, the few examples given by participants had little to do with WhatsApp, and more to do with other social media platforms, such as Twitter, or just mainstream media.

INTERVIEWER: Did you feel that people were manipulating news around, that maybe that there were disinformation around political parties?

UFS STUDENT 1: Always! I feel like because of Twitter, we are more exposed to a lot of things. So we can actually, cause also, if there's like a false reporting or false news, sometimes you can just go through the comments, and there'll be someone who will just tell you that this is the actual story and where you can actually verify to check that it's the actual story.

INTERVIEWER: So we have elections coming up the year after next, big
elections. So where will you go to get information?

UFS STUDENT 2: I think for me I will try and get my information
from somebody who's directly affiliated with politics and has maybe
technical knowledge about issues surrounding politics, that would
maybe give me a clearer sense of which party to vote for based on the
party's manifesto.

UFS STUDENT 3: On eNCA there's a broadcaster, Prof. JJ Tabane, he
also interviews a lot of politicians. His show is usually very political.
So, I think, also just listening to some of those because I'm sure
closer to the time he'll have the different leaders on the show, just
talking about their different manifestos.

When it comes to foreign political disinformation, opinions were
divided, with some referring to recent developments in Ukraine as examples, like these students from CPUT:

CPUT STUDENT 1: Yeah, I think with Russia, with Russia . . . Ukraine
fake news coming from. I don't know, maybe like they're gonna attack South Africa. Yeah. Yeah. Soldiers. Social media. It was just the
time it was starting. I was scared so I would search it up just to see.
[Participant: First time hearing about that.] [Another participant:
What is that? Fact checks?] I always go to Google (inaudible).

CPUT STUDENT 2: But then Google puts everything there (inaudible).
When you sick, right, you have a flu and then you google the symp-
toms then you self-diagnose, which is bad (inaudible). When you do
that, you think of a whole bunch of things . . . you think you have a
heart disease? If you search something on Google, it will give you a
bunch of different things then you don't know which one is true.

However, pressed about their opinion about the war in Ukraine, students at UFS showed some degree of fatigue with the topic:

INTERVIEWER: So the war on Ukraine . . . Do you follow news around
that?

UFS STUDENT 1: I stopped! Not anymore. I lost interest! Because also I
think we realize that it, okay, it affects us, but it's not gonna physically

affect us. I feel like all of us were on it because we were scared about "Oh my gosh! these people are gonna attack, these people are gonna whatnot." And then as soon as we saw that, it's sort of a them affair. And we were just like "Okay!"

UFS STUDENT 2: Because also I think we realize that it, okay, it affects us, but it's not gonna physically affect us. We're not gonna (inaudible).

Overall, foreign propaganda is, to some, an issue to consider, but, in their views, this type of disinformation is not necessarily that coming from China, Russia, or Iran, as a lot of current scholarship posits, but to many students, a cause of concern is the United States, and the variety of American content, from news to entertainment that reaches South Africa. In our interviews with young people, participants indicated the United States as a source of foreign influence and propaganda:

INTERVIEWER: You said that you also think that maybe foreign powers might influence? What do you have in mind? Which country is it?

UFS STUDENT: USA! I think because obviously like they are the kings of marketing and advertising and journalism, so whatever agenda they feel that they want to drive, that's what they'll push, regardless of how it affects other countries, as long as they get to push a narrative that suits [them].

China, Russia, and Iran came up very rarely in the focus group discussion. In one of the discussions, with honors [postgraduate] students at UKZN, the culprit was a totally different one: India.

UKZN STUDENT 1: Um, so quite a few times, I've seen that. Like, India links come up. And sometimes the language will be different, or this . . . you can see it's like basically comes from India, maybe in the, in the URL or . . . but that's what pops up quite often.

UKZN STUDENT 2: I agree. I think the India links do pop up quite often.

INTERVIEWER: Is it with some of these scams and false vouchers when India links come up?

UKZN STUDENT 1: Yes, I've seen that.

Conclusion

Research on political disinformation and extreme speech is currently dominated by scholarship on North America and Europe. The focus on the impact on disinformation in US elections has meant that disinformation studies currently lack geographical, cultural, linguistic, and geopolitical diversity. We still know comparatively little on how disinformation is impacting African elections, the strategies of foreign influence operations, and the use of disinformation as tactics by domestic political actors. Previous research has not only emphasized the relative shortage of research on political disinformation in Africa, but also stressed the importance of contextual knowledge for understanding how disinformation spreads in particular settings. This knowledge gap is important because of the high contextuality of the "information disorder" in most African nations, including South Africa. Even in countries where ostensibly democratic systems are in place, there has been an authoritarian creep, as governments have clamped down on media criticism and citizen protest by shutting down the internet for prolonged periods, creating the space for rumors and unverified information to thrive. The fragilities of government institutions and the frayed social fabric in African settings mean that efforts to respond to disinformation has often given rise to censorship and the infringement of freedom of expression. A key issue for media development on the continent is that the complex relationships between geopolitics, internet governance, legal systems/journalism, and free expression remain severely understudied. This chapter aims to contribute to understanding these issues within a particular African context.

Our research among South African youth users shows that WhatsApp is much more than a source of news: it is also a tool for social connections, which influences the ways individuals engage with the content (accurate or inaccurate) that reaches them through the platform.

As far as using WhatsApp as a source for news is concerned, WhatsApp is not among the top sources of "news" for the majority of the young adults we interviewed. Nonetheless, as the focus group data indicates, information does reach them via the app, which forces them to decide whether to engage or disengage with the content. This decision creates a dilemma for the users, as they have to decide whether to confront family members, which may create problems in contexts where respect for elders is high. The decision to disengage may also just be because of apathy or lack of interest in matters of political or societal importance. Our findings show that youth users of WhatsApp in South Africa have little interest in political information, such as messages pertaining to elections. While WhatsApp may potentially hold a risk for spreading disinformation that may threaten electoral integrity, or promote extreme speech that weaponizes political competition around elections, such risks could be mitigated by apathy toward political issues among users. Alternatively, disengagement by certain groups of users, such as university students, from political discussions may be a missed opportunity for more robust discussion, fact-checking, and correctional practices around politics, which in turn may allow disinformation narratives to go unchallenged.

These decisions to disengage from political content on WhatsApp are informed by specific social considerations in the South African context, such as the importance of familial relations, which may differ from such motivations among users in other contexts. This finding points to the importance of understanding the affordances, limitations, and risks of WhatsApp within particular contextual conditions and lived experiences.

7

Discourses of Misinformation in the Russian Diaspora

Building Trust across Instant Messaging Channels

YULIA BELINSKAYA AND JOAN RAMON RODRIGUEZ-AMAT

The new wave of restrictive laws banning Facebook, Instagram, and Twitter on the Russian internet (Sauer 2022), beyond the immediate implications for access to information and freedom of expression, has played a significant role in audiences shifting away from "public platforms like Twitter" (Machado et al. 2019, 1013) toward more private spaces for news, political debates, and entertainment content. In this context, Telegram and WhatsApp groups have become vital environments of support for the fleeing Russian migrant population in the aftermath of the Russian invasion of Ukraine in February 2022.

For many Russian migrants, separated from friends and relatives remaining in their homeland or settling in other countries, familiar platforms such as Telegram and WhatsApp have emerged as tools, offering a safety net of a community amidst uncertainty. Moreover, for newcomers who are often lacking fluency in the local language, the new online communities become invaluable resources of information and support. The participants were not only seeking guidance on practical matters such as visas, housing, employment, but also within the unfamiliar environment, these digital networks provided a sense of belonging and empowerment. These platforms also facilitated the formation of new connections, networks, and communities based on shared interests and fostering the cultural exchange.

Telegram and WhatsApp are not studied particularly well, especially in the nondemocratic contexts, despite being particularly popular in such countries as Iran, Brazil, Uzbekistan, and Russia (Salikov 2019). This chapter aims to address this gap by exploring the practices of Russian expatriate communities on instant messaging platforms and shapes a conceptual map to enhance our understanding of how these communities operate within digital spaces and their role in the dissemination of misinformation. By investigating the dynamics of these communities, this chapter seeks to contribute to available literature on misinformation practices and provide insights on the online interactions, and the specific behaviors contributing to the spread of false information.

Using the community of Russian expats in Austria as a case study, this chapter also offers a methodological strategy to better understand the extension of the community of Russian expats connected through websites and across chats and information channels as a complex assemblage of citizens, devices, and platforms that work together to form a civic community.

In this chapter, we explore the discourses of misinformation spread with the help of WhatsApp and Telegram in the specific case of the Russian-speaking community based in Austria. This is one of the first attempts to access these closed WhatsApp chats and to compare the methodological strategy with the approach to Telegram that has been already previously tested (Rodriguez-Amat and Belinskaya 2023). One of the contributions of this chapter involves implementation of ethnographic methods to the studying of WhatsApp as a potential resource. This combines time-consuming, and overt and covert observation opportunities while posing relevant ethical challenges. In this chapter, we refer to the stories that were spread with no aim or intention to deceive the other participants. We thus refer to them as "misinformation" as an encompassing umbrella term. The chapter looks at the specific language and the strategies of legitimation of the information being spread and how trust and credibility are built within closed online communities.

In particular, we look at the following examples of misinformation spread in the chats during the time of observation: the alleged damage to Gustav Klimt's painting *Death and Life* by eco activists; misinterpretation of the news of a closure of a pedestrian crossing between Narva and Ivangorod on the Russian-Estonian border; rumors around mobilization in Russia; and misleading information regarding Russian citizens' bank accounts in the EU.

This chapter begins with an exploration of methodological and ethical complexities inherent in conducting research of WhatsApp and Telegram. In the next sections, an elucidation of the Russian diaspora context, sample selection, and analytical approach is provided. Following this, the chapter delves into the presentation and examination of results, which are further contextualized through discussions. Finally, the chapter culminates with a reflexive analysis and concluding remarks, consolidating the key findings and insights derived from the study.

Studying WhatsApp and Telegram: Methodological and Ethical Challenges

In spite of the assumption that WhatsApp is "a real goldmine for scholars" (Barbosa and Milan 2019, 50), and that it may be the key to the spread of misinformation, major methodological and ethical challenges kept the platform under-researched. This has opened a knowledge gap that contrasts with the major social relevance of the platform.

First, there are several technical issues. These sites cannot be accessed using application programming interface (API) access provided for academic research, as for several years was the case for Twitter. API access allowed the researchers to request and retrieve public tweets and related data, such as user profiles, follower counts, retweets, and mentions in an automated and systematic manner. Furthermore, in the case of WhatsApp, gaining access is particularly difficult, even if it can seem like a quite standard and well-researched issue in qualitative ethnographic research as participants are often recruited through the researcher's personal networks through the process of "negotiating access" (Atkinson 2007). However, chats cannot be found through a word search, and the groups cannot be joined without a personal invitation from the administrators. Telegram public channels can be searched online; however, the search is quite limited, and closed groups also remain inaccessible. The absence of technical tools available that could help with scraping the data is also problematic. As WhatsApp stores all data from the chats on the local device, technically it could be periodically extracted; it is, however, necessary to decrypt the messages from the database. As the number of permitted members in a group chat constantly increases, the immediacy and spontaneity of communication make ethnographic research rather impossible, as it may require 24/7 observation and data collection.

Pang and Woo (2020) suggested that the scrollback interview method is more appropriate for research on WhatsApp, as participants could describe the changes in their media usage. The method was earlier employed by Robards and Lincoln (2017, 721) who developed it to "engage participants in the research process as co-analysts of their own digital traces." While scrolling back through their own Facebook timelines, participants reflect on the content and context of their disclosures, unfolding the development of their online identity (Robards and Lincoln 2017).

The ethical dilemmas include the potential risks that researchers, who are exposed to presumably intimate conversations, may bring or amplify. As noted by Barbosa and Milan (2019), end-to-end encryption and other characteristics of WhatsApp mentioned earlier provide group members with a false sense of security. In the case of semipublic or fully public channels, such as those provided by Telegram, the data is shared in an open public domain, which lifts the dilemma of access. The issue of perceived trust and security, however, remains. Conversations and personal stories brought to a trusted group of people generally should not be eavesdropped on and, furthermore, published, even if the chats and participants are anonymized, as it may create risks for participants. One of the solutions was to avoid text messages and focus the study only on links and media files. The question of revealing the researchers' identity is also crucial for such studies. It is highly problematic to obtain consent from a large group of participants, which could exceed a thousand (Barbosa and Milan 2019).

Due to these reasons, the most popular method used by researchers to understand WhatsApp are ethnographic methods (Barbosa and Milan 2019), and the research is mostly done on publicly accessible WhatsApp group chats.

Russian Diaspora Messaging

In 2022, emigration from Russia reached the level of a massive wave (Domańska 2023), marking the largest brain drain since the collapse of Soviet Union (Kamalov et al. 2022). Despite this noteworthy phenomenon, accurate data on the scale of new Russian emigration remains elusive and fragmented. Various estimates suggest that within a year after the onset of the war, anywhere between six hundred thousand to over one million people left Russia, with the majority seeking refuge in countries that have provided aggregated statistics (Anastasiadou, Volgin, and Leasure 2023). The primary reasons behind this exodus were rooted in the threats posed by the Russian invasion of Ukraine: political persecution, security concerns, fear of military mobilization, and economic risks. As a result, and especially after the announcement of mobilization, many individuals from the affected regions found solace in former USSR republics, such as Georgia, Armenia, and Kazakhstan, and beyond, including countries like Turkey, UAE, Thailand, Serbia, and Argentina, where visa requirements were less stringent (Anastasiadou, Volgin, and Leasure 2023). European countries, where the visa barriers are much stronger, received comparatively fewer immigrants. Conversely, the EU accepted approximately eight million Ukrainian refugees, with Austria alone hosting more than one hundred thousand newcomers, many of whom spoke Russian as their first or second language (United Nations High Commissioner for Refugees n.d.).

Understanding the size, demographic makeup, and socioeconomic characteristics of the Russian diaspora proves to be a complex task. In scholarly discussions, the Russian diaspora is often equated with the Russian-speaking community, while census data from host countries typically counts Russian passport holders or those who identify "Russian" as their nationality. According to census data, the largest Russian diaspora is still to be found in European countries, comprising over nine hundred thousand people (Maximova et al. 2019). Austria, particularly Vienna, has historically experienced a steady growth in its Russian-speaking community, thanks to several waves of immigration before and after the Soviet Union's collapse. Presently, Austria is home to approximately thirty-five thousand permanent Russian residents (Statistik Austria 2023). However, this data only includes first-generation immigrants holding Russian passports. The Russian diaspora in Austria exhibits a diverse nature in terms of nationality, economic status, political affiliations, and generational backgrounds, leading to its high fragmentation, as observed in media usage. Traditional diaspora forums such as Новый Венский Журнал (New Viennese Journal) and Dawai! (last updated in December 2021), which initially thrived as print publications and later online websites, have witnessed a decline in their audiences due to the proliferation of numerous Facebook groups and, more recently, WhatsApp and Telegram chats and channels. With the recent influx of immigration, Telegram and WhatsApp have emerged as crucial sources of information, especially for newcomers seeking guidance and support. The popularity of these encrypted messengers can be partially attributed to their widespread usage in both Russia and Ukraine.

Methodology

In an effort to address both the methodological challenges and misinformation discourses, we worked with two different datasets collected, one from WhatsApp and one from Telegram data. By including data from Telegram, we aimed to provide a broader context for our analysis. This diversification allows us to draw comparisons and contrasts between the two platforms, shedding light on the nuances of misinformation propagation and reception in different digital environments. We accessed three Telegram and three WhatsApp chats and collected the textual data involving the thirty days of November 2022 (as the data from WhatsApp chats could not be tracked from earlier) with the help of self-written Python scripts. The analysis focused on messages that contained some forwarded information.

We also marked several messages as misinformation, approaching it quite broadly: we included here information that contained incorrect facts, or partly misleading information, and rumors. Group chats in the sample were either known to the author beforehand or found following a snowball method that consisted of tracking personal connections as well. Table 7.1 provides information about the chats in the sample, including the number of participants and the number of posts

Table 7.1: Chats in the sample

Chat code	Platform	Number of participants	Number of messages in November
A	Telegram chat	1652	2789
B	Telegram chat	2477	706
C	Telegram channel	99	18
D	WhatsApp chat	722	301
E	WhatsApp chat	75	232
F	WhatsApp chat	83	147

in November 2022. Following the ethical guidelines of the Association of Internet Researchers (Franzke et al. 2019) in order to protect the privacy of the participants, the titles of the chats were anonymized and coded with letters.

The analysis identified topics and discourses of misinformation spread in the chats on the apps across the messages with particular interest in the specific language and strategies of legitimation of information spread and on forms of building trust and credibility. To do so, we applied Van Leeuwen's (2007) framework involving four main categories: authorization, moral evaluation, rationalization, and mythopoesis. Authorization requires validation through an institution, political leader, expert, or some kind of other authority, including conformity (if everyone does it, they cannot be wrong). Moral evaluation is associated with moral values and could be accompanied by adjectives such as "healthy," "normal," or "natural." Rationalization is based on goal- or effect-orientation and purposefulness including eventually theoretical rationalization, under the form of "the way things are." Mythopoesis or storytelling legitimizes action through common myths, symbols, and narratives, and could be presented in the form of moral tales (good is rewarded) or cautionary tales (bad is punished).

Chats analyzed for the chapter were generally devoted to the exchange of useful information and search for certain services, products, explanations of regulations, events, and locations, and mostly were thematically organized. In this sense, for instance, a specific chat devoted to parties in Vienna did not seem to carry any specific cases of misinformation.

One of the analytical difficulties included the fact that several topics were discussed simultaneously. In Telegram, there is a possibility to see the thread separately. In addition, at the end of 2022, Telegram introduced "Topics"—different thematic sections within one large group chat. In WhatsApp, however, there are no such features. The majority of the discussions were also abrupt, as participants were entering with questions, disrupting the ongoing discussion that then would often fade out.

We have identified several topics that could be marked as misinformation, rumors, or conspiracies on WhatsApp and Telegram groups. To identify them, the analysis designed a two-round sampling process. The first round consisted of sorting messages that contained links to other sources and messages that provided any kind of news-related content.

The second round consisted of labeling all messages. Most of them had been questioned by other participants and proved to be false or misrepresenting some parts of the information. During the month of November 2022, four transversal discussions emerged across most of the chats. The four topics were eco activism, border closure, mobilization, and banking.

The Language of Legitimation across Topics

One of the captivating examples of partly incorrect information that was actively spread and discussed in the chat on November 15, 2022, was the case of two eco activists who poured black oily liquid over Gustav Klimt's painting *Death and Life* in the Leopold Museum in Vienna and the news was forwarded to the group from another Telegram channel. The painting was not damaged; however, the information was received and then reproduced incorrectly by the group members. This may be also explained by the message being accompanied by a video of the action, the black stain on the painting clearly visible in the video. The following discussion not only involved numerous insults and heated discussions about the activists' action building on moral imperatives of what is good or bad; the news also led to the spread of misinformation on climate change with messages including the infamous and proven misleading graph titled "Climate History over 9,500 years" that shows the "present climate" point as 1885, while misrepresenting the average Earth temperature. This discussion and the use of the graph is a great example of authorization by the "expert" data visualization.

Another interesting example was a message about the closing of the pedestrian border crossing between Narva and Ivangorod on the Russian-Estonian border. The border crossing for all other means of transportation was to remain open. This fact was also mentioned in the original piece of news forwarded from the Russian online newspaper Fontanka.ru; however, in this case as in the previous one, the information was not correctly understood, and the discussion followed with a debate on border crossing and visa issuing with statements such as: "The Poles argued that it [closure of the borders for Russians] was because Russia opened direct flights with who knows who, like Afghanistan and Iran. And to keep refugees from flooding in." This case shows the emergence of an argument to explain the case, serves as an example of rationalization—the rumors seem logical, but there is also a reasonable explanation of why such a step could be necessary.

In times of war and mobilization in the Russian Federation, one of the prominent topics was connected to the rumors surrounding the rules of mobilization and border crossing. Many messages were forwarded from other channels or copy-pasted to the groups without identifying the source. Some of the messages were accompanied with sentences like "Urgent! Spread the information among your friends!," "Please, help spread the word!," and "Important news for everyone who is currently abroad—send to family and friends if they are there" or similar wording. These findings align with Feng et al. (2022) according to whom the diffusion of misinformation is often well-intentioned, as family members or other people from close networks warn as a form of caring. Also, in order to draw more attention to the urgency of the news, many messages forwarded from other chats started with emphatic emojis as "❗," "⚡ 🌐," and "⚡⚡⚡." These messages also increase the perceived importance of the news and contribute to the feeling that the information is relevant and originated in some hidden trustful source. Such rhetoric also boosts the forwarding of unproven messages (similar to what Belinskaya [2023] discussed as "insider news" on Telegram); and the construction of proximity between source and readership through the myth of exclusive access and of relevant information. Precisely, some messages spread in chat F referred to insider sources close to power: "Urgent news from our informant from the Federal Security Service in Moscow and Moscow region: most likely full mobilization will be introduced with a total ban on men leaving the country." Other messages using the legitimation based on authorization by quoting the higher-rank officials: "There won't be any additional restrictions said Chairman of the State Duma Defence Committee Andrey Kartapolov," "said the head of the Federal Security Service in Moscow and the Moscow region, FSB General Alexei Dorofeyev," or "⚡ Member of the Human Rights Council [consultative body to the president of the Russian Federation] Kabakov: there is no legal ban on leaving the place of residence during partial mobilization; it takes effect from the moment you receive the draft notice."

Mobilization clearly was one of the most frequently discussed concerns among the participants. The announcement of full mobilization in September 2022 caused several waves of arguments about its consequences, in the first place, in terms of restrictions on leaving the country and of the rules of deferment of military service: "Borders can be closed. And they can also announce full mobilization. That means they can also take students." Some participants were searching together for the trustworthy sources of information. For instance, the shared piece of news titled "Russia Proposed Banning Citizens of Draft Age from Leaving the Country" was accompanied by a discussion started by the author of the comment whether the source "m.ura.news" is known to the public and can be reliable. Another participant stated that they called to Aeroflot and asked directly if there were any restrictions at the moment on selling plane tickets. While many individuals were trying to find reliable information among the avalanche of fake news and general uncertainty about the outcome for individuals who evaded mobilization or attempted to flee, other participants showed more confidence in how the process of mobilization would be structured and what rules were proposed by the Russian government and other countries: "If you served, are fit, are of military age—they won't let you leave" or "They [Austrian and German governments] will only issue them [humanitarian visas] to those who evade mobilization. I will only be considered a draft dodger when I am declared wanted. Being wanted, I won't be able to leave Russia."

Also, there were several instances of anecdotal evidence of individuals illegally crossing borders and carrying more money than was allowed, or people being denied entry to the country despite possessing the required documents. When confronted by other chat members, the author of one of such evidence that was presented as a new universal rule, stated: "Some very specific people wrote this. I hope it's a hoax and a lie, but that's the reality. You don't have to believe me, I understand that too."

In general, the participants demonstrated a lot of doubt regarding the accuracy of information circulating and asked the authors to post more factual information and less emotional responses: "But I sincerely don't understand why throw all these discussions of news here into the chat. (P.S. I'm not talking about the fact itself, but specifically about discussing the nuances of the chaos)."

On the topic of banking, there were several examples of spreading misleading information; for example, that all bank accounts owned by Russian citizens in the EU would be blocked. Such messages, often exaggerated by anxiety and uncertainty, circulated widely across both platforms, triggering further speculations. There were explanations accompanying these messages that attempted to rationalize the purported actions of the banks, citing reasons such as money laundering laws and sanctions compliance: "It is based on money laundering law and sanctions," "Any bank can just reject anyone in their services, it is normal," or "They evaluate their risks and in order to avoid them . . ."

This kind of goal-oriented legitimation is a typical tool for rationalization. In this example, even more personal anecdotes were presented as evidence to support broader claims. The participants shared stories, allegedly based on personal experience, suggesting that their accounts or those of acquaintances had been frozen or rejected by banks in the past. These anecdotes, although not necessarily grounded in verifiable fact, were extrapolated to establish a new universal rule.

This kind of "firsthand experience" triggers dangerous dynamics in the trusted environment, fostering a sense of collective urgency and concern among recipients. The participants of the chat, when confronted for generalizing their experiences, justified it in the chat by stating: "That was a couple of years ago though, maybe they've changed it now. I apologize if I misinformed."

Similarly, messages that shared such firsthand experience were often cautiously framed with the phrases: "I heard," "As far as I know," "Someone told me," "At some point, someone posted," "They say," "I read somewhere," "(as far as I know again) 100 percent receive (as far as I know, I can't vouch for the words)." These phrases served as markers of uncertainty, signaling that the information being shared was not necessarily verified or reliable and thus, the apology becomes permission to deny responsibility. When one of the participants was accused of spreading deliberately incorrect information that may cause panic, they stated: "I would only be glad if I accidentally spread a fake [instead of this being truth], and in fact, nothing of the kind is planned." This casual attitude toward the veracity of information may reflect a broader trend where the pursuit of engagement and attention often eclipses concerns about accuracy and truthfulness.

The Role of the Community

Chats with a big number of participants organized a system of message moderation. The rules of one of the chats were described as follows: "NO to: insults, politics, spam, advertising, trolling." If any of the rules were violated, participants called for administrators to delete the messages or even to ban the participants. The rules of another chat prohibited the use of languages other than Russian: "Here we communicate in Russian. Exceptionally, in English. If other languages are used, the member will be removed from the group."

Participants were also sensitive to the issue of polarizing topics, such as the reasons of the Russian invasion or politics regarding refugees in Austria and the EU, and asked authors of such messages to remove the conversation from the public domain and continue, for example, in a private chat. In several cases, the community demonstrated a high level of self-regulation regarding the spread of misinformation or of misleading messages. The authors of the posts were asked to provide reliable sources: "It makes sense to attach a source to such news," "We need proof" and confronted in cases of not providing the source: "Why write what you don't know? This is how misinformation is born." For some news, participants cooperated to find the initial source together and to disclose the truth. Often, the conversations about visas and residence permits aroused and one of the pieces of news from a Russian source was found to be about the new tourist visas and not about residence permits: "It seems to me that this was TASS [Russian News Agency], who 'translated' [the news in such a confusing manner]."

Participants showed certain levels of awareness about the consequences of posting unproven information from nonreliable sources: "Then you'd better delete your first post before you actually figure it out. Otherwise, people will faint, ask about it in all the chats, and tomorrow RIA Novosti and then TASS will publish it as information obtained from their sources." Also, when it came to the most controversial and often discussed topics, participants often were recommended first to check available official sources: "[In order to avoid rumors around the topic] I have collected basic and reliable information in a file."

We have also observed a tendency to label differing viewpoints as misinformation, which reflects a cognitive bias, when individuals are predisposed to seek out information that confirms their preexisting beliefs while dismissing contradictory evidence. However, the discussions also showed a certain level of reflexivity about this issue: "So it turns out to be an amazing discovery when you don't like others' opinions, then it's lies, misinformation, and provocation, silence!" This quote shows that participants not only recognize the potential harm caused by spreading misinformation, but also can call out others on lack of critical engagement.

In one of the WhatsApp chats, a user posted repeatedly about the predictions of Nostradamus, and how they "surprisingly foretold reality." Other members reacted in a more violent manner: "Can you at least not post this crap here?" Some commented in a more ironic way: "What did Vanga predict? What is the horoscope for tomorrow?" Some, however, defended the author and asked others not to insult them: "Everyone has a different protecting mechanism."

Sources

Garimella and Tyson (2018) have reported that in their dataset, 39 percent of messages on WhatsApp chats contained links to websites. In contrast to those findings, only 4 percent to 5 percent of the messages on Telegram contained links to other sources. The straightforward explanation is that these are just peculiarities of the platforms, but this requires further investigation. In the study by Garimella and Tyson (2018), the datasets showed how all the groups were openly accessible and thematically organized, suggesting that people interested in politics, for example, were more interested in discussing the news about politics. In this dataset, the groups were devoted to expat related issues and chats were also quite strictly moderated. It can thus be assumed that certain links that provide deceptive information had been deleted. Figure 7.1 shows the sources linked in chat A.

As Figure 7.1 shows, the news takes a relatively small percentage, compared to links to various services, products, events, and locations or links to official sources, such as Austrian immigration offices, for example. The links to social media, mostly to other Telegram chats or specific posts, were aimed to help the participants to find the solution to their specific problems, for example, links to chat about real estate.

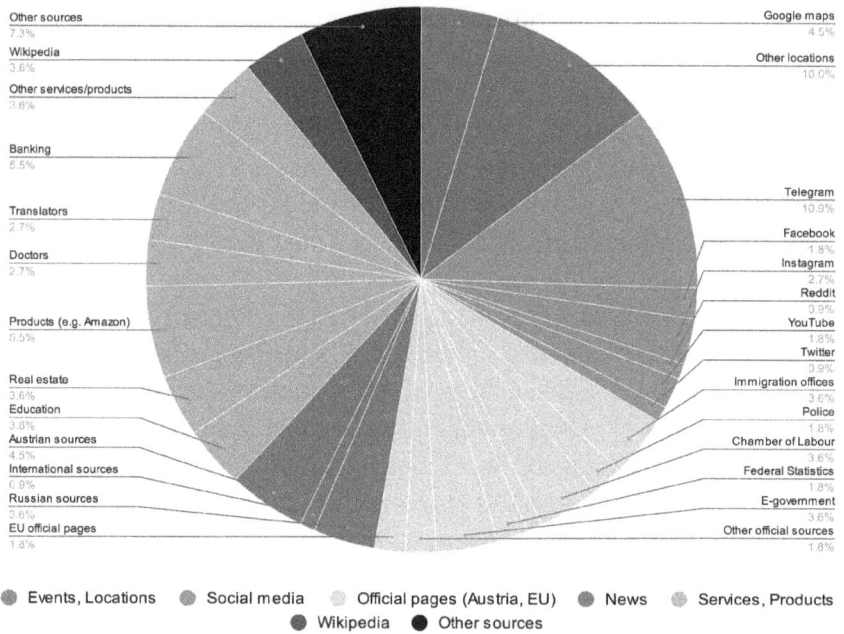

Figure 7.1: Detailed structure of the sources posted in the Telegram chat A. Source: Authors

WhatsApp chats, compared to Telegram, had fewer participants and fewer messages in general. Within the span of only one month, the participants of the Telegram chat A exchanged almost 2,800 messages. Conversations in relatively small WhatsApp chat E with seventy-five participants during the same period consisted of just 232 messages. Among those messages, 24.6 percent contained links, mainly to Instagram, YouTube, and Facebook. Table 7.2 shows the number and percentage of links in all chats in the sample. The large-scale Telegram chats had a 4 to 5 percent range of links to outside sources, while smaller WhatsApp chats had relatively higher percentages: 6.3 percent, 9.5 percent, and 24.6 percent.

Telegram channel C exemplified an interesting approach, consisting of accompanying some of the news with the disclaimer "Information confirmed by our readers." The channel often referenced official sources, such as the Russian Embassy, official press releases of the Council of the EU, various regulations from the official EU webpage, or executive orders

Table 7.2: Number of links in all chats in the sample

Chat code	Number of messages in November	Number of links	% of links
A	2789	110	4%
B	706	37	5.2%
C	18	6	33.3%
D	301	19	6.3%
E	232	57	24.6%
F	147	14	9.5%

of the president of the Russian Federation. As indicated in Table 7.2, the channel had the highest number of links; however, this is explained by the nature of public channels that mostly publish relevant news. However, in twelve cases out of eighteen, despite the official language and the accompanying images, the source of the information was not indicated.

The Perceived Online Safety and Trust

The perceived privacy of group chats contributes to the impression among users that such spaces are trustworthy and safe environments. The feeling of closeness among users who form tight-knit networks and communicative spaces facilitates the building of social communities (Brantner, Rodríguez-Amat, and Belinskaya 2021). These affordances also contribute to the perception of privacy and anonymity, fostering a culture of openness and disclosure. In our example, group chat participants often felt free to share personal anecdotes, real names, and even details of illegal deeds from the past or planned in the future, under the assumption that their communications were safeguarded. This phenomenon was particularly pronounced on Telegram, within publicly accessible group chats. For Telegram, however, the perception of privacy is deeply ingrained in the brand identity and actively promoted by its creator, Pavel Durov. The platform's end-to-end encryption and emphasis on user anonymity create an illusion of invulnerability, leading many to believe that their conversations are impervious to surveillance or scrutiny. However, this perceived privacy can also have unintended consequences, as individuals may inadvertently expose themselves to legal risks or exploitation by malicious actors, especially in the case of publicly accessible chats.

Simultaneously, many chat members expressed appreciation for the support they received, reflecting on the purpose and atmosphere of the chat group. They expressed gratitude for the chat being a place where they could escape from heavy topics: "For me, this little chat was, on the contrary, a breath of fresh air for reading :)) there's mobilization, whether the borders will close or not, whether the money in banks will burn or not :) but here you could listen to which store has the best selection and to whine and grumble about schnitzel and coffee :)."

Conclusion and Reflections

As many studies have suggested, the role of instant messaging platforms in misinformation diffusion is crucial. However, due to ethical and technical problems, there is still little evidence coming from private WhatsApp chats. This work, being one of the first attempts to access the hidden WhatsApp chats and compare them to Telegram groups, shows several differences. One notable distinction between WhatsApp and Telegram lies in their handling of links to external sources. Garimella and Tyson (2018) reported that a significant portion of messages in WhatsApp chats contained links to websites, which our finding partly confirms: the chats contained a relatively higher percentage of links (up to 24.6 percent), often leading to discussions around shared content from external sources. However, in the Telegram dataset, only a small percentage of messages included such links, indicating a lower prevalence of external source sharing. The reason behind this discrepancy warrants further investigation, but it is likely influenced by the thematic organization and strict moderation of the Telegram groups in our chapter. Regarding the source attribution, Telegram channels often referenced official sources, such as the Russian Embassy, official press releases of the Council of the EU, and WhatsApp users often occasionally posted information without clear attribution to its source, leading to discussions that may have lacked proper context or verification. The degree of message moderation and self-regulation within these platforms is also noteworthy. Telegram groups, particularly those with a substantial number of participants, instituted message moderation rules that restrict insults, political discussions, spam, advertising, and trolling. WhatsApp chats appeared to have fewer participants and messages, potentially leading to less stringent moderation.

Within both WhatsApp and Telegram, all the described cases of misinformation spread were present, and rationalization and legitimation played crucial roles in the dissemination of that misinformation. Participants often provided goal-oriented legitimation to rationalize the information, attempting to explain its logic even when it was based on unverified sources. The use of phrases like "I heard," "As far as I know," or "Someone told me" indicated an awareness of the lack of concrete evidence, enabling individuals to distance themselves from potential misinformation consequences. Also, the legitimation by authorization was more often employed by the participants of the WhatsApp chats. By applying tactics of quoting high-ranking officials, the participants enhanced the perceived importance and reliability of the messages, often urging recipients to forward unproven information.

It should be mentioned that at the beginning of the war in February 2022 and within the following months, a majority of participants had left many private chats that had been available to the authors due to polarizing political discussions that were starting to arise. Many participants moved their communities to other platforms, mainly Telegram. Also, the new groups applied the moderation system as participants became sensitive to political discussions. Furthermore, as discussed earlier, participants showed a high level of awareness about the spread of misinformation.

Several limitations of this chapter should be pointed out. First, several chats that had been accessible to the authors are not active anymore or were simply deleted. Second, there is a certain bias in the chats' selection—only those available to the researcher could be included. Also, as the chats' history was archived in January 2023, there is no way to track back which changes have been made and which messages have been deleted. We also could only access the chat history starting from November, when we began the data collection. In the case of Telegram chats, all the history was accessible and was indeed collected; this can be also used for future research. However, as already mentioned, the WhatsApp chat history could not be accessed upon joining the chat, which is why the decision was made to address only one month of an archive that was available across all chats. Another limitation is that images were not part of the analysis; however, there are several cases in which news has been posted in the form of a screenshot and not the link to the source.

Due to these listed limitations and other reasons discussed, the number of cases of misinformation being spread was lower than expected at the beginning. Future research may look for smaller WhatsApp groups with fewer participants, that have higher levels of trust and, arguably, more possibilities for misinformation to spread. The biggest challenge is still access, as family chats, for example, could not be joined or monitored.

8

WhatsApp in the United States

The Political Relevance of Subversive Platforms

INGA KRISTINA TRAUTHIG

When platforms like Facebook and Twitter entered the media fray and disrupted existing structures of authority, the hope was that a more diverse set of voices would emerge—and that this would ultimately lead to a more inclusive public sphere (Fuchs 2015; Hampton, Livio, and Sessions Goulet 2010; Schrape 2016). In the United States, popular social media platforms have largely disappointed in this regard. Facebook, YouTube, Instagram, and Twitter, among the most popular social media platforms by absolute user numbers and frequently relied on for news consumption (Auxier and Anderson 2021; Newman et al. 2021; Statista 2022c), have proven to privilege some users' voices over others (Oversight Board 2022). They also frequently promote hazardous content (Picardo et al. 2020) and weaken fringe voices by "over penalizing" the speech of marginalized communities (Yee et al. 2023). However, there are another set of platforms that have developed into main avenues for communication, news sharing, and content creation globally that are understudied in these terms. Their impacts upon the flow of communication also have repercussions for U.S. public life: encrypted messaging apps, such as Telegram, WhatsApp, and WeChat (Kuru et al. 2022; Rossini et al. 2021; Scherman et al. 2022; Sun and Yu 2020).

Given the fundamentally different platform features of WhatsApp in comparison to more studied "mainstream" platforms such as Facebook or X (Twitter), their potential functions for the public sphere are also markedly different. Most significantly, encryption and the consequent intimacy between users on encrypted communication platforms combined with nonalgorithmic traffic opens up opportunities for minority communities to converse, share, and discuss various contested issues, such as the COVID-19 pandemic or U.S. election results outside of dominating rationales or discourses.

Simultaneous to the rise in U.S. usage of WhatsApp, a "crisis of public trust," (Dalton, 2004) related to "a crisis of democracy" (Crozier, Huntington, and Watanuki 2012) has taken hold of many Western democracies, including the United States—a decline that may usually be associated with more fragile democracies in the Global South (Hetherington 2005). The erosion of legitimacy traditionally conferred to state institutions and office holders by the majority population has been partially attributed to social media—digital platforms have also regularly been blamed for societal ills, such as antidemocratic behavior or health risks, such as low vaccination rates (Gaber 2009).

While democratic institutions have always needed support and trust from the majority population to survive (Almond and Verba 1963; Habermas 1975), American democracy has repeatedly disenfranchised and/or marginalized minority communities (Judis 2001). The "crisis of public trust" can therefore be challenged as a crisis of authority that had relied on certain parts of the population but disregarded others (Richardson 2020). Furthermore, this breakdown of faith in the state and the corporate sector has been accompanied by the dependence on other, less formal structures such as the family, neighborhoods, and other communities that provide much needed support during times of crises (Hutchison and Johnson 2017). For some diaspora communities in the United States, WhatsApp can offer (perceived) safe spaces for political communication away from the mainstream and for hence discussions that allow a different tenor as well as topics—without fearing judgment from other Americans as WhatsApp communication is seen as closed and intracommunal.

This chapter aligns with critical media theory, which has argued that community-owned media have empowered marginalized groups by offering them an alternative environment for inclusion and representation (Appadurai 1990; Srinivasan 2006). Our research on WhatsApp and the diaspora reveals a critical shift in the way diaspora communities in the United States engage with news and, therefore, with public life. I aim to explain how WhatsApp affects diaspora communities' democratic engagement, with implications for the United States' future. Crucially, any functioning democracy needs a space between the market and the state to thematize, problematize, and address the challenges of society—and resilient public spheres are characterized by multiple opinions. I argue, considering this, that the inherent subversiveness of encrypted communication nurtures potentials for democratic inclusiveness against the backdrop of multiplicity, by allowing for a plurality of contesting publics (Fraser 1992; Mouffe 2000). These understandings also have implications for a media practice approach as practice theory helps understand how negotiations allow different actors (single actors, communities, or social movements) to participate, articulate themselves, and challenge dominant viewpoints (Lünenborg and Raetzsch 2018).

While WhatsApp is less widely used per capita in the United States than in other parts of the world (Bengani 2019), the Pew Research Center reported outsized importance of the app among U.S. Hispanic users (46 percent use it) compared to White users (16 percent) (Auxier and Anderson 2021). In August 2021, a survey of 1,010 adults in the United States asked if they used WhatsApp in the last six months, and the following breakdown of people answering "yes" emerged: 27 percent of Hispanic communities, 21 percent of Asian communities, 9.5 percent of people who identified with two or more racial/ethnic categories, 8 percent of Black respondents, and a mere 4 percent of White people (Gursky, Riedl, and Woolley 2021). In August 2022, a survey of 1,544 adult WhatsApp users who belong to either the Cuban American, Mexican American, or Indian American community asked them to answer questions about their perceptions of the platform, their news and information consumption behaviors, and their encounters with false information. It found that a significant percentage used WhatsApp for political discussion (Riedl et al. 2022).

WhatsApp is often discussed in terms of its convenience of use for text messaging. It is also popular because users can communicate with friends and family in countries around the world where WhatsApp is immensely popular, such as India. For this chapter, however, broader sociological parameters explain the success of encrypted platforms. Among diaspora communities in the United States, these environments offer an alternative to what is considered the majority-dominated public discourse. In the words of Fraser (1992), they have the potential to cultivate "counterpublics," which can inspire more confidence or trust among communities that have been historically marginalized from the public sphere.

Contestations over "the Public Sphere": Minority Communications and "the Media"

Among the strongest criticisms of early conceptualizations of the public sphere was its normative overtone, which structurally privileged certain parts of the population over others (Eley 1990; Fine 2010; Fraser 1989; Ikegami 2000) since the emphasis on consensus finding ultimately attributes power to majorities (Fraser 1992, 128). While the "rational-critical debate" was supposed to be radical by removing hereditary social status, Habermas himself recognized the cruciality of diverse identities and, therefore, the legitimacy of multiple forms and sites of deliberation with varying power, partially removing the priority of consensus (Bohman 2004). Most importantly for this study, however, is the continuity of striving for a political compromise and solutions against the backdrop of disagreement and various different interpretations of the environment (Lunt and Livingstone 2013). Crucially, however, this study provides food for discussion about how subversiveness (via the inherent privacy afforded by end-to-encryption) and alternative forms of news consumption and political talk can be conducive for an inclusive democracy and hence challenge existing convictions that ascribe challenges/dangers to the former (Downey and Fenton 2003).

Thanks to prolific scholarship exploring the multiplicity of publics and counterpublics, the "public sphere" is now commonly understood as contested, evolving, and undergoing challenges. Specifically, this affects minority communities, who often differentiate themselves from the "rational-critical debate" of dominant publics through contrastive dispositions, varying styles, and tactics to influence public attention. "Counterpublics" is used to refer to partially organized publics of individuals and/or groups whose individual or group identity consigns their public contributions to an "inferior position vis-à-vis the wider and dominant public" (Breese 2011, 136). Hence, counterpublics are marginalized, subaltern, subordinated publics (Fraser 2019; Squires 2002; Warner 2021).

Media organizations, news consumption, and the resulting formation of political opinion are pivotal in this regard since differing public spheres contest for both the immaterial legitimacy and material resources that help them to organize, show up in the wider public, and make demands to the dominant public and the state (Breese 2011). Non-mass media, sometimes referred to as "small, alternative, non-mainstream, radical, grassroots, or community media," (Downey and Fenton 2003) represent a potpourri of a myriad cultural backgrounds of production and often emphasize citizen participation (O'Sullivan 1993). They have received a large boost with the advent of the internet (Downey and Fenton 2003). While scholars initially derided alternative media as unsuccessful and hence unimportant (Atton 2007; Curran and Liebes 2002; Verstraeten 1996), the internet came to be hailed as the savior of alternative media and politics generally, or democracy more specifically (Sawchuk 2000). However, scholarship in this area is still undertheorized and simultaneously has become overshadowed by the securitization of alternative media and platforms—especially after the explosion of right-wing online presences (Downey and Fenton 2003) and deleterious effects of social media upon democracy due to disinformation and other factors (Hindman 2008).

However, for minority communities, experiences and interactions with the national media environment have always defined communities' experiences of inclusion and representation. Fraser (2019) elaborates on two dynamics in this regard: one related to the stigmatization of stereotypical media frames under the imperative of diversity in the context of multiculturalism and the second the vulnerability to populist rhetoric in different time periods. With regard to Mexican Americans, for example, Aguirre et al. analyzed how the cultural production of Mexican identity in the U.S. media produces a Mexican threat narrative in the American public's mind emphasizing the alleged criminality or foreignness of those "others" in U.S. civic culture (Aguirre, Rodriguez, and Simmers 2011). With regard to Hispanic communities more generally, Retis (2019) explains how the formation of Hispanic diasporas in recent decades spurred racial frameworks and discriminatory discourses in North America that tend to equate the "Latino otherness as a homogeneous peripheral group." This continues to marginalize the fastest growing part of the U.S. population.

Contrary to existing assertions that argue that alternative media sites tend to polarize (Bentivegna 2006; Hill and Hughes 1998; Sunstein 2017), we reassess the potential of alternative media and communication channels (such as WhatsApp) for political talk. We argue that a prerequisite of a safe space needs to be created for these communities before they can engage in political talk in a relatively equal manner.

WhatsApp and Its Communicative and Societal Role

Existing scholarship on WhatsApp, with relevance to this argument, is clustered around two main themes. Firstly, Scherman et al. (2022, 78) define WhatsApp as "strong-tied social media which are generally homogeneous [and] formed by close people." By building on work by Granovetter, a sociologist who contributed to the burgeoning scholarship of "network societies" that define modern states, they show that the use of social media with strong ties, specifically WhatsApp, is related to perceived political polarization and nonconventional political participation (Scherman et al. 2022). With this, they contribute to scholarship that addresses the impact of social media on polarization (Stroud 2010; Tucker et al. 2018) but also political activism (Agur and Frisch 2019; Gil de Zúñiga, Ardèvol-Abreu, and Casero-Ripollés 2021; Valenzuela, 2013) by focusing specifically on WhatsApp. Most importantly for this piece is the line drawn between access routes to political information (or "news") and differing impacts depending on whether strong or weak networks are involved. For many, regularly accessing the information needed for forming a political opinion and making decisions on (non) action can be overwhelming (Pang and Woo 2020). What is more, because political discussions are not of interest to all citizens, "short cuts" like receiving political information from a more or less well-informed friend or family member (Ladini, Mancosu, and Vezzoni 2020) can be of convenience as access is easy and the source is seen as "more reliable than mass media and messages from politicians" (Huckfeldt, Mendez, and Osborn 2004).

The impact of close ties is particularly relevant for potentially resulting political actions because it leaves less room for political disagreement and hence diverging political actions (Sinclair 2012). This study acknowledges existing research complicating the relationship between social network use and political polarization (Muddiman and Stroud 2017; Tucker et al. 2018) and sees itself in this line of thinking but advances those efforts by transposing the approach as we connect the "crisis of public trust" to a majority-dominated challenge that inadequately acknowledges democratic efforts from the margins, which can come from more closed, and potentially more polarized, spaces such as WhatsApp.

This requires a theoretical break with the normative tradition of Habermas as described earlier because consensus, or even compromise, is not necessarily the end goal—nor does it need to be. This argument is made possible because we focus specifically on political talk over WhatsApp among minority communities in a consolidated democracy like the United States.

Secondly, some communication scholars emphasize that these apps are designed for more private communication. They argue, therefore, that they enable communication between existing, trusted networks of people with close ties to one another. This, in turn, enables the spread of mis/disinformation and broader political manipulation efforts (Rossini et al. 2021), targeted forms of disinformation (Melo et al. 2019) and/or hate speech (Giusti and Iannàccaro 2020).

Overall, WhatsApp has moved into scholarly focus in the political communication field as a new "semipublic space," due to the above-described dynamics of increased popularity—and hence quantitative success when it comes to user numbers. Unique features of these apps also offer novel means of access to news and interpersonal political discussion due to their more fluid conversational settings, where exchanges can include texts, audio, videos, images, and/or links (Matassi, Boczkowski, and Mitchelstein 2019). In a recent Reuters Institute *Digital News Report*, Newman et al. (2021) show that, while Facebook continues to be the main social media source of information, users are more likely to take part in private discussions about the news through WhatsApp. This is important for the purposes of this chapter because diaspora communities exhibit high levels of trust in political institutions—but those trust levels are still relatively low (Pew Research Center 2023). As a result, strong(er) social networks, embodied by WhatsApp, which connect family, friends, or neighborhood communities are central to the discussion of politics.

Methodology

This chapter relies on fifty-six interviews with members of three different diaspora communities in the United States: Cuban Americans,

Indian Americans, and Mexican Americans. These communities were selected since (1) their WhatsApp user numbers have been reportedly much higher than for most of the U.S. population (Auxier and Anderson 2021) and (2) the political importance of these communities for future U.S. politics/U.S. democracy as they have been identified as crucial demographics to potentially "swing" election results (Cai and Fessenden 2020).

Interviews were conducted from June to November 2022 during research trips to Houston (Indian Americans), Miami (Cuban Americans), and San Antonio (Mexican Americans) as well as via Zoom for some follow-up interviews following established trust and resulting snowballing after the research trips. We started contacting community leaders, such as presidents of campus associations, and then relied on snowball sampling (Galletta 2013) to expand beyond our original compilation of prospective contacts. All interviews were done by team members of the Propaganda Research Lab at the University of Texas at Austin.

For the purposes of this work, we use the term "diaspora communities" to include users who told us that they regularly use WhatsApp to communicate with people in their country of origin or where their family is from, with individuals who share their cultural context, and with people living in the United States identifying with the same community. While this approach risks overincluding individuals who are not part of identical communities, the connecting thread for our research is the usage of WhatsApp. Instead of basing the study on sociologically deterministic inclusion/exclusion criteria, such as a certain age or nationality, this chapter follows well-established conceptualizations by Anderson that subvert the determinist scheme in which any nation is portrayed as a product of specific sociological conditions; instead, the nation is an "imagined political community" (Anderson 2006). When asking about "false news," "misinformation," or "disinformation," we privileged field perspectives as we explained academic understandings of these terms but let interviewees define what they considered falling under those terms, and which examples they wanted to share.

The qualitative data analysis of interview memos relied on NVivo to organize and structure the data into thematic clusters that share explicit or implicit similar meanings (Wildemuth 2016). The coding is rooted in grounded theory, whereby the analysis (thematic clusters) is guided by the gathered data and inductively constructed, instead of preconceptualized thematic clusters (Flick 2013). The themes elaborated on in the findings section of this chapter evolved out of both explicit content (surface content) as well as implicit content (underlying meaning).

Findings

Three main themes, each containing several codes that fall under the overarching theme, emerged from the thematic cluster analysis—each describing how contemporary engagements with mis/disinformation on WhatsApp conjugate with larger social, political, and cultural parameters: (1) WhatsApp as a protected news space, (2) WhatsApp as a reverse news sharing mechanism, and (3) WhatsApp as having increased importance for "older" diaspora members (forty-plus).

WhatsApp as a "Protected" News Space and Reverse News Sharing Mechanism

Sharing news and discussing political developments on WhatsApp has grown into a daily activity for most interviewees. News is shared in a variety of formats and can range from alternative sources such as links to small news outlets or memes or screenshots of articles by mainstream news outlets. However, the interviewees initially expressed WhatsApp as one of the least reliable platforms to obtain factual information. Digging deeper into the topic, we found an alternate reality—when we continued the conversation discussing habits of news consumption, most interviewees agreed that while WhatsApp was not their main go-to source for news, nevertheless the platform is where they come across the most news.

The perceived intimacy diaspora community members experience when using WhatsApp translates into how they consume political news and engage in political discussions. This perceived intimacy forms due to (1) the platform design, which forefronts individual chats (one-on-one or group chats), but also because (2) diaspora community members associate WhatsApp with their community and actively engage with their community while relying on communication via iMessage with noncommunity members. This pattern is fortified among the Mexican Americans we interviewed. A fifty-two-year-old interviewee explained:

"Via WhatsApp you define your community." To underscore the importance of Spanish for WhatsApp communications among Mexican Americans in San Antonio, a twenty-one-year-old student added: "My Spanish would never vanish because I use it on WhatsApp every day."

Due to the perceived intracommunity communication on the different apps, diaspora community members expressed their conviction that they can talk about topics on those apps that they would not address on other social media. For example, a twenty-one-year-old student in San Antonio explained that on WhatsApp, she only talks with people she is close to, that immigration is always a big issue, she feels that people misunderstand Mexicans, and her community needs to protect themselves against a Mexican threat narrative. This suggests our research agenda must appreciate that despite their stated news and information habits, people cannot avoid seeing political and news content on WhatsApp numerous times throughout the day. While they might not accredit WhatsApp with the highest importance for news consumption, we need to continue studying what latent news consumption and passive exposure to misinformation via messaging apps has on their political lives. Furthermore, interviewees noted the importance of sharing and discussing news on WhatsApp in a protected manner among family and friends—and removed from the majority-dominated public discourse in the United States.

In our data the second-largest theme showed that guilt of "leaving your country behind" amongst diaspora community members in the United States translated into heightened attendance into sharing news related to the community members' countries of origin. This dynamic was reciprocated by friends and family abroad sharing news pertaining to the United States. This "reverse news sharing mechanism" developed into a cradle of misinformation as interviewees felt inhibited in commenting even if particular news items were seen as dubious—exacerbated by an underlying dynamic of increased trust in news shared on WhatsApp as the platform is seen as run by friends and family (not algorithms and/or fake accounts).

Priscilla[1] described that political conversations in her small family group chats are often triggered after sending funny memes or stickers of

1. All names used in the text are chosen pseudonyms protecting the interviewee's identity.

Trump, which are often left without further comment or context. Small group chats seem to be the main organizational feature of the WhatsApp conversations and multimediality, such as the use of voice messages, videos, and memes, and accompanying humor is crucial. Some group members are more invested than others. A twenty-two-year-old Mexican American student in San Antonio showed us a group her grandfather, who lives in Mexico, started with all his grandchildren that he uses to share news almost daily: "He tries to keep us up-to-date with Mexican politics."

Mainly through the high frequency of news sharing, misinformation also enters the fray such as a popular fake video of Barack Obama being voiced over in Spanish uttering nonsense, which "pretty much everyone in my WhatsApp groups believed," a thirty-year-old Cuban American said. Several interviewees pointed out how they are inhibited to point out when they think something is fake—one Venezuelan interviewee summed it up: "This is our main way to be connected with our family abroad (. . .) [and] I don't want to make them feel like I don't care about Venezuela anymore (. . .) whenever they share something about Venezuela, I let it be." Reversely, interviewees expressed how this is similar with American news they would share.

This reverse news sharing mechanism can be interpreted as part of "political remittances" of diaspora communities as well as "infopolitics" (power contestation over information), which have previously proven to be innovative ways of using digital technologies to subvert state power, such as with the Eritrean diaspora and the state of Eritrea (Bernal 2014) or more broadly as part of diaspora activism in multiple interfaces (Udupa and Kramer 2023).

WhatsApp Has Increased Importance for "Older" Diaspora (Forty-Plus)

Finally, our interview data pointed toward increased importance of WhatsApp for political news sharing as well as political discussion among "older" diaspora members. Most "younger interviewees"— mainly college students between twenty and twenty-five—expressed their conviction that they get most news on WhatsApp from older generations. Ivette, a twenty-one-year-old Mexican American, summed this up: "I swear if it weren't for my older family members, I would get all news on Instagram (. . .) [but] these family group chats are where my uncle and grandma and also my parents just share random news all the time." Juan added: "Since my parents understood how to forward things on WhatsApp, they have not stopped (. . .) I explained to them that forwarded news can also be bad, but they think it's all from their friends and family, so it must be legit news."

However, when asked about their news consumption and (critical) engagement with news and political content on WhatsApp, our "older" interviewees vehemently disagreed with the described behaviors. Instead, they emphasized their advanced understanding of news consumption and point toward their children who—in the words of fifty-year-old Victor—"believe everything on social media [including WhatsApp]." What has become clear overall, however, was the increased time "older" interviewees engaged with WhatsApp generally, including political news and discussions. How far these higher amounts of time translate into more misinformation being spread and consumed, however, needs to be studied in future work (Wasserman and Madrid-Morales 2023).

Conclusion

Given the described dynamics in the (perceived) safe spaces of WhatsApp, I argue that political discourse on WhatsApp among diaspora communities in the United States can turn into a form of subversive speech practice that is central in defining lived forms of political discourse and meaning.

This has important implications. Firstly, this research suggests connecting the "crisis of public trust" to a majority-dominated challenge that inadequately acknowledges democratic efforts from the margins, which can come from more closed, and potentially more polarized, spaces such as WhatsApp as they have to develop as a counterpublic first before they can seek broader political change due to minorities' structural disadvantages in societies, including the United States. Basically, WhatsApp is important for building confidence, which can be a first step for marginalized communities to then participate (more) in mainstream discourse and U.S. politics. Future research could investigate how these dynamics are used to subvert state power (or not) in their countries of origin (Cuba, Mexico, and Venezuela).

Secondly, academics, journalists, and policymakers need to work harder to understand the significance of WhatsApp for minority communities to create more inclusive democracies. Addressing this question is a consistent task for political systems, many of which are defined by histories of exclusion or marginalization, such as the United States. Putting agency and context into responses to disinformation is central in this regard. Also, the three studied communities showed differences with regard to which topics were discussed on WhatsApp/removed from the mainstream (e.g., communist fears for Cuban and Venezuelan Americans or mafia developments for Mexican Americans, to mention only two). As such, more nuanced responses that grow through community-led initiatives are therefore vital.

Therefore, this research has implications for conceptualizations around countering mis- and disinformation. Singling out diaspora communities and political conversations into which misinformation creeps should not be viewed as an effort to "educate" some parts of the population. Instead, ways need to be found to counter problematic content on WhatsApp—without infringing on the described protected news spaces. One could argue that diaspora communities are sometimes more vulnerable to populist rhetoric precisely because they demonstrate a higher degree of skepticism and a critical stance with regard to official news sources. In other words, marginalized communities are often well placed to apply critical thinking when consuming the media. Any policy recommendation intended to strictly reassert the legitimacy of official news sources fails to address the challenge of trust and inclusion. In the long term, we need to design digital literacy initiatives that initiate an inclusive dialogue that can be truly interactive, so as to begin the process of (re)building this public trust.

Acknowledgments

The authors thank the interview participants for their willingness to speak with us. I am sincerely grateful to Mirya Dila, Katlyn Glover, and Kayo Mimizuka for their research support, and thank Sahana Udupa, Herman Wasserman, Samuel Woolley, and Ruth Moon for their feedback to previous versions of the manuscript. This study is a project of the Center for Media Engagement at Moody College of Communication at the University of Texas at Austin.

Funders

John S. and James L. Knight Foundation

PART III

Infrastructure

9

Contextualizing WhatsApp as Reporting Infrastructure

RUTH MOON

Democracy is on the decline globally; as of 2021, the democratic gains of the past thirty years had been practically erased, with the average person experiencing a level of democracy on par with 1989 (Bachelet 2022). At the same time, covert tools of authoritarianism are on the rise, and implications extend even to seemingly secure and private platforms like WhatsApp. Many contemporary authoritarian regimes rely heavily on publicity and other image-management techniques to bolster their reputations among global policymakers and aid organizations (Dukalskis 2021). As a result, autocratic control increasingly incorporates covert strategies, often wrapped up in democratic-seeming symbols and infrastructure of governance like regular elections and laws guaranteeing freedom of the press (Michener 2011; Tripp 2004). This chapter examines the ways journalists in the authoritarian East African country of Rwanda use and are shaped by WhatsApp as an element of infrastructure: a "boring thing" that distributes justice and power in the background so that visible work can proceed (Star 1999). My core argument is that, while the security affordances of WhatsApp, such as end-to-end encryption, seem to support information sharing and access, these affordances are interpreted within a local context of far-reaching surveillance that leaves journalists reluctant to freely share information, especially when it is unverified and might be false. In this chapter, I explore the ramifications of a strong, central, authoritarian government on the uses and limitations of WhatsApp. My findings draw from observations and interviews with journalists in Rwanda, but they suggest some important factors that could affect WhatsApp's information-spreading capacity in other contexts as well.

The Infrastructure of Journalism

Infrastructure is integral to the everyday practices of work and daily life, but the same features that make it essential can make it a challenge to study. In everyday language, "infrastructure" is often defined to include easily ignored and simple-to-use tools that facilitate movement, including power grids, water, and the internet (Star and Ruhleder 1996). In other words, infrastructure constitutes "the physical networks through which goods, ideas, waste, power, people, and finance are trafficked" (Larkin 2013). It can include objects, knowledge systems (such as classifications and definitions, like the Dewey decimal system), and people—anything that enables movement of matter across some sort of space can be infrastructure (Larkin 2013).

This classic definition, however, glosses over some important complications. As Star and others point out, one person's infrastructure can be another person's problem: to a scholar, the internet is a tool; to the IT department, it is a source of challenges (Larkin 2013; Plantin and Punathambekar 2019; Star 1999). In addition, the affordances that make a tool easy to use and attractive in one context can pose challenges in another. The chapters in this volume illustrate this, as the affordances that make WhatsApp an attractive site of misinformation (anonymity and secrecy) can be weaponized against marginalized populations (as Kim Schumann shows in their chapter). They are, in other words, socially contingent—reinforcing the importance of understanding context when trying to understand the uses and implications of WhatsApp in regulating and anarchizing speech practices (Wasserman and Madrid-Morales 2022).

The infrastructure of news gathering, then, includes the set of tools that enable the movement of information to journalists who then select, arrange, and repackage it for audience consumption. This includes, of course, closed social messaging platforms like WhatsApp, Signal, Telegram, open networks like Twitter and Facebook, traditional communication tools such as telephones and email, and the infrastructure that enables physical travel to gather information and conduct interviews. Infrastructure is of course embedded in social contexts and as such both shape and are shaped by those existing contexts, resulting in sometimes-hidden systems that can guide and shape the social world, including processes like news production. Infrastructural systems are embedded in their social context; transparent in use; persist beyond single uses and sites; used in routine ways by group members; linked to conventions of practice; intersect with other tools in standardized ways; and change iteratively from the bottom up within existing local contexts (Bowker and Star 2000; Star and Ruhleder 1996). As mentioned, infrastructure is a "boring thing" that distributes justice and power in the background so that visible work can proceed (Star 1999).

Infrastructural discussions related to communication tend to focus on specific tools, unpacking the complexity and related power dynamics of the physical infrastructure that enables interpersonal communication or mass communication practices. Studies in this vein highlight the way that news flow and information access depend on, for instance, working mobile networks and geotagged information, both especially problematic in disaster zones where reliable news updates are crucial (Sheller 2015). Scholars have examined how journalists encounter technological tools, including proprietary and often-clunky content management systems designed to enable faster, easier reporting but actually creating frustration and yet enduring because of path-dependent systems (Rodgers 2015). While journalists in the field more readily adopt new tools, especially as those tools enable faster and easier on-location reporting, the adoption includes a lot of "tinkering" as they learn on the fly, thus encumbering the process (Guribye and Nyre 2017). Platforms and actors enable certain flows of information in the process of news circulation for digital news (Carlson 2020b).

Infrastructural studies have the potential to contribute to important discussions around power distribution within journalism practice. Some of these discussions are happening already though not tied explicitly to infrastructure; for instance, Creech (2018) critically analyzes the rhetoric of risk around foreign correspondence to unpack the increasingly uneven precarity of global labor that contributes to stories about world events for local audiences. This study builds on current approaches to understanding journalistic tools by focusing on the ways that WhatsApp intersects with practices of eyewitnessing and verification, two practices central to journalistic legitimacy and journalists' abilities to hold powerful actors accountable. Eyewitnessing is a fundamental component of journalistic practice and contributes to journalistic legitimacy (Carlson 2017). The concept is so important that it has taken on "keyword" status, marking the collective cultural identity of journalists (Zelizer 2007).

And, while digital technologies increasingly enable long-distance "eye-witnessing," journalists are still expected to immerse themselves in dangerous and precarious physical situations to obtain information (Palmer 2018). On-the-ground footage satisfies audience demand and confirms that journalists were physically present (Allan, Sonwalkar, and Carter 2007; Cottle 2013). A specific kind of eyewitnessing is the in-person interview, which satisfies multiple journalistic goals, including allowing the reporter to verify identity (Parks 2022). Verification is itself an important routine, increasingly a way journalists combat mis- and disinformation (Graves 2016). It is also a "strategic ritual"—a way journalists symbolically (but not always actually) ensure that information is accurate before publication (Shapiro et al. 2013). This practice increasingly takes the form of fact-checking, with dedicated fact-checking roles springing up across the world, from the United States to Zimbabwe and South Africa (Cheruiyot and Ferrer-Conill 2018; Graves 2016; Mare and Munoriyarwa 2022). I contribute to scholarly understanding of these practices by examining the ways that infrastructure aids, hinders, and changes their implementation in the everyday work of journalism in the Global South, with specific reference to Rwanda.

What Good Is WhatsApp as Infrastructure?

Following from the definition above, WhatsApp is not exclusively infrastructure, but can be part of an infrastructure to the extent that it is used in a particular time and place to move information between actors or locations (Larkin 2013). It is part of a network that includes digital, physical, and human actors that function and break down to varying degrees in everyday journalistic routines. Electricity breakdown, for instance, shapes the everyday process of journalism. Power outages are eight times more common in sub-Saharan Africa than the global average and play a major role in hampering business processes (Cole et al. 2018; Kaseke and Hosking 2013). I watched electricity failures wipe out nearly finished news articles from desktop computers and old laptops when power outages swept through Rwandan newsrooms, leaving journalists to work longer days and rewrite stories from memory. While mobile phones are common, computers are relatively rare and expensive compared to global price indices, making computer access another challenge of newswork (Da Silva 2020, Grover n.d.). As a result, newsroom computers tend to be old and may not be standardized; in a newsroom I visited, one reporter typed away on an Arabic keyboard while another used a keyboard set up for the French language. People tend to take care of and repair technological tools that might be discarded in more technologically rich environments; one journalist postponed an interview with me because he had to take his mouse to be repaired by a repairman in another town. Physical infrastructure, like road quality and transportation access, also influences news production, as I have documented elsewhere (Moon 2022). Journalists adopt and adapt WhatsApp within this network of news production tools, and like the other tools, it proves useful in some ways and challenging in others.

As an element of communication infrastructure, several WhatsApp features are potentially relevant in shaping the way journalists use the platform to share and gather information. WhatsApp claims to offer a secure interface, using end-to-end encryption to ensure that "your personal messages stay between you and who you send them to" (WhatsApp n.d.). In some contexts, journalists rely on this encryption to gather information under surveillance (Belair-Gagnon, Agur, and Frisch 2017; Pang and Woo 2020). However, in others, WhatsApp security breaches have led journalists to doubt the platform's security (Di Salvo 2022; Moon 2022). The WhatsApp platform easily facilitates the spread of dis- and misinformation (defined respectively as false information shared with or without harmful intentions) (Brenes Peralta, Sánchez, and González 2022; Udupa 2023). It extends newsroom space and editorial coordination for newswork, especially relevant in places where physical access is challenging because of built infrastructure, unpredictable weather, or other elements (Moon 2022). It provides a platform for sharing information and disinformation with large groups (Kligler-Vilenchik 2021; Kligler-Vilenchik and Tenenboim 2020). Finally—and important for journalists, who rely on phone calls and other message exchanges to gather information for their reports—WhatsApp reduces the cost of voice calls and reliance on airtime, which tends to be more expensive than data in many African countries (Pindayi 2017). Digital tools, especially WhatsApp and similar messaging platforms, became increasingly important for journalistic work during the COVID-19 pandemic, and some uses have persisted since (Ndlovu and Sibanda 2022). And for journalists, a key factor motivating WhatsApp use is its popularity—as one of the most widely used social messaging platforms across Africa, it enables connections with a wide range of information sources and contacts. WhatsApp along with WeChat dominates the global messaging market (Goggin 2020). Thus, while some of WhatsApp's specific features are clunky or weak—for instance, its purported data security—path dependency ensures that it remains quite popular for journalists; sources continue to use it, so journalists must use it too (Business Insider Africa 2022; Ryfe and Kemmelmeier 2011).

Method

These observations about the use of WhatsApp are based on interviews and extensive fieldwork I conducted through East and Southern Africa since 2017. Direct quotes and observations come from interviews conducted in early 2023 with journalists in East Africa (primarily in Rwanda). To understand the role WhatsApp plays in eyewitnessing and verification, I used process-tracing and reconstruction approaches to interview journalists about their use of various tools in their reporting work (Reich and Barnoy 2016; Tansey 2007). The study used an intentional sample of journalists to understand how these experts make fast-paced decisions about the quality and trustworthiness of information obtained over WhatsApp. This user-centered approach is particularly valuable given that WhatsApp messaging data is inherently private and difficult for researchers or other outsiders to access. In addition, this methodological approach focuses on a select quantity of high-quality data to draw conclusions about an important group.

Qualitative interviewing is an ideal research method for understanding how socially situated actors participate in processes that are not easily observed or understood firsthand, and particularly to understanding the actor's experiences, knowledge, and perspective (Lindlof and Taylor 2017). Process-tracing and reconstruction interviews are specifically designed to explore the steps by which a cause leads to an outcome in the presence of multiple complex variables (Reich and Barnoy 2016). This approach is particularly useful to identify and describe new social phenomena (Collier 2011) such as the rapid judgments of truth, falsehood, and quality required of journalists and others assessing messages on

WhatsApp as they consider whether to act on or forward the messages. In this chapter, process tracing will serve the same purpose, revealing the steps journalists take to assess the relative information quality of WhatsApp messages. I recruited journalists using a snowball sample method, appropriate to access a hard-to-reach population, such as busy journalists (Dosek 2021).

Individual and small-group interviews are an effective way to study WhatsApp, which is difficult to study directly because of its closed nature (Masip et al. 2021). Social desirability bias is probably a main drawback of this method—for instance, subjects tend to report that they are very cautious and able to avoid misinformation but worry that other people cannot do this (Masip et al. 2021). As a result, this chapter is limited in that it unpacks what journalists *say about* WhatsApp and their engagement with it, which may be quite different from how they *actually* engage with the platform. However, this is a valuable way to begin understanding infrastructure, since emotions and other user reflections constitute part of the embeddedness of infrastructure and help us understand the impact of tools on their users (Larkin 2013).

For this chapter, I used Zoom, WhatsApp voice, and Signal to interview six journalists in several rounds. In the first-round interview, I collected demographic information and general perspectives on journalism. In the second-round interview, I asked journalists to reconstruct specific news articles or packages they had written or produced. I wrote memos during or immediately after the interviews, cataloged themes and observations, and saved the transcripts for further coding and evaluation with MAXQDA. Table 9.1 lists demographic information for the journalists I interviewed.

Table 9.1: Interview subject demographics

Journalist	Location	News outlet(s)	Medium
J1	Kigali	International wire	Print
J2	Kigali	Local news	Television
J3	Rubavu	Local sports	Print/web
J4	Kigali	International wire	Radio
J5	Kigali	Freelance economist	Radio
J6	Kigali	Local news	Print

Findings

So, what does WhatsApp do, exactly? I focus here on three major uses and the ways these might influence the spread of information and misinformation among journalists and between journalists and the public. First, WhatsApp allows journalists to gather eyewitness testimony without being physically present. Second, WhatsApp enables journalistic collaboration, enabling verification through a network of trust. Third, it encourages journalists to gather in physical locations with internet, thus enabling collaboration and government oversight.

Eyewitnessing without Presence

Journalists use WhatsApp to substitute for in-person encounters when constraints—deadlines, distance, or cost—prevent them from getting to a place to gather information. For print journalists, quotes collected on WhatsApp could be used in news articles. For J4, a successful story pitch for a radio spot would have to include audio of someone else's voice, and WhatsApp voice clips provided the ideal tool. Since the COVID-19 pandemic lockdowns, he said, he conducted 90 percent of his work over WhatsApp. "As a radio person, I rely on voices," he said. Before COVID-19, "I would go meet them with a recorder. I realized I didn't need to do that, to move up and down . . . WhatsApp voice notes produce a good sound for radio, so I usually try to convince my sources that we can use that." WhatsApp provided a way for J4 to produce radio stories for his employer while using fewer of his own resources (time and money), which was important because his employer did not reimburse for personal expenses incurred while reporting.

J1, a Kigali-based correspondent for a transnational print wire service, used WhatsApp to gather on-the-record quotes for his reports. While his byline promotes the fact that he is an on-location foreign correspondent (it usually appears as NAME, reporting from Kigali), he conducted all the interviews for two of his recent stories—both of which we discussed in detail through a reconstruction interview—via WhatsApp from a correspondence office in Kigali. Story 1 centered around an interview, conducted over WhatsApp, with the leader of an important rebel group in a neighboring conflict. Tracking down this person's contact information took a network of five or six other journalists, who J1 reached out to one by one, receiving phone numbers that led to dead ends before he finally got the correct contact for his main source. Story 2—breaking news coverage—revolved around local reactions to a French court decision about a genocidaire on trial. The process to report story 2 uncovered some limitations of WhatsApp; J1 said he prioritized talking on the phone and called each source directly because he was reporting on deadline and sources would take too long to read and respond over WhatsApp, but phone calls would get immediate responses.

Routine-wise, then, WhatsApp is an accepted way of gathering information for a news product. Editors see it as legitimate, and reporters gathering news use it to reach sources in lieu of physical presence. This is sometimes seen as a last resort, but in other cases (like that of J4), it is a newly preferred way of collecting information for news pieces because it saves time and money, both of which many freelance journalists must guard closely as they are not reimbursed by employers. As a stand-in for eyewitnessing, WhatsApp has the potential to reinforce existing networks and lead to insularity in news reporting; to interview someone over WhatsApp, one must have that person's contact information, implying a level of relationship (even if it is several degrees removed). Interviewing a stranger on the street doesn't necessarily need a relationship of any kind (though in places like Rwanda, where journalists are generally treated with distrust, getting a man-on-the-street interview is a challenge on the best day). In this use, then, WhatsApp might reinforce existing power networks. Insofar as these networks reinforce the already closed system of communication on the platform, they could either help journalists resist misinformation or reinforce its spread, since WhatsApp users tend to trust information they receive from strong ties and social in-groups—leading, in fact, to my second finding (Pasquetto et al. 2022).

Verification by Collaboration

Journalists rely on their close networks—often consisting of other journalists—to verify information collected over WhatsApp. J4, for instance, said he would not interview a person he had never met before over WhatsApp exclusively; he would try to meet the person face-to-face to see their working environment and situation, or verify their identity through colleagues. "I rely on people I know," he said. Rwandan journalists also do not think of WhatsApp as a particularly secure platform. Instead, it is a place to collect information that they will use on the record. If a journalist needs to talk to someone in a way that will not be traced, they switch platforms; J1, for instance, uses WhatsApp to collect quotes and uses Signal for background information from sources.

In general, my sources used WhatsApp to collaborate with each other to solve reporting problems. When J1 was trying to reach an elusive source across the border, he followed a trail of reporters, via WhatsApp, that took him to a regional bureau chief (for a different network) in Nairobi before he found the information he needed. They assume that information shared on WhatsApp might be visible to others (whether because the platform can be hacked, someone's phone has spyware on it, or the recipient just shows messages to their colleagues). As a result, WhatsApp does not facilitate a space for privacy, but instead fosters a platform of collaboration (for better or worse).

WhatsApp as a Gathering Space

WhatsApp facilitates gathering, both physically and virtually. In the physical sense, journalists—especially those with freelance gigs—would congregate around spots with free and consistent wireless internet, which enabled consistent WhatsApp access along with other internet-linked tools. These centralized spaces were often government-sponsored; in Kigali, the Rwandan Journalists Association provides a newsroom space with wireless internet for freelance reporters to gather, but journalists routinely speculate that some of the "freelance journalists" posted in the office are in fact government spies, watching out for especially critical enterprise reporting and other noteworthy developments. Outside of Kigali, consistent wireless internet could be found at local government offices, with J3 for instance walking to his local sector office every day to access the free internet and work on his reporting assignments. These gathering spaces of course make it easier to keep an eye on journalists, important in a place with an authoritarian image- management scheme and democratic-seeming media policies.

Finally, WhatsApp provides a platform for virtual gatherings of professionals, which facilitates resource sharing but also facilitates the spread of misinformation. J5 said he was part of several WhatsApp groups that he would scan for updates every morning when looking for economic story ideas: "People dump publications on the networks," he said. "Without wasting my time to look for it, someone has shared it. These networks facilitate my research work." He would be added to groups by contacts after finishing contract projects—for instance, after he finished a project with the Australian Broadcasting Corporation, a member of the team added him to a distribution list, where he now gathers information regularly. J4 noted that he is a member of five different WhatsApp groups for journalists and he used those spaces to gather helpful information but also had to be on guard for misinformation, which would often spread through these groups. "Everyone wants to break a story, wants to pretend to be on top of their game," he said. As a result, journalists might share ideas or tips before they had taken time to verify the information, with the goal of impressing their colleagues and seeming "on top of their game," but with the actual outcome of spreading sensational, unverified information. J4 had learned to check up on story ideas himself and verify through other contacts before he would pitch them to an editor. "If it is a true story, in Rwanda it will not take me one hour to verify," J4 said. "You'll get in trouble . . . with the government, with the editor. Your editor will not trust you if you share stupid ideas two or three times."

The gathering spaces encouraged by WhatsApp thus serve as a double-edged sword. On one hand, they encourage collaboration and information sharing, both by gently maneuvering journalists into physical locations with internet access—often shared spaces funded by government actors—and by creating virtual information-sharing forums. However, gathering brings downsides: physical gatherings are more easily intercepted by outsiders, who might steal ideas or just keep an eye on possible misbehavior. Virtual gatherings provide spaces for information and misinformation to circulate, as journalists eager to look in-the-know share ideas and rumors before verifying them for themselves.

Conclusion

Across sub-Saharan Africa, physical infrastructure that usually functions smoothly in the background often breaks down, and the tools available to journalists require adaptation and attention to function smoothly. This is evident in transportation and technology infrastructure such as roads, electricity, and fixed-line telephone networks, which have developed slowly and remain limited and poorly maintained (Calderón and Servén 2010; Platteau 1996). Infrastructure services are expensive and contribute to poor access; paved roads, reliable electricity, and telephone access have become less accessible across the continent in recent decades (Ajakaiye and Ncube 2010). New information communication technologies like mobile phones cannot overcome all the infrastructure challenges facing residents (Alzouma 2005). The limited availability and poor quality of material infrastructure influence news production; journalists in Rwanda (for instance) routinely sidestep stories that would incur exorbitant transportation costs compared to similar travel in the United States (Moon 2022). WhatsApp has many affordances that situate it to excel in this context of expensive electricity and airtime, limited technological availability, and inadequate wired information networks. It also benefits from the path dependency that comes from millions of users. For a journalist gathering information, the best tool is the one that the largest number of people are using. So while WhatsApp has technical limitations that other platforms have updated—for instance, Signal is seen as a more secure space—WhatsApp is often the newsgathering tool of choice. Journalists often haven't selected it intentionally so much as started using it because it's the tool that their sources use to contact each other and thus makes it easiest to get in touch easily and quickly. This was evident in the way that many journalists talked about using it—they didn't have specific reasons for adopting this tool but could talk about its benefits and the ways they incorporated it into their work. J1 specifically said that, as a journalist, he felt compelled to use all the messaging platforms and tried to reach sources where they were most comfortable (often on WhatsApp).

This chapter explores the ways the infrastructural capacity of WhatsApp shapes journalistic practice, especially in the areas of verification and information gathering. As an element of journalistic infrastructure, WhatsApp facilitates information access without physical presence; to generate collaborative networks, often across geographic distances and national borders; and to encourage communal gathering at internet access points. Each of these activities has potential ramifications for WhatsApp's relationship to misinformation and extreme speech.

In some ways, WhatsApp elicits increased skepticism from journalists in ways that might lead them to temper, rather than perpetuate, misinformation. When WhatsApp replaces physical presence, it disrupts journalistic reliance on eyewitnessing as a signal of authority. Journalists have long treated "being there" as an authoritative signal of authenticity and truth in their work (Zelizer 2007). When journalists, or other professionals, face routine disruptions, they tend to work especially hard to justify their behavior and to show how they meet professional standards (Coddington 2019). Journalists I spoke with insisted that they had established routines for verifying information they received via WhatsApp before they passed it along to editors or audiences. And indeed, the incentives of journalism in Rwanda make this likely: journalists accused of spreading false information are often subject to vague laws and punitive legal measures (Moon 2023). The result is that journalists work especially hard to verify information before publication if it came from nontraditional sources. Journalists are, of course, a special kind of information workers with more experience and training than most in sorting and verifying information. However, this finding suggests that one of the affordances that makes WhatsApp a virulent source of misinformation—the ease of sharing anonymous messages—has context-dependent ramifications that may slow the spread of disinformation in certain situations. By fostering virtual networks and physical gatherings among Rwandan journalists, WhatsApp contributes to spaces where information can be both shared and verified. The result is both an increased ability to share information quickly—a key factor in the spread of misinformation—and an increased ability to verify that information with a broadened network, which could help slow its spread.

Overshadowing all these affordances is the specific context of Rwanda, where an atmosphere of constant surveillance is reinforced by physical manifestations like identity cards and mandatory group activities (Purdeková 2015, 2016). The Rwandan state has used supposedly private WhatsApp data to prosecute legal cases, leading many journalists to conclude, legitimately, that the platform is not a safe space for rumors or private thoughts (Srivastava and Wilson 2019). The use of WhatsApp in this way is unusual, but the surveillance state tactics are not unique; surveillance in Zimbabwe, China, and elsewhere leads journalists to adapt their reporting techniques to the fact that "big brother" might be watching (Munoriyarwa and Chiumbu 2019; Palfrey 2020; Waters 2018). Thus, in the Rwandan context—and potentially in other surveillance-oriented authoritarian states—while journalists and others use WhatsApp frequently and find it useful in many situations, the safety and privacy of WhatsApp that often encourage the rapid spread of unverified misinformation are negated by the infrastructure of surveillance, the reminders that nothing is private.

WhatsApp is increasingly popular and complicated in Africa (Bailur and Schoemaker 2016; Pindayi 2017). It thus represents an important site of infrastructural innovation, symbolizing the tension of journalism practice in Africa's hybrid authoritarian states, which combine symbolic nods to democracy with power centralization and practical authoritarian constraints (Tripp 2004). Understanding the role of WhatsApp in journalistic newsgathering routines in this context illuminates the ways that infrastructural elements work together to shape the adoption and use of individual tools in particular situations.

10

Beyond Algorithms

How Politicians Use Human Infrastructure to Spread
Disinformation and Hate Speech on WhatsApp in Nigeria

SAMUEL OLANIRAN

During general elections in Nigeria in 2023, WhatsApp emerged as a prominent platform for communication, information sharing, and political discourse. With a user base of over nineteen million people in the country (Datareportal 2023), WhatsApp has become deeply ingrained in the social fabric and political landscape of Nigeria. Its ease of use, wide accessibility, and ability to reach large audiences quickly have made it a preferred channel for political campaigns, grassroots mobilization, and the spread of both genuine information and disinformation (Oyebode and Adegoju 2017).

In Nigeria's vibrant democratic environment, political parties and candidates have recognized WhatsApp's power as a tool to engage with voters directly. They form dedicated campaign teams and deploy them to create WhatsApp groups that serve as virtual hubs for party supporters and volunteers. These groups become bustling spaces where campaign updates, party manifestos, and candidate profiles are shared. Volunteers actively participate in discussions, organizing rallies and coordinating door-to-door campaigns. WhatsApp's interactive nature enables politicians to establish direct connections with voters, fostering a sense of personal engagement and accountability (Chagas 2022; Dvir-Gvirsman et al. 2022).

However, WhatsApp's use during elections is not limited to official campaigns (Hitchen et al. 2019; Sahoo 2022; Santini et al. 2021). Grassroots movements, civil society organizations, and concerned citizens also leverage the platform to raise awareness, advocate for issues, and mobilize support. Social and political movements, such as the

#BringBackOurGirls campaign and #EndSARS protest (Ogbonnaya 2020), have effectively utilized WhatsApp to disseminate information and rally public support for their causes. By sharing news articles, videos, and personal accounts, these movements created a network of engaged citizens who could drive change and influence public opinion.

Yet, alongside its positive impact, WhatsApp has also been associated with the spread of disinformation and hate speech during election periods (Cheeseman et al. 2020). As the most popular social media platform in Nigeria (Statista 2022a), it easily becomes a hotbed for political disinformation. The ease with which messages are created and shared makes it challenging for researchers to track harmful content or the users spreading it. These limitations also force political actors to rely on deliberate human action to create and spread disinformation on the app during campaigns, what Nemer (2021) termed as the "human infrastructure" of disinformation.

The actions of tech-savvy users online aided by the algorithmic affordances of platforms premised on sharing content through mechanisms such as status updates and broadcast make cross-platform migration of content possible. The integrated nature of contemporary information ecosystems globally makes it crucial to now move beyond single-platform approaches to better understand how harmful content travels *in-between different social media channels*. This is especially relevant in countries like Nigeria where the information ecosystem is rapidly changing—such as the current challenge posed by TikTok for the spread of disinformation (Madung 2022)—or during critical events such as elections or the EndSARS protests that elevate the risks of inauthentic and coordinated behavior (Adekoya 2021). There are several reasons why it is important to study cross-platform behavior of sharing disinformation content on Twitter and WhatsApp. Research on mis/disinformation and social media has focused on automated content, coordinated behavior, networked propaganda, or problematic content spread by regular users on Facebook and Twitter) (Chadwick, Vaccari, and O'Loughlin 2018; Valenzuela et al. 2019). More people are turning to messaging apps to consume news or participate in politics online, and given the link between social media use, news/information consumption, and political participation (Gil de Zúñiga et al. 2019; Newman et al., 2023), it has become imperative to shift attention to the largest private messaging app in Nigeria and understand how its unique affordances make it endearing to propagators of disinformation.

By adopting a human infrastructural lens, this chapter brings together narratives, user elements, and tactics to investigate how disinformation is coordinated across platforms by agents. This chapter focuses on WhatsApp groups as networks to advance the argument that offline and online structures are interlinked, and that they reinforce and build on each other in interesting ways. As a result, in many respects WhatsApp amplifies the significance and influence of networks that already exist within the Nigerian political space and broader society.

The chapter is divided into five parts. The first part examines existing scholarship on WhatsApp, emphasizing its framing as human infrastructure. It also explores Twitter's role in amplifying content and its interconnectedness with WhatsApp in disseminating disinformation and hate speech. The second part delves into the historical context of disinformation campaigns in Nigeria. Part 3 outlines the methodological approach for researching cross-platform flow of disinformation, and part 4 focuses on human-networked dissemination of problematic content and the exploitation of WhatsApp's versatility to amplify messages. Finally, part 5 presents the chapter's conclusions.

WhatsApp's Human Infrastructure

Infrastructures, as defined by anthropologist Brian Larkin, are "built networks that facilitate the flow of goods, people, or ideas and allow for their exchange over space" (2013). This framework is used to understand how politicians tap into the interpersonal connections and communication channels facilitated by human networks to disseminate messages. What happened during the 2019 presidential election in Nigeria debunks the idea that WhatsApp is a level playing field (Hassan and Hitchen 2022). WhatsApp's peer-to-peer encrypted architecture may give users a sense of security and privacy, since there is no algorithm intervening in their messages and content shared is part of deliberate human interactions. It may also give them a sense of spontaneity since the app allows anyone to produce and share content. However, the 2023 campaign relied on disinformation that was systematically created and spread by a human infrastructure that orchestrated a targeted campaign. For example, some messages were circulated that the All Progressives Congress candidate, Bola Ahmed Tinubu's, choice of running mate, Ahmed Shettima, a Muslim like Tinubu but from the North East of Nigeria, was a plot to Islamize Nigeria. Peter Obi was described in some disinformation messages as being a supporter of the Indigenous People of Biafra (IPOB) and should not be voted for. It is hard to verify the exact impacts that digital populism had on the 2023 presidential election. The technical and human infrastructures behind social media platforms are not enough to guarantee electoral victory (Olaniran and Diepeveen 2023). While these infrastructures may facilitate the dissemination of political messaging (including disinformation and hate speech), their effectiveness in influencing election outcomes depends on various factors beyond the structural components of social media platforms. WhatsApp's design prioritizes human interaction and relies heavily on personal contacts and group dynamics for content dissemination. Unlike Twitter, where the technological structure, such as algorithms and trending topics, plays a significant role in determining content visibility and virality, WhatsApp's closed-group nature places greater emphasis on interpersonal relationships and individual interactions facilitated by dedicated volunteers, party activists, and sympathetic influencers.

Disinformation messages that originate from WhatsApp can take on a life of their own as they move across other platforms. However, controlling how a message spreads on WhatsApp requires an effective human structure, unlike on Twitter where the technological structure takes precedence. As Hassan and Hitchen (2022) noted, political parties in Nigeria like the All Progressives Congress (APC), Peoples Democratic Party (PDP), and Labour Party (LP) and their support groups go the extra mile to co-opt existing nonpolitical WhatsApp networks into their structure to ensure that their messages reach the widest possible targets.

Examining disinformation from an infrastructural lens reveals how disinformation operations extend well beyond the use of social media and the construction of false narratives. As Pasquetto et al. argued (2022), while disinformation continues to evolve and adapt, the overarching (dis)information infrastructure through which "epistemic evidence" is constructed and constantly updated is rather stable and has increased in size and complexity over time.

Disinformation Campaigns in Nigeria

Disinformation campaigns in Nigeria didn't start with social media. Before the proliferation of social media, propaganda was spread via radio, street leaflets, and newspapers (Okoro et al. 2018). The absence or lack of information about important issues in the Nigerian society or media space encouraged the appearance and spread of rumors on them. These rumors not only filled the informational vacuum but also contributed to shaping perceptions of events. They circulated among individuals seeking to understand ambiguous or potentially threatening situations.

Modern disinformation has its origin in black propaganda (typically used to vilify an enemy or opponent through misrepresentation) (Becker 1949; Jowett and O'Donnell 2012). Thus, deception and manipulation might take new forms in modern disinformation, but the underlying roots and causes would remain the same. The APC spread disinformation during the 2015 election to vilify Goodluck Jonathan and misrepresented data in the form of creative deceit to claim the PDP was corrupt and looting the national treasury of the country. The PDP also responded with counterpropaganda that the APC was a party of strange bedfellows made of aggrieved politicians who lacked a common political ideology and credibility.

To spread the false narrative that another party was dishonest and plundering the nation's treasury, APC stalwarts and supporters circulated unemployment statistics and photos of uncompleted or poorly executed projects by the PDP administration that turned out to be fake (Oyeleke 2021). The PDP, through its agents, spread countermessages, some of which targeted the presidential candidate of the APC, Muhammadu Buhari, claiming he possessed a fake academic record and was sympathetic to the cause of Islamic fundamentalism in Nigeria.

The success of any disinformation campaign stems from thorough research and meticulous planning by its instigators (Freelon 2017; Marwick and Lewis 2017), in addition to context, tone, and timeliness of the message. Propagators of disinformation also leverage their understanding of the "rhythm" of communication, knowing when to disseminate certain messages and when to abstain (Wardle and Derakhshan 2017). For instance, when Cambridge Analytica attempted to interfere in Nigeria's 2015 election, their aim was to undermine confidence in the election's legitimacy and erode trust in the nation's democratic processes (Akindipe 2023; Bradshaw and Howard 2018; Dowling 2022; Obodo 2022). This campaign sought to exacerbate ethnic and religious tensions by exploiting existing or perceived divisions within the country, inadvertently impacting Nigeria's democracy.

Relationships and power networks are integrated in digital political information. For instance, George's (2016) research of case studies of "hate spin"—manufactured vilification or outrage—used as a political tactic to enlist supporters and intimidate opponents discovers that propagating hate is a common political tactic employed by ruling governments. Udupa (2018a) also noted a combination of the following elements in situations where disinformation appears to have had a significant impact: democratic structures are weak; democratic institutions have been depleted; instruments of the state sponsor and circulate disinformation; mainstream political parties/actors support the disinformation campaign; and audiences perceive an electoral crisis. Nigeria's electoral management body, the Independent National Electoral Commission (INEC) has suffered a bad image following several flawed past elections (Saliu and Ifejika 2017). Its reputation was so negatively affected that many feared the 2023 general elections would not yield anything different from the norm of delivering a flawed election. INEC was targeted by disinformation campaigns during the voting exercise, with some claiming it was aiding rigging, voter suppression, and disenfranchisement to favor the ruling party.

Disinformation is used as a political weapon alongside other tactics such the repression of opposing viewpoints, the erasing of accurate accounts, and the use of physical force (Dutta and Gangopadhyay 2019). For instance, when political disinformation that exploits majoritarian politics is spread in Nigeria by networks and organizations associated with the country's political party in power, it is swiftly propagated through digital media (Twitter, WhatsApp, Telegram, and Facebook) with collaboration from offline networks. Ruling parties are often accused of funding violent incidents while simultaneously undermining democratic outlets for alternative voices (Hassan and Hitchen 2022).

This chapter is guided by two interlinked questions:

- First, it asks if the increasing use of WhatsApp offers insights to how the app's informal design is leveraged to spread disinformation during the 2023 presidential election in Nigeria.
- Second, it explores whether WhatsApp groups replicate existing supporter and commercial customer networks on Facebook and Twitter, thereby providing a platform for a range of actors to enter the political arena.

Method and Data

Building on the works of Krafft and Donovan (2020), Marwick (2018), Marwick and Lewis (2017), Starbird, Arif, and Wilson (2019), and Tufekci (2014), this chapter is grounded in the premise that disinformation is better understood as a process that is actively designed and curated by key individuals, via different degrees of coordination rather than a machine-controlled activity.

Since chat groups on WhatsApp are mostly private, they are much harder to monitor than Facebook or Twitter discussions. Because of that, I contacted the group admins to get access to messages posted in the private WhatsApp groups dedicated to the presidential election. The groups are made up of senior party officials, communications consultants, campaign strategists, and campaign coordinators. The data obtained was used to investigate the structure of private WhatsApp groups and compare its characteristics with a synthetic social network like Twitter.

I collected 56,087 tweets published between September 28, 2022, and December 31, 2022, from the Twitter Academic API using selected keywords ("bola ahmed tinubu," "BAT," "drug dealer," and "tinubu documentary"). The time period was selected to consider how a popular disinformation campaign that targeted one of the presidential candidates, Bola Ahmed Tinubu and Peter Gregory Obi, evolved and spread between WhatsApp and Twitter leading up to the election. A total of 24,264 unique users were identified in the dataset.

The Twitter data was first processed as "interaction network," which captures all the interactions of Twitter users (retweets, mentions, replies, and quotes). I visualized the networks in Gephi to gain further insight into how different clusters interacted and evolved. To facilitate this exploration, the Leiden algorithm was used for community detection and ForceAtlas 2 for setting up the layout of the networks. The aim of this first stage was to map the users, relationships, and content related to the disinformation on Twitter during the study period. Interactions are analyzed and visualized across the entire dataset as well as temporarily focusing on peak periods (see Figure 10.1).

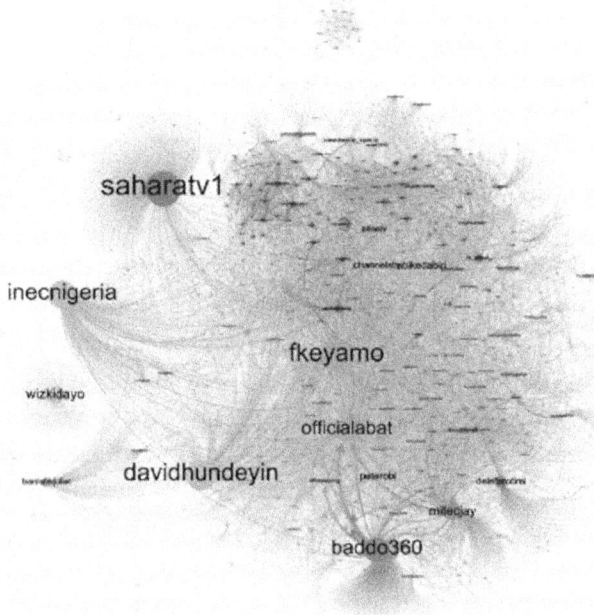

Figure 10.1: Interaction map of the "Tinubu the drug dealer" campaign. The peaks were primarily driven by reactions to allegations of Tinubu's involvement in drug trafficking, criticism of his age and physical stamina, and discussions regarding his suitability for the presidential office. Source: Author

Organizing Human Networks

WhatsApp emerged as a crucial tool for political parties and campaign groups to coordinate events and disseminate messages promoting candidates' credentials while criticizing opponents. The organizational structure within WhatsApp groups facilitated top-down and bottom-up communication, ensuring the efficient dissemination of campaign materials and instructions. Human networks, comprising supporters, volunteers, and influencers, played a pivotal role in amplifying political messaging across geographically dispersed locations. The parties formed at least one WhatsApp group in each of the thirty-six states to mobilize supporters and coordinate campaign groups. Leaders of these groups coordinated smaller support groups and campaign activities within their domain. The coordinators were managed through another WhatsApp group headed by a "director of campaign support groups." Messages from the national campaign group went through them and before cascading to the state/LG/ward level groups they micromanaged.

In a politically charged environment, contesting parties leave nothing to chance when it comes to soliciting votes. The blame game and mudslinging become prevalent, and such conditions provide fertile ground for creation and spread of mis/disinformation (Farooq 2018; Garrett 2017). The spread of disinformation and fake news became rife as the election date approached. In November 2022, Bola Ahmed Tinubu, the presidential candidate of Nigeria's ruling All Progressives Congress (APC), was mentioned in an alleged drug trafficking case (Majeed 2022). Viral media reports linked Tinubu to a drug trafficking case and a $460,000 forfeiture to the American government. It soon became a political weapon and one of the popular disinformation campaigns that resonated with the election. Although Tinubu and his party denied any such claims, short video clips and documentaries were shared widely on WhatsApp.

WhatsApp groups serve as virtual spaces where supporters congregate to discuss political issues and share content, thus facilitating the exchange of ideas and information across geographically dispersed locations. As Larkin (2013) demonstrated in his study, politicians construct and utilize human networks comprising supporters, party members, volunteers, and even paid influencers. These networks are strategically built to reach specific demographics of the target audience within the electorate. Rather than relying solely on technological infrastructures, politicians tap into the interpersonal connections and communication channels facilitated by human networks.

The study revealed the prevalence of four types of private WhatsApp groups—approval group, policy group, media team, and a general communication group—each serving distinct purposes in disseminating campaign content. Save the general communication group, which had a membership of twenty, others were limited to seven people. Campaign messages go through a vetting process by the approval group before the media team sends it to the general communication group. Where there is information (for example publicity content or response to viral negative publicity) to be circulated, the director of the support groups sends a link (the message or a list of target accounts) to a supporter's group to be sent to their Twitter counterparts to ensure virality, using their own initiative. However, the shift in disinformation tactics was evident, with the reliance on machine coordination diminishing due to platform updates and regulatory measures. The spread of manipulated narratives highlighted the emotional potency of disinformation and its impact on public discourse during elections.

Meta's introduction of a Nigeria-specific elections operations center was among several measures introduced to identify potential threats in real time and speed up responses to flagged cases during the election (Ikenze 2023). As corroborated by one of the group administrators, the microtargeted campaign executed on WhatsApp (the sending of bulk SMS to voters based on their location and ethnicity) in 2019 was not possible in 2023. Instead, they developed more campaign messages in local languages, relying on human consultants. Other Meta-owned platforms (Instagram and Facebook) were used as alternative channels, though they were not as effective. Word-of-mouth spreaders, known as "canvassers," were dispatched to the interiors to distribute merchandise (T-shirts, caps, fabrics, packaged food items, etc.) with local inscriptions. WhatsApp was used to monitor such activities, respond to safety issues, and evaluate the campaign exercise.

Exploiting WhatsApp's Versatility and the Influence of Fear

The choice of WhatsApp as a primary communication tool stemmed from its widespread adoption and accessibility across diverse demographics in Nigeria. Despite challenges such as infiltration by malicious actors, its end-to-end encryption instilled confidence in users, fostering a sense of security in communication. However, updates to the platform's privacy features and content moderation mechanisms signify a shift toward greater accountability in combating disinformation. One of the WhatsApp group administrators affirmed:

> WhatsApp was the most active platform in Nigeria. Every demographic is on WhatsApp, and it is easier for coordination of large groups. Unlike email chains that require follow ups to ensure people checked their email on time, WhatsApp is handy. The end-to-end encryption increased our trust in the app. We felt safe with communication, save a few cases of infiltration by human moles. The sense that our communication would not be breached because it was encrypted gave us confidence. It was efficient in that we could communicate instantly or coordinate meetings at the click of a button without the stress of creating links. We could call the group on the spot to agree on an urgent action.

Other measures include user reporting of content as well as deceleration of viral messages (Farooq 2018; Reis et al. 2020). This has also enhanced the app's ability to track the source of a message (Pennycook and Rand 2021). The reliance on machine coordination that sent out automated messages to ten thousand voters every thirty minutes in 2019 is no longer available for party digital influence operators. One of the administrators asserted they used INEC's verified voters' register to bulk message voters, up to ten thousand numbers every thirty minutes, without having them as individual contacts by using third-party software. Other updates have switched the cards, forcing the parties to rely on human infrastructure.

Instances such as the dissemination of a fabricated viral message (see tweets below) alleging Bola Ahmed Tinubu is sick and unfit to run for the office of the president or the claim that likened Bola Ahmed Tinubu to Colombia's Pablo Escobar gained momentum due to the power of emotion (Horner et al. 2021).

Figure 10.2: Time-based distribution of the data with peaks. Source: Author

@MrRefor So u are agbado eran boy dat he send to ask our incoming president Peter Obi question about IPOB abi. Shame on u, u are a disgrace to Nigeria Youth's. Is only a ppl lk u with half brain dat will be supporting a drugs Lord with fake identity to be ur President.

@Ife_dayo760 @TheNationNews Really? I never knew that! The same IPOB that isn't supporting Peter Obi? Tinubu that is a drug lord & Shettima is a sponsor of Boko Haram but u all close ur eyes to that. And we have shown u pple proof. Useless argument. Show us proof Obi is IPOB leader. Mumu all of u.

The fear factor in politically charged environments amplifies the spread of disinformation, as supported by Pocyte (2019) and evidenced in the heightened engagement during peaks of emotionally charged narratives (see Figure 10.2). However, not all misinformation is rooted in fear; ideological motives and the pursuit of social reform also drive the creation and dissemination of distorted information. The delay in platform intervention exacerbates the challenge of mitigating the impact of disinformation, as human infrastructures evolve and adapt to circumvent technological filters. As Udupa (2015) noted, ideology is a formidable factor behind the use of the internet. In some instances, users suspend critical thinking and verifying, and cross-checking news or information becomes minimal. The delay in platform owners intervening exacerbates the challenge of maintaining a safe online ecosystem. As time passes, human infrastructures evolve, establishing new layers and migrating across platforms. Furthermore, they leverage existing relationships, making it increasingly difficult to mitigate their impact.

Disinformation is also a source of money for many who operate multiple or fake accounts or strive to monetize their online popularity and influence. Lack of ideology and use of partisan media characterize this type of electoral disinformation (Guess et al. 2021; Gupta et al. 2023), with disinformation spread by incumbents, further muddying the waters of institutional trust. Trust in institutions significantly influences beliefs regarding electoral disinformation, with varying effects produced by different institutions. Past instances have seen official government sources flagged for disseminating disinformation claims, particularly targeting the judicial system during previous elections. The results of this chapter add nuance to the notion that politicians exploit the trust, social ties, and influence wielded by individuals within human networks to amplify their messaging.

Politics and ideology are not the only reasons people create and spread disinformation. Operators of many fake/multiple social media accounts understand that certain people and issues get more attention and therefore more hits. Using religion, nationalism, patriotism, and gender, these users share manipulated or harmful content on various platforms, including WhatsApp. Not only does this help them earn money, it creates an atmosphere where criticism of certain political actors or ideologies guarantees message virality. Some of these accounts are operated by trolls and paid social media influencers, who not only share their content but also defend it and confront voices that are critical to their thinking.

A significant finding of this chapter is that disinformation shared on WhatsApp cascades to other platforms like Twitter and gets spread by channels other than news sources. As shown in Figure 10.1, although the news media (@channelstv, @arisetv, and @saharareporters) recorded high engagement with the "Tinubu is a drug dealer" allegation turned disinformation campaign, there was a strong presence of nonmedia accounts in the gray cluster (@vawulence_space, @chudemedia, @moreish7, @wutang464, @fs_yusuf_, @obatojo, and @steelwrld1). The peak recorded in Figure 10.2, triggered by this campaign, which saw @fkeyamo and @officialbat leading Tinubu's side of the narrative, was corroborated by one of the WhatsApp group administrators in the interview. Other users like @baddo360 and @mitecjay, despite their small followings, were among the loudest on the topic. Most of the top-performing accounts were created in 2022. Other users, like @baddo360 and @mitecjay, despite their small following, were among the loudest on the topic. Most of the top-performing accounts in the dataset were created in 2022, suggesting that a surge in new accounts was strategically timed to amplify specific narratives ahead of the presidential election.

As political parties organized Twitter Spaces almost as daily discussions, hosted by hired social media influencers or party activists, efforts were made to delegitimize opponents, boost the profile of their own candidate, and galvanize supporters and would-be voters according to "WhatsApp instruction." Spaces are a public, ephemeral, live audio conversations feature that allows for open, authentic, and unfiltered discussions on any topic, from small and intimate to millions of listeners. During the election, this feature gave users the opportunity to express themselves without restriction on the app and reach more audiences since anyone can join as a listener, including people who do not follow the host.

One of the biggest Twitter Spaces was hosted by Peter Obi (@Peter-Obi) on May 25, 2023, which had forty-one million live listeners. Twitter Spaces featured prominently in the 2023 election and was used in spreading awareness of politics in Nigeria.

Similarly, with its reach in terms of number of users, WhatsApp remains useful to coordinate election campaigns and circulate multimedia content. Voice notes remain critical, especially in local languages, with content regularly played and replayed to a voting audience that has no direct online access, and is less literate or visually impaired.

Conclusion

The widespread dissemination of disinformation on platforms like WhatsApp, driven by financial incentives and political agendas, presents a significant threat to democratic processes. The link between WhatsApp usage and the spread of electoral falsehoods underscores the necessity for a nuanced comprehension of private messaging dynamics and their implications within political contexts.

Politicians strategically leverage loyal supporters and influencers to amplify their messages within personal networks and WhatsApp groups, thereby enhancing their credibility and reach. The platform's accessibility, particularly for individuals with limited literacy skills, coupled with its audio-visual features, facilitates cost-effective dissemination of content, raising concerns about the ease of spreading misinformation.

The increasing use of WhatsApp during elections supplements the roles of other social media platforms, broadening the channels through which disinformation proliferates. Political parties exploit WhatsApp for various purposes, including organizing discussions, promoting candidates, and mobilizing supporters, leveraging its expansive user base and multimedia capabilities.

Voice notes, particularly in local languages, play a significant role in reaching voters with limited online access or literacy, further amplifying the impact of disinformation dissemination. Politicians capitalize on human networks as infrastructures to propagate disinformation and hate speech, exploiting interpersonal relationships to circumvent technological filters and shape political discourse.

Understanding the diverse functions of platforms like WhatsApp is crucial. While direct promotion of electoral disinformation may not be inherent to WhatsApp's news consumption, engagement in political groups correlates with belief in such falsehoods (Evangelista and Bruno 2019). This underscores the imperative of addressing user motivations and behaviors to effectively combat disinformation proliferation.

11

Dis/Misinformation, WhatsApp Groups, and
Informal Fact-Checking Practices in Namibia,
South Africa, and Zimbabwe

ADMIRE MARE AND ALLEN MUNORIYARWA

This chapter contributes to our understanding of organic and informal user correction practices emerging in WhatsApp groups in Namibia, South Africa, and Zimbabwe. This is important in a context where formal infrastructures of correcting and debunking dis/misinformation have been dominated by top-down initiatives. These formal infrastructures include platform-centric content moderation practices and professional fact-checking processes. Unlike social platforms such as Twitter and Facebook, which can perform content moderation and hence take down offending content, the end-to-end encrypted (E2EE) infrastructure of WhatsApp creates a very different scenario where the same approach is not possible. This is because only the users involved in the conversation have access to the content shared, shielding false and abusive content from being detected or removed. As Kuru et al. (2022) opine, the privacy of end-to-end encryption provides a highly closed communication space, posing a different set of challenges for misinformation detection and intervention than with more open social media, such as Facebook and Twitter. In this regard, false and misleading information on WhatsApp constitutes "a distinctive problem" (Kuru et al. 2022; Melo et al. 2020). As Reis et al. (2020, 2) observe, "the end-to-end encrypted (E2EE) structure of WhatsApp creates a very different scenario" where content moderation and fact checking at scale is not possible. Fact-checking WhatsApp groups, which have been flagged as the major distributors of mis- and disinformation is equally difficult. This means that alternative fact checking occurring at the margins needs to be investigated on their own terms. Although this problem is not unique to

WhatsApp, given the popular usage of this platform in the Global South where professional fact- checking initiatives are often disconnected from the grassroots, it becomes imperative to examine informal fact-checker practices and cultures.

Given the ubiquity of false and misleading information on WhatsApp, there have been loud calls for necessary and proportionate regulatory, policy, and technical interventions (Mare and Munoriyarwa 2022; Reis et al. 2020). In response to these calls, Meta, the parent company of WhatsApp, has put in place interventions aimed at the production and distribution mechanisms (Mare and Munoriyarwa 2022). These measures include inserting warning labels that the message has been forwarded more than five times and limiting the forwarding of viral messages. WhatsApp has introduced a feature allowing users to change their settings so others cannot add them to groups, or so only certain people can do so. Meta has also partnered with fact-checking organizations around the world. These measures have been criticized for being inadequate and halfhearted given the enormity of the information disorder in the platformized ecosystem. In some jurisdictions, national governments have proposed the breaking of encryption as a way to enable moderation and law enforcement on WhatsApp (Wasserman and Madrid-Morales 2022). For instance, in 2018, the Zambian government made international news headlines when it announced that all WhatsApp group administrators would be required to register their groups and set up codes of ethics or risk being arrested if there was a breach. It is within this regulatory vacuum coupled with halfhearted platform-specific interventions that "the burden of correcting misinformation [has tended to] rest on users" (Ng and Neyazi 2023, 426). WhatsApp, therefore, relies heavily on active participation from its users to reduce misinformation. However, these informal and everyday user correction practices and cultures on WhatsApp are yet to be systematically explored in Africa.

Southern Africa is one of the most mobile-centric regions on the continent. The influx of mid-to low-cost smartphone brands from Asia has boosted access to these devices in Southern Africa. Countries such as Namibia, South Africa, and Zimbabwe have an average of 90 percent smartphone penetration rate (Internet World Stats 2023). Because of these high smartphone penetration rates, WhatsApp has emerged as one of the most popular messaging platforms in Southern Africa (Mare

2023). In Namibia, South Africa, and Zimbabwe, telecommunication companies have launched social media bundles. WhatsApp data bundles are some of the most popular in that bouquet. This has contributed significantly to the ubiquitous presence of WhatsApp in the everyday lives of most citizens in these three countries. As of 2022, 58 percent of South African mobile phone owners used WhatsApp Messenger, making it the most popular social media app (Statista 2022b). As far as Namibia is concerned, WhatsApp accounts for 98 percent of all instant messages sent on the country's major telecommunications operator, MTC (Communications Regulatory Authority of Namibia 2022). In Zimbabwe, WhatsApp accounts for half of the country's internet traffic (Postal & Telecommunications Regulatory Authority of Zimbabwe 2022). This means that the app is the default channel for accessing the internet for most citizens in these three case nations. Yet, there are no studies on how everyday users of the app negotiate exposure to false and misleading information.

In view of research lacunae, this chapter critically examines the extent to which informal fact-checking practices on WhatsApp groups hold promise (or not) for sustainable organic and informal user correction practices in Southern Africa. As Ng and Neyazi (2023, 426) observe, "Instant messaging platforms like WhatsApp have limited systems in place to reduce misinformation." Thus, an investigation of nascent forms of informal fact-checking manifesting at the margins of the formalized user correction practices has the potential to illuminate interesting practical and regulatory insights relevant for platform companies and policymakers in the Global South. Less is known about informal fact-checking practices and cultures manifesting on WhatsApp groups in the Global South. In this chapter, we explore the following questions: (1) How do administrators and members within WhatsApp groups experience and respond to the sharing of misleading and false information in a group context? and (2) What do nascent forms of fact-checking look like in WhatsApp groups, which are generally inaccessible to professional fact checkers?

Conceptual Frameworks

We draw on the concepts of *informality* and *provisionality* as well as Bruns's (2011) gatewatching theory in order to analyze the informal user correction practices and cultures within WhatsApp group contexts. By foregrounding concepts like *informality* and *provisionality*, we intend to shed light on how everyday users of the app negotiate the production, circulation, and consumption of false and misleading information, outside of the reach of platform companies and professional fact-checking organizations. Building on Obadare and Willems's (2014) insightful observation that *informality* is central to the way both state officials and citizens exercise agency in Africa, we argue that everyday users of WhatsApp are not waiting for external salvation from professional fact checkers to clean up their digital public sphere. Instead, they are using local resources, knowledge, and skills to push back against the avalanche of false and misleading information. The notion of *informality* is often associated with unregulated, unplanned, and unsystematic ways in which people make do at the margins of formal laws, regulations, and procedures (Banks, Lombard, and Mitlin 2020). We deploy the term "informal" to underscore the everyday *post-hoc* user correction practices and cultures deployed by those at the margins as a way of coping with information disorders. We define "informal user correction" as the unofficial, uncoordinated, and unorganized process of verifying the factual accuracy of questionable information, reports, and statements circulated on social media platforms. There are no written rules, codes of ethics, or systematic methodology involved in the process of informal user correction and fact-checking. The informal fact checker relies on their intuition, epistemic capital, social networks, media literacy skills, and investigative skills to verify the veracity of information circulated in closed WhatsApp groups. This is slightly different from "self-correction," which refers to correcting oneself after sending untrue information (Arif et al. 2017). However, it has similarities with "social correction," which denotes correcting others who have shared untrue information and with whom one has a social relationship (Bode and Vraga 2018). These episodic acts of authentication and verification may appear disorganized, fragmented, and unscientific at the surface but represent everyday forms of user correction by certain groups that lack media and information resources.

Our chapter also draws on the analytical strengths of the concept of *provisionality*. This concept represents the temporary, flexible, and open-to-change arrangements prevalent in the lives and interactions of ordinary people. Thus, informal user correction on WhatsApp groups represents *provisional* practices and cultures, exploring the inherent tensions between the formal and informal, the professional and unprofessional, the truth and falsehoods, and the acceptable and unacceptable. It denotes the ways of doing and being, and both a mode and method of verifying the authenticity of news and information in a crowded information environment. As a *way of doing*, informal fact-checking are everyday practices and routines that are arranged or created for the time being. They are intended as temporary, flexible, and stopgap measures (Simone 2018). It is a way of navigating the volatile, unpredictable, complex, and ambiguous information and communication environment. Unlike formal processes, provisional practices are predicated on making do for the present. As Simone (2018, 13) aptly observes, there are often "a wide range of provisional, highly fluid, yet coordinated and collective actions [that] are being generated that run parallel to, yet intersect with" professional fact-checking practices and cultures. Hence, provisional fact-checking practices focus on the ephemeral, emergent, and cross-cutting forms of verifying the authenticity of information shared in a group context.

Another concept that we deploy in this chapter is *gatewatching* (Bruns 2011). In its latest reincarnation, it is constructed as a replacement of gatekeeping processes in the digital age.[1] Noteworthy to highlight is that gatekeeping as a practice (Lewin 1947) was fundamentally born out of an environment of scarcity (of news channels, and of news hole space within those channels). Unlike gatekeepers, gatewatchers on social media platforms gather and curate information from a wide range of sources based on their commonsense understanding of what is in the public interest. The process of gatewatching as a social practice has always been part of journalism. However, it has assumed added importance in the digital age. Conceptualizing WhatsApp administrators and self-appointed informal fact checkers on WhatsApp as *gatewatchers* helps us to view them as information intermediaries and mediators. This voluntary or user-initiated form of fact-checking on WhatsApp

---

1. "Gatekeeping" refers to the process by which selections are made in media work, especially decisions whether or not to admit a particular news story to pass through the "gates" of a news medium into the news channels.

groups represents the everyday and informal ways of dealing with the scourge of mis/disinformation. Although WhatsApp group administrators and members have no ability "to keep—to control—the gates of any of these channels . . . however, what they are able to do is to participate in a distributed and loosely organized effort to watch—to keep track of—what information passes through these channels" (Bruns 2011, 121). In this context, informal user correction on WhatsApp groups often manifests as a crowdsourcing effort involving a number of group members with diverse media and information literacy skills and competencies. Thus, some group administrators and members engage in a form of internal gatewatching by tracking, following up, and verifying the veracity of externally sourced information and news circulated on the app. These varied yet unorganized and informal ways in which WhatsApp administrators and users exercise agency within the context of the group constitutes the main focus of this chapter.

Methods and Materials

We contribute to new knowledge on informal user correction practices and cultures by using Namibia, South Africa, and Zimbabwe as illustrative case studies. These three countries were chosen because of their heavy use of WhatsApp in everyday communication (Internet World Stats 2023). They were also chosen because of increased cases of mis- and disinformation in the three African countries (Wasserman et al. 2019). Several researchers have acknowledged that WhatsApp is very hard to research largely because of its encrypted nature, which means that one cannot easily track how messages are shared, or who shares them (Cheeseman et al. 2020, 146). In order to circumvent these challenges, interviews with fifty WhatsApp group administrators and users were conducted to explore their understanding of informal fact-checking practices and cultures and to gauge their reactions to emergent forms of social correction in a group context. WhatsApp group administrators were chosen because of their personal and professional roles as gatekeepers, intermediaries, and mediators of group interactions. They have control primarily over who enters the group and also have the privilege to remove participants. WhatsApp hosts a diverse range of groups. These include religious, sporting, family, work, school, and professional associations. These spaces are constructed as ideal for communicating care and phatic exchanges. In order to have access to the conversations on purposively selected WhatsApp groups in the three case nations, we sought permission from administrators.

However, there were challenges because we realized that there are some WhatsApp groups with settings that give all users administrator rights. Some of the admins allowed us to participate and observe group interactions. Others refused to grant us permission. Before starting our participant observation phase, we briefed all the group members who were present at the time we were added to their WhatsApp group about the purpose of our research. In the end, we worked with those who were receptive to our research agenda. In Namibia, we managed to join three WhatsApp groups constituted by religious, football fans, and alumni communities. In South Africa, we participated in three religious, football fans, and academic professional WhatsApp groups. In Zimbabwe, we immersed ourselves in three alumni, football fans, and journalists WhatsApp groups. We joined most of these WhatsApp groups in March and April 2022. Most of these groups had 256 members. Very few groups had fewer than one hundred members. We also realized that academic and football fans groups were mostly dominated by men. The only groups where women were episodically visible were religious and alumni associations. This suggests that these WhatsApp group interactions and discussions are gendered, and classed.

We conducted a deep-dive participant observation in nine WhatsApp groups in three case nations for a period of twelve months. This extended period of time allowed us to engage in background listening, interacting, observing, and archiving of data on WhatsApp groups. Observation of WhatsApp group interactions enable researchers to witness and monitor users' self-reported activity in real time, from real-time data. Through structured participant observation, we were able to observe conversations, interactions, posting, and commenting behaviors of WhatsApp group administrators and members. Most of the conversations were conducted in English. However, there were cases where local languages such as ChiShona, IsiNdebele, Oshiwambo, Herero, Damara Nama, IsiXhosa, Sotho, Tswana, and isiZulu were used. Taking into consideration ethical guidelines, we decided to observe and document cases where false and misleading information was shared, debated, and fact-checked by group members. Directly observing how group administrators and members from the three case nations interacted with, negotiated, and navigated false and misleading information on WhatsApp enabled us to get insights into the nascent forms of fact-checking emerging at the margins of professional fact-checking initiatives.

We complemented our participant observation with thirty-two in-depth interviews with WhatsApp group members. We also interviewed eighteen WhatsApp group administrators. In each group, we interviewed two administrators on how they dealt with false and misleading information shared by the members. In total, we interviewed fifty respondents. Fifteen of these respondents identified as female while thirty-five identified as male. These interviewees were purposively sampled from the nine WhatsApp groups. All the interviews were conducted on WhatsApp. We made use of voice notes, chat, and calls (voice and video) to communicate with our research participants. On average, our WhatsApp interviews lasted between twenty and thirty minutes. In cases where the participants had no data bundles, we offered to buy the bundles for them. These interviews were conducted between August 2022 and July 2023. Our interviewees were scattered all over the world, although they belonged to WhatsApp groups in Namibia, South Africa, and Zimbabwe. Most of these research participants were between the ages of fifteen and seventy years. The respondents were asked to recall and reflect on specific incidents or "social media events" where false and misleading information was shared, debated, and pre- or debunked in the group context.

All the interviews were recorded and transcribed soon after the data collection phase. All the data from participant observation and interviews was anonymized. In cases where we use names, we rely on pseudonyms to protect the identities of our participants. In order to ensure transparency, we sought informed consent from the group administrators and members before commencing with the data collection. Everyone present during the time we joined the WhatsApp groups was briefed about the purpose of the research. We also guaranteed full anonymization for our research subjects. In this endeavor, we were guided by South Africa's Protection of Personal Information Act (POPIA) (2013), which came into effect in July 2021. In Zimbabwe, we paid special attention to the Data Protection Act of 2021 in Zimbabwe. Data was thematically analyzed in line with Braun and Clarke's (2006) suggestion on how to code data, to search for and refine themes, and to report findings in qualitative studies.

Responses toward the Sharing of Dis/Misinformation in WhatsApp Groups

WhatsApp groups provide an important infrastructure for informal user correction practices and cultures. They also provide a safe space for administrators and participants to debate about the veracity of information shared in group contexts. Respondents from the three case nations revealed they reacted differently when pieces of false and misleading information were shared in WhatsApp groups. The nature of the misinformation, rules of the group, quality and nature of relationships, severity of the falsehoods, intention of the spreader, and the group contexts were flagged as some of the key determinants. Based on these determinants, respondents explained that they reacted differently to cases of mis/disinformation in groups. Our respondents indicated that it depended on the relationship they had with the spreader of the dis/misinformation. In family WhatsApp groups, it was revealed that the dynamics were very complicated when the mis/disinformation was circulated by the elderly and highly educated members of the families. Participants also raised concern especially when the group administrators who were socially expected to be the gatekeepers become habitual spreaders of mis/disinformation. This created a complex situation because calling them to order and socially correcting them often led to conflict with group administrators. Our participants observed that they feared being removed or blacklisted from vibrant WhatsApp groups, especially those where football and religious issues were regularly discussed. Some pointed out that instead of publicly embarrassing the group administrator(s) for spreading falsehoods, they often resorted to alerting the concerned person through the chat (inbox) and call functions. This was confirmed by most of our participants in South Africa and Zimbabwe as follows:

> From my own experience, people try as much as they can to correct the viral spread of falsehoods. These include political mis/disinformation and even videos which are recycled where people will be warning

others about an impending cyclone (such as Cyclone Freddy). You find that users in a WhatsApp group are quick to correct the sender that this is an old video from 2018. Even if it is political falsehoods, people correct you there and there. They don't wait for the message to be read by most people. In that case, they are very proactive (R21, Zimbabwe).

Sometimes when a very argumentative post has been shared, something that's considered outrageous and inaccurate, people just keep quiet to avoid inflaming the issue out of control. But increasingly in most groups I am part of, I have realized that there are people who are willing to stand up to the truth. They don't allow misleading and false narratives to be peddled without challenging it. There is some fact-checking going on (R12, South Africa).

It is evident from the foregoing that WhatsApp users in South Africa and Zimbabwe are employing different tactics and strategies to deal with false and misleading information in group contexts. Most of the responses centered around ignoring, pushing back, fact-checking, conducting their own investigations (Googling) as well as publicly and privately reprimanding the sharers/spreaders of mis/disinformation. It is noteworthy to emphasize that these responses vary from group to group and country to country. For instance, in politically polarized contexts like Zimbabwe, most of the users tended to ignore political misinformation for fear of reprisals, and surveillance by the state. This was not evident in Namibian and South African WhatsApp groups, although ethnic and racial differences also played themselves out. In Namibia, for instance, we noticed that ideologies of respect and ubuntu mediated the ways in which participants interacted with each other in group contexts. Instead of publicly humiliating the spreader of misinformation, participants preferred gentle chiding and respect for social coexistence. South Africa was somewhat different because of its liberal democratic traits and human rights orientation.

One of the findings of this chapter is that group dynamics also shaped the ways in which participants dealt with dis/misinformation. We observed that the situation was less complicated among football fans and alumni associations where most people were associated along weak ties. In strong tie networks like family and church WhatsApp groups, participants indicated that they avoided confronting the spreaders of misinformation to avoid straining social relationships. This confirms recent studies by Ng and Neyazi (2023) in Singapore, where similar conclusions were reached. These findings also dovetail with Granovetter's (1973) argument that strong ties are characterized by emotional and familial support while weak ties tend to be based on distant and casual relationships. This means that such relationships can be broken without collateral damage to the individual. In cases where participants felt that socially correcting the spreader of falsehoods on the group would strain personal and social relationships, they often resorted to silence or speaking with the concerned person offline. In a weak ties context, informal user correction could cause a backlash that strengthened misinformation beliefs due to a desire to avoid publicly admitting a mistake (Ng and Neyazi 2023). In football fandom and alumni WhatsApp groups, we observed that such convivial outlets tended to diversify the free flow of information and promote robust discussions when compared to church groups, which are generally insular and conservative. This further supports the idea that groups composed primarily of weak tie networks (De Meo et al. 2014) have the potential to engage in crowdsourcing fact-checking (which we will discuss below) when compared to religious WhatsApp groups.

Participants also explained that they responded strategically to dis/misinformation depending on its severity and context. Health-related misinformation associated with the COVID-19 pandemic was generally described as "severe" and hence required strong debunking before it got out of hand. Most of the respondents indicated that severe types of misinformation had the potential to cause loss of life and livelihoods. In most WhatsApp groups, natural disasters-related dis/misinformation was quickly and effectively debunked. There was a lot of heated discussions when such kinds of falsehoods were circulated in WhatsApp groups in Namibia, South Africa, and Zimbabwe. This is in contrast to "less severe" types of mis/disinformation where people would choose to remain silent instead of making their opinions known in a group context.

Media literacy was identified as one of the most important determinants of whether group members would volunteer to fact-check false and misleading information. Media literacy is generally understood as the ability to sift through and analyze the messages that inform, entertain, and sell to us every day (Malik 2008). It's about asking pertinent questions about what's there and noticing what's not there. It was evident from our observations that there are WhatsApp groups dominated by media literate users. There are also groups where most members are less media literate. Media literate members of a WhatsApp group are able to question, analyze, and evaluate the information circulated in groups. They do not believe information at face value. These WhatsApp users possess higher-order critical thinking skills. In groups where media literate users are switched on, it was easier to observe quick and timeous debunking of mis/disinformation. Similar to studies (Nemer 2021, 2022) from Brazil, we found that WhatsApp users with lower levels of educational attainment and limited digital literacy skills were less inclined to engage in informal fact-checking practices. In reflecting on the importance of media literacy in fact-checking information on WhatsApp, respondents observed that:

> So, some groups are more media literate than others. Some groups are not very active. The news center is very attractive and it's sort of a crowdsourcing model but with some rudimentary journalism skills. But they do not seem to be journalists, but they seem to understand country dynamics, topical news (R3, Namibia).

Respondents described the complex ways in which administrators and members navigate the spread of misinformation in WhatsApp groups. While it is expected that the wisdom of the crowd will help in pushing back against this scourge, it was noted that:

> In some groups even though people refute claims, people are hyperpolitical, hypersensitive, and generally angry. As a result, some people downplay the truth, trivialize issues, or get personal especially on issues that have to do with religious leaders or politicians (R12, Zimbabwe).
>
> Toxicity, especially on politics and religious issues, divide groups to the extent where people who are less exposed to information are left more confused and at times dread to take part or air views. Political conversations are characterized by threats, at times something that deters further debate. In some instances, serious issues are humorized so much that they lose traction and value (R10, Zimbabwe).

When asked about whether members of the WhatsApp group were upfront in debunking false and misleading information, this is what one of the participants from Zimbabwe said:

> Yes, but it's scary coz (because) they get even police dockets and so am not sure what is happening. And they are up-to-date. Like the story of the Central Intelligence Organisation (CIO) alleged murderer. They would tell you where he is. Church groups are different. People get away with fake news because people are more relaxed in those groups.

It is evident from the above that WhatsApp users consider group norms and dynamics when deciding whether to fact-check something. It is generally more relaxed on religious and alumni WhatsApp groups.

Typologies of Informal User Correction on WhatsApp Groups

Fact-checking is the systematic assessment and publication of claims made by organizations or public figures to assess their validity (Markowitz et al. 2023). It encapsulates methods of verifying the factual accuracy of questioned reporting and statements. It can be conducted before (*ante hoc*) or after (*post hoc*) the text or content is published. In the context of WhatsApp groups, informal fact-checking starts when the group administrator or member cross-checks the veracity of the news and information before sharing with others. It can also take the form of group members commenting on posts contaminated with dis/misinformation. During our data collection phase, we observed that in most religious and alumni WhatsApp groups, members often took it upon themselves to fact-check information shared by their former classmates and church mates. This is because WhatsApp groups are made up of participants with diverse knowledges, skills, and competencies. It was easier for knowledgeable members of a WhatsApp group to counter misinformation with truthful and credible information. We found out that most of the WhatsApp groups were made up of experts (such as journalists, lawyers, medical doctors, researchers, public health experts, engineers, teachers, academics, and students). These experts assumed roles of episodic or volunteer fact checkers. Experts in this study are conceptualized as "individuals who possess significant general knowledge and information, know a lot about their field" such that they are able give informed opinions on issues of common interest (Goldman 2001, 91).

We also found that there are some WhatsApp group administrators who are very alert and proactive in identifying, correcting, and debunking false and misleading information. They often used their position as gatewatchers to delete, question, cross-check, and flag something as false or problematic. For instance, some administrators would ask sarcastically: "Is this not tomato sauce?" The question was meant to probe the sharer of the information to verify the authenticity of the document or information shared. Because most professional fact checkers struggle to access false and misleading content shared on WhatsApp, we observed nascent forms of user-initiated fact-checking sprouting in various groups. Our chapter identifies three main types of informal fact-checking practices emerging on WhatsApp groups in Namibia, South Africa, and Zimbabwe. These are volunteer or self-appointed, crowdsourcing, and amplification of fact-checked information; we will discuss each of these in turn.

1. Volunteer/Self-Appointed Fact Checkers

The first type of informal fact checkers we identified in the nine WhatsApp groups are what we refer to as "volunteer/self-appointed fact checkers." This refers to users within a (WhatsApp) group context who take it upon themselves to verify and debunk any type of false and misleading information shared by other members. These users rely on their own media literacy skills, media resources, and social networks to fact-check the authenticity of various information. Because of their role performance as "unofficial fact checkers" within WhatsApp groups, they are expected to correct dis/misinformation when it rears its ugly head. There were situations where controversial and questionable information was shared in some of the WhatsApp groups. Before many group participants commented on it, we realized that either the administrator(s) or anyone else would ask the "de facto fact checker" to verify content. Questions such as "Is this true?," "Has this been verified?," "How legit is it?," and "Believers, is this source credible?" were a recurring feature in most of the WhatsApp groups we observed in Namibia, South Africa, and Zimbabwe. In most cases, these informal fact checkers had journalistic skills, access to a wide range of media sources, access to reliable uncapped Wi-Fi, and had contacts in the media industry. Below is an excerpt from one WhatsApp group from a volunteer fact checker:

> Well, this was posted on someone's Facebook page and not on official media outlets. If you look for this story, it's only appearing on the said Facebook post and comments confirm that it's fake and photoshopped (field observations, Zimbabwe, March 18, 2023).

That most WhatsApp groups had "a go-to person(s)" in cases of shared problematic and inauthentic information clearly demonstrated that users of these apps in Namibia, South Africa, and Zimbabwe are not waiting for professional fact checkers and platform companies to intervene as "saviors" in their information and communicative ecosystems. Instead, they are busy coming up with their own everyday, informal user fact-checking practices that are context specific. It is interesting to note that very few volunteer fact checkers relied on professional fact-checking organizations (such as Africa Check, ZimFact, and Namibia Fact Check) to debunk fake news. Rather they used content from mainstream media, comments from public officials, press statements, and tweets from trusted social media influencers to corroborate their arguments. In Zimbabwe, we realized that people like Hopewell Chin'ono, Fadzayi Mahere, Gift Ostallos Siziba, Freeman Chari, and Blessed Mhlanga were believed to be bearers of verified information. It was also observed that most football WhatsApp groups in the three countries had certain sources that they associated with factual reporting. For instance, during football transfer news in Europe, most participants considered something to be true if it was published by Sky TV news, ESPN, and Goal.com. Most of the participants relied on social media handles of the clubs and athletes to fact-check rumors and conspiracy theories shared in WhatsApp groups. They also had lists of sports journalists such as Fabrizio Romano, Gerard Romero, Sid Lowe, Guillem Balague, and Kaveh Solhekol, whom they believed were purveyors of factual news and information.

As scholars have shown, informal user correction leverages on the nature of social media, network structure, and the efforts of volunteer fact checkers. In this context, a "collaborative system" that involves volunteer and professional fact checkers in identifying and debunking dis/misinformation has the potential to address the elitism associated with professionalized and institutionalized fact-checking. These group-specific fact checkers work as individuals tapping into their deep and wide social networks.

2. Amplifiers of Fact-Checked Information

The second type of informal fact-checking that we found has to do with amplification of fact-checked information. We observed that there are also some WhatsApp group members who have access to recent fact-checked information from Africa Check, Namibia Fact Check, Fact-CheckZW, and ZimFact. Most of these users were members of WhatsApp groups administered by these professional fact checkers. Some of them used their access to broadband internet to source fact-checked information on trusted websites, social media handles, and mainstream media. These amplifiers assumed the role of informal fact checkers because they were interested in making sure group members had access to verified information. Amplification through WhatsApp groups served the role of popularizing professional fact-checking practices. We also observed that because most people are not aware of the role of professional fact checkers, they regularly questioned the credibility of fact-checked information. Our interviews also confirmed that most WhatsApp group participants are not aware of the existence of professional fact checkers. They still relied on mainstream media for fact-checking the veracity of information shared on WhatsApp groups.

3. Crowdsourced Fact-Checking

The last typology we observed on most WhatsApp groups relates to crowdsourced fact-checking. This entailed a situation where members with different media literacy skill sets, resources, and social networks leveraged on them to verify the authenticity of false and misleading information. *Crowdsourcing* refers to "the act of a company or institution taking a function once performed by employees and outsourcing it to an undefined (and generally large) network of people in the form of an open call" (Howe 2008). In the context of journalism, crowdsourcing has been associated with a model for distributing reporting function across many people (i.e., crowd) (Kelly 2009, 18). When applied to the field of fact-checking, it means that WhatsApp group members engage in collaborative fact-checking of various information circulated by others. This often involves a large group of dispersed group participants contributing or producing evidence, and ideas about the truthfulness or lack thereof of any particular information. These group members work as volunteers. Unlike self-appointed fact checkers, crowdsourcing fact-checking taps into the *wisdom of the crowd* (and *ignorance of the crowd*). This taps into the African concept of *humwe* or *nhimbe* (festival of work). In the chiShona language, *humwe* or *nhimbe* means working together to achieve a shared goal. Thus, in a WhatsApp group context, members come together to verify the authenticity of information generally considered false. This type of communal work is faster, more flexible, scalable, and relatively cheaper. The role of fact-checking misinformation in a group context is left to one or two people. Group members utilize their skills, networks, and resources in finding out the truth.

Based on virtual ethnography and interviews, we found that crowdsourcing fact-checking was the most popular form of pushing back against the spread of false and misleading information. This chimes with Kligler-Vilenchik's (2021) notion of collective social correction, which refers to an ongoing practice of information verification occurring within group contexts. It was generally used by group participants to make sense of controversial and questionable information. Some of the research participants mentioned that they had muted most of their WhatsApp groups as a way of managing the flow of content, and they still cross-checked the veracity of certain information before sharing it with other groups. Our virtual ethnographic vignettes revealed that informal fact checkers often cross-check with various news sources, bring fact-checked information to the table, and conduct their own investigations before approving the circulation of information. As *modern-day gatewatchers*, informal fact checkers on WhatsApp groups are playing an invaluable role in cleaning up the contaminated digital public sphere. Most of these informal user correction practices are contributing immensely toward pushing back against the normalization of information disorders in the digital ecosystem. Our findings also show that informal fact-checking on WhatsApp groups is aided by members' internal connectedness. This is largely because members of

WhatsApp groups belong to different groups at the same time. This allows them to crowdsource information and news from a wide range of groups, opinion leaders, and social networks.

Conclusion

This chapter has examined the phenomenon of informal user correction practices on WhatsApp groups, which is emerging in the shadows of professional fact-checking practices and cultures popularized by Africa Check, FactCheckZW, Namibia Fact Check, and ZimFact. It has demonstrated that focusing on what platform companies and professional fact checkers are doing to push back against the spread of mis/disinformation on mainstream social media platforms often leads to analytical and empirical blind spots. It obfuscates our understanding of what happens in infrastructures such as WhatsApp groups where diverse members from different sociocultural and demographic backgrounds converge and negotiate the sharing of dis/misinformation (Parreiras, this volume; Udupa, this volume; Wasserman and Madrid-Morales, this volume). Drawing on empirical data from Namibia, South Africa, and Zimbabwe, the chapter has discussed three types of informal fact-checking that are emerging on WhatsApp groups. These are volunteer/self-appointed, crowdsourcing, and amplification of fact-checked information. Our analysis has demonstrated that the severity of the mis/disinformation, nature of social relationships, ideology of respectability, and group dynamics influenced how the administrators and members responded to false and misleading information. Three main conclusions may be drawn based on the findings.

It has shown that informal fact checkers often cross-check with various news sources, bring fact-checked information to the table, and conduct their own investigations before approving the circulation of information. Thus, although professional fact checkers and platform companies have limited access to what happens within semipublic infrastructures like WhatsApp groups, this chapter has shown that provisional and informal user correction practices are emerging. While these practices are not professionalized and formalized, they offer everyday forms of fact-checking in group contexts. Our findings also show that administrators and participants in WhatsApp groups often send their fact-checked information to everyone in the group, and also to the sharers/spreaders of misinformation. There are also instances where no one fact-checked information because of absence of contrary evidence.

This chapter contradicts popular assumptions that WhatsApp users are helpless victims of mis/disinformation. While the WhatsApp infrastructure provides fertile ground for the spread of false and misleading information, it also creates an outlet for informal fact-checking within group contexts. This suggests that WhatsApp as a communicative infrastructure enables and also disables different kinds of possibilities—with regards to being a conveyer belt of information disorders while at the same time allowing informal fact checkers to rise to the occasion as "fire fighters." From our findings, it was evident that the administrators and members are not waiting for "external support." Rather they are busy trying to find context-specific solutions to the challenge at hand. These findings go against popular belief that people have no agency to negotiate the misinformation headwinds buffeting WhatsApp groups. In cases where most of the users had limited media resources to cross-check the veracity of information circulated within a group, they often relied on volunteer/self-appointed fact checkers. In some cases, they leaned heavily on crowdsourcing fact-checking. All these context-specific strategies and tactics show that ordinary people have agency even in situations of resource constraints.

In the shadows of top-down initiatives (such as platform-centric content moderation and professional fact checking), informal fact-checking practices and cultures are emerging in group contexts. These informal ways should not be misconstrued with unorganized, unplanned, and unproductive tactics and strategies mounted at the margins. Instead, these informal fact-checking practices represents what Giddens (1984) calls intentional activities by situated actors who are focused on satisfying their needs and goals within structural barriers. They symbolize *self-help and improvisation* on the part of administrators and members relying on WhatsApp for every communication, communion, and sociality. Instead of waiting for aid through prebunking and digital literacy campaigns, WhatsApp administrators and users are finding their own solutions to the polluted information ecosystem.

12

How to Approach Speech Regulation on WhatsApp

Lessons from Regulatory Experiments in India

AMBER SINHA

In several large countries in the majority world, messaging services like WhatsApp have emerged as the default service to connect with people and share information. Simultaneously, it has become a primary vector for disseminating problematic speech in countries like India.

In May 2017, Divya (name changed), a sixty-five-year-old woman, was traveling to a temple town in the southern Indian state of Tamil Nadu. During a stop to get information on road directions, an elderly woman became suspicious witnessing Divya's family offering chocolates to children playing nearby. She immediately contacted her son, which led to an escalation of the situation. Upon reaching the next village, Divya and her family found a large mob waiting for them. These villagers, influenced by viral WhatsApp videos depicting child trafficking, doubted the family's intentions despite their explanation of visiting their family temple. The mob remained unconvinced, resulting in a horrifying ordeal for Divya and her family. They were subjected to a violent attack, including being stripped naked, brutally beaten with iron and wooden sticks, and the attackers also recorded the assault on their cameras. Tragically, Divya lost her life due to her injuries, while the rest of her family managed to survive (Jayarajan 2018).

This incident is not an isolated case but emblematic of numerous instances of lynching that occurred throughout the country in 2017–2018. In my research, I discovered fifty-four documented cases through news reports between 2017 and 2018 where unconfirmed suspicions of child abduction led to mob violence. These figures were based solely on English language news reports, suggesting that the actual number of incidents could be higher. The mob sizes ranged from small groups of fifteen to crowds as large as two thousand people. The victims were diverse, including innocent individuals like Divya, as well as tourists passing through villages, people unintentionally entering certain areas, transgender people, and travelers seeking road directions. These recorded incidents took place in sixteen Indian states: Madhya Pradesh, West Bengal, Bihar, Karnataka, Assam, Tamil Nadu, Chhattisgarh, Gujarat, Tripura, Maharashtra, Odisha, Telangana, Andhra Pradesh, Jharkhand, Kerala, and Rajasthan, encompassing more than half of India's area and population. The videos responsible for inciting these incidents were challenging to trace due to the end-to-end encryption used by WhatsApp. They often depicted alleged abductors, sometimes exploiting religious or regional biases, such as featuring a Muslim woman wearing a burqa (Reuters and Hindustan Times 2018). Other videos showcased graphic images of mutilated children, raising a warning about supposed organ harvesting schemes. These messages exploited people's concerns for their children's safety, utilizing horrifying visuals to stoke fear among viewers. A common thread among these videos was the intentional manipulation of existing fears, biases, and prejudices toward migrants, tourists, and those seen as outsiders, suggesting that the narratives tapped complex social divisions and anxieties. They were carefully edited to mobilize community prejudices, creating a climate of unease for visitors. When someone in the vicinity was suspected of being a child abductor, these false messages were frequently shared in large WhatsApp groups with hundreds of participants, resulting in mob gatherings that left local law enforcement with limited time to respond. The authorities had made efforts to educate and caution residents about such disinformation in some instances, including in Tamil Nadu prior to Divya's visit. However, debunking and preventing the rapid spread of rumors on WhatsApp proved extremely challenging. These videos predominantly circulated in areas with low literacy rates, where people heavily rely on visual and video content received through messaging apps (Census of India 2011).

This series of lynchings brought nationwide media and regulatory attention to WhatsApp as a source of misinformation. The conversation on the spread of digital misinformation was until then focused on social media platforms driven by recommendation algorithms. However, the lynching episodes placed WhatsApp at the center of the misinformation discourse in India.

Based on analysis of news reports and policy documents and interviews with fact checkers, this chapter examines different aspects of the spread of problematic speech through WhatsApp in India, and possible regulatory responses. By "problematic speech," I refer to two forms of content: the first type is misinformation and disinformation, which involves intentional or unintentional creation of false information; and second, dealt with in lesser detail, is the practice of hateful speech, abusive language, and online harassment, which I refer to as "extreme speech" (Udupa and Pohjonen 2019). In the following section, I will describe how WhatsApp is used to spread information and mobilize groups in India, with a specific focus on its technical architecture and the governance systems that are in place, and how they enable and impede the spread of information. In the third and fourth sections, I will provide a brief overview of learnings from the domains of communication and propaganda studies to understand how we respond to misinformation and extreme speech. In the final section, I will look at three responses to misinformation and extreme speech on WhatsApp: fact-checking, traceability, and design friction, and assess their relative merits and pitfalls.

Platform Features of WhatsApp

Misinformation and extreme speech can be found on both social media and messaging platforms. However, there is a difference in user behavior between these two kinds of platforms. Facebook's algorithms control the way we come across and engage with the content, while WhatsApp displays all messages in chronological order. On Facebook, users are expected to interact with a post by leaving a comment, sharing it, or reacting by using emojis. In a WhatsApp group, a stream of messages from different members is displayed, and it is up to users how they engage. There is no personalized algorithmic training on WhatsApp, and norms and practices evolve through user exchange. This feature alongside the preponderance of groups on WhatsApp makes it reminiscent of tightly monitored forum discussions in the first decade of the Web.

Messages and forwards sent to one group do not necessarily find easy mobility to other groups, and people are acutely conscious of which messages belong where. Groups might have their own codes of conduct, but

WhatsApp offers little to no recourse to avoiding or reporting abuse or flagging misinformation in countries like India. In September 2018, they appointed a single grievance officer for India, who could be contacted for concerns and complaints (PTI 2018b). WhatsApp did not make it straightforward to reach the grievance officer. They could not be contacted via WhatsApp, and a digital signature is required to reach them over email.

While messages do become viral on WhatsApp, it is much harder to manage or monitor their virality than on social media platforms. It is this aspect of WhatsApp that makes it a suitable platform for mobilization of people, especially in the same group in a short span of time. Using WhatsApp to spread rumors that prey on community prejudices and lead to local sentiments turning into violence is thus achieved in a surprisingly easy way.

Even though WhatsApp was intended as a private messaging service, it is difficult to think about it today as anything other than a hotbed of group conversations. Groups on WhatsApp are built around common interests or associations, ranging from personal (extended family, friends, weddings, or holiday planning) and work-related (company-wide, department- and project-related) to hobbies (cricket, cinema, or quizzing) and other communities (housing complex, alumni groups, new parents). This helps in achieving homophily, or the drawing together of people in tight networks of like-mindedness (Chakrabarti 2018). Shared identity, association, and beliefs lead to group members suffering from a confirmation bias. This homophily provides an ingenious method of microtargeting—not at an individual level, but at a group identity level. Coupled with the minimal oversight WhatsApp provides in countries like India, group dynamics have resulted in an unchecked, free flow of misinformation that vested interests may wish to perpetuate.

Additionally, on WhatsApp, sender primacy is one of the keys to understanding why people share things. BBC's research found that this was the key heuristic that users in India relied upon when deciding whether to share content or not, and if it was credible. If the sender is influential and respected, there is a greater chance of their messages and forwards being consumed and shared further. On the other hand, if someone is perceived as an irritant, their messages are more often ignored (Chakrabarti 2018).

Platform Measures and Regulatory Responses

WhatsApp's current approach to addressing the problems on its platform are restricted to training its algorithms to detect "how" messages are shared and made to go viral, rather than "what" messages are shared (Bhushan 2019). These could include paying attention to "spam farms" that play a role in disseminating misinformation or hate speech. It has also reportedly sent "cease and desist letters" to marketing firms on mass messaging (Indo Asian News Service 2019a). More notably, in 2018, it announced restrictions on forwarding. Earlier, the platform allowed a user to send a hundred forwards in one go. WhatsApp reduced that number to five in India and to twenty for the rest of the world. It also introduced a "forwarded" label on messages to help people identify that the message is not directly from the sender, and that they are only circulating a message shared by someone else. The company also disabled the "quick forward" option next to media messages (photos and videos) and introduced a "suspicious link" label for URLs that its algorithm detected as containing unusual characters. There has been limited development on this end, and progress on analysis of suspicious URLs for misinformation or hate speech has been slow.

As the limitations of platform measures suggest, regulation of extreme speech poses significant challenges. Unlike platforms like Facebook and YouTube, where the regulatory focus has been on tightening data protection practices or accountability of recommendation systems, an algorithm-free platform like WhatsApp poses extremely different regulatory challenges. Importantly, for most of India, particularly rural India and even smaller towns are largely served by 2G internet connectivity (Mukhopadhyay and Mandal 2019). This infrastructural context, alongside the popularity of mobile phones as the primary devices to access the internet, positions WhatsApp as a unique platform with a wide network and affordance.

Over the last few years, more dedicated regulatory developments have emerged globally in direct response to the perception that platforms are negligent in addressing hate speech and misinformation on their networks. The Network Enforcement Act (Netzwerkdurchsetzungsgesetz, NetzDG) came into effect in 2018 in Germany (Bundesministerium der Justiz 2017) and imposed high fines on platforms, which, despite being alerted to "manifestly unlawful" content, failed to remove them in a timely manner. The European Union also voted in 2019 to mandate websites accessible in the EU to "remove any content deemed 'terrorist' content by a 'competent authority' within three hours of being notified" (Masnick 2022).

In India, the Information Technology (Intermediary Guidelines and Digital Media Ethics Code) Rules, notified in 2021, made significant changes to the roles of intermediaries in actively monitoring and removing content (Ministry of Electronics & Information Technology 2023). It aimed to ensure traceability of communications on their platforms. Specifically, intermediaries would be obliged to proactively identify and remove or disable public access to illegal information or content. This means that intermediaries would be responsible for screening user speech to determine if it is illegal, instead of waiting for notification, such as a court order, to remove it. The rules further specify that this proactive responsibility should be fulfilled using technology-based automated tools or appropriate mechanisms.

In relation to WhatsApp, three possible regulatory responses are pertinent in exploring ways to combat extreme speech.

Fact-Checking and Its Limitations

In recent years, the rise of misinformation on the internet and mainstream media has prompted the emergence of various fact-checking platforms. Founded by Pratik Sinha in 2016, Alt News is a team committed to verifying viral stories on social media and WhatsApp, authenticating photos and videos, and exposing misinformation in media reports (Desai 2019). Similarly, SM Hoax Slayer, initiated by Pankaj Jain, gained prominence for debunking a viral news story about an alleged "nano GPS chip" in India's new 2,000 rupee note (Doshi 2017). Shammas Oliyath manages Check4Spam, actively debunking false forwards during his spare time and operating a helpline for reporting hoaxes on WhatsApp. By 2017, Oliyath was already receiving over sixty forwards daily (Shekhar 2017).

This trend has led to the proliferation of fact-checking websites. In the lead-up to the 2019 general elections, several television channels dedicated segments to debunking viral hoaxes and misinformation. Even ideologically driven websites like OpIndia.com now include sections for fact-checking content shared by opposing groups. The valuable work carried out by these fact checkers initially represented individual efforts by committed individuals to engage rigorously with extreme speech and dispel myths, hoaxes, and propaganda. However, both Sinha and Oliyath acknowledge that fact checkers alone cannot effectively combat the misinformation ecosystem. Further, even the initiative launched by Facebook in partnership with BoomLive, an independent Mumbai-based fact-checking organization that was certified by the International Fact-Checking Network, suffered from limited support from Facebook (PTI 2019). While recognizing the challenges faced by fact checkers in terms of scale and the organized machinery they contend with, it is crucial to critically examine the underlying assumptions regarding the efficacy of fact-checking.

While fact-checking is essential and underscores the need for reform within the journalism industry, considering it as a sustainable solution for countering misinformation beyond mainstream media may be misguided. The challenges of scale have been acknowledged, but it is equally important to question the foundational premise. Firstly, it assumes that individuals will alter their opinions when presented with evidence debunking a political falsehood. Moreover, it presupposes that online discussions serve as a deliberative process where people engage with others to inform, persuade, and debate. However, these assumptions may not hold true for the consumption and dissemination of news online.

In a sense, viewing fact-checking as an effective solution, on its own, would be tantamount to falling into the trap of considering recipients of misinformation as passive actors who will automatically change their viewpoints upon being presented with evidence that refutes the information they rely on. In reality, the public is far more complex, responding to information based on their own identities, biases, and preferences.

Reports have highlighted the use of private companies, such as Sarv Webs Private Limited, by political parties to spread messaging on WhatsApp (Sathe 2019). These companies maintain numerous SIM cards and use multiple numbers to send messages across various WhatsApp groups. They closely monitor the number of groups each phone number is part of and the volume of messages sent, received, and read, as well as the number of replies and engagement from recipients

(Indo Asian News Service 2019b). This practice extends beyond a single company or party, as other political groups have recognized the effectiveness of WhatsApp as a communication medium. WhatsApp also enables group members to collect the mobile numbers of all other group members, presenting opportunities for ruling parties to identify both supporters and detractors based on their presence in ideologically aligned WhatsApp groups or by analyzing messages from users. Indian laws mandating the registration of mobile numbers facilitate this identification process.

Shortly before the 2019 general elections, WhatsApp introduced a fact-checking service in India, allowing users to forward messages to the Checkpoint Tipline. A team led by local startup Proto would evaluate and label messages as "true," "false," "misleading," or "disputed." However, users noted that the verification process took a considerable amount of time after reporting a message. Proto acknowledged that the fact-checking service would have limited impact in combating the misinformation ecosystem prior to the election (PTI 2018a). Its effectiveness relied on users voluntarily submitting messages for review and did little to address consumption by individuals already inclined to believe the information and unlikely to report it. The primary goal was to study the phenomenon of misinformation on a large scale, assisting WhatsApp in identifying the most affected regions, languages, and issues. However, these efforts were considered inadequate and untimely.

The Bogey of Message Traceability

The traceability requirement in India's new Information Technology regulations mandates identification of the first originator without specifying how this may be technically implemented (Kumar 2022). Encryption involves scrambling plaintext messages to render them unreadable except to those with the secret key. One of the most commonly used technologies for securing and transmitting information over the internet is end-to-end encryption (E2EE). E2EE relies on hardware embedded in phones and computers to generate random locks and keys that only work on the devices involved in the conversation (Deeks 2016).

Unlike open standardized communication protocols such as Extensible Messaging and Presence Protocol (XMPP) or Internet Relay Chat

(IRC), most instant messaging (IM) protocols are centralized. This means that users of each application can only communicate with each other through that specific application. As a result, users cannot choose the most trustworthy provider but instead need to fully trust the one provider that develops both the protocol and application. Once the application is installed, keys are automatically generated and encryption is enabled. WhatsApp is a closed-source instant messaging protocol that uses the Signal protocol for key exchange and encryption. However, it is independent of Signal's messaging and group communication protocols.

Over the past few years, the Indian government has made a series of demands urging WhatsApp to implement traceability measures in order to identify the sources of misinformation and problematic content. The Ministry of Electronics and Information Technology has issued notices to WhatsApp, expressing concerns about the circulation of irresponsible and inflammatory messages on the platform, which have been linked to instances of lynchings. The Ministry directed WhatsApp to use technology to prevent the spread of such messages and take immediate action (Sridhar and Choudhary 2018).

While it is unwise for the Ministry to solely blame a technology company for lynchings without considering the underlying social issues and the complicity of the government and political parties, it is worth exploring the actions WhatsApp can take. In one of the notices, the Ministry demanded that WhatsApp develop effective solutions to facilitate law enforcement and incorporate traceability (The Hindu Businessline 2018). WhatsApp has resisted this suggestion, citing potential compromises to its end-to-end encryption and threats to user privacy.

There are broadly three ways in which traceability can be built. The first way would be to do away with E2EE, which would have a deleterious effect on both security and privacy of communications. The second would be to store hashes of all messages, which has accuracy challenges as a motivated individual can easily modify hashes. It also has confidentiality challenges, as the content of the messages could be reverse engineered using the hashes and facilitate censorship and profiling. The third would be to attach originator information to messages as metadata. This may have limited effectiveness and lead only to the identification of relative originators and not absolute originators (Grover, Rajwade, and Katira 2021).

Having a clear traceability mandate is a direct restriction on the right to privacy. For it to be legal, the restriction must be reasonable and satisfy the necessity and proportionality test. As described above, each type of implementation of traceability encounters significant limitations that hinders its ability to achieve the intended goal, while also creating operational challenges for messaging services. The "necessity" requirement articulated by the Supreme Court as a precondition for restrictions of privacy also requires that an assessment is made for the availability of alternatives with a lesser degree of privacy restriction that can achieve the same purpose (Puttaswamy 2017).

Reliance on metadata instead of originator information is worth assessing as a regulatory measure. WhatsApp takes proactive measures by scanning all unencrypted user data to identify and prevent instances of child sexual exploitation and other forms of abuse. Additionally, the messaging service Matrix provides guidelines for users and administrators on content moderation and the application of specific rules based on metadata.

Such measures that offer alternative means to address extreme speech on WhatsApp highlight the point that the legal mandate for traceability is misguided and creates new issues for privacy and surveillance harms without adequately addressing the primary issue.

Experience Friction

The other set of regulatory options discussed much less are design solutions that can work toward addressing misinformation and extreme speech. The user interface of platforms is currently designed to facilitate quick and effortless content sharing. This unhindered usability contributes to the perception of lack of consequences when sharing content. There is an increasing body of scholarly research highlighting the potential decline in cognitive abilities among humans as various tasks and processes become automated (IEEE 2016). This phenomenon can be observed in simple examples such as map reading, spelling, and memorizing phone numbers. Furthermore, this concern extends to the consumption of content, as the proliferation of personalized content has led to a diminished ability within society to discern and actively seek out genuine information. Extensive research already exists regarding the impact of the "filter bubble" (Allred 2018) and the manner in which content is disseminated through networks and communities, resulting in users placing trust in content based on its source rather than engaging in critical evaluation (Udupa 2019).

The key question is whether the spread of extreme speech can be controlled through design solutions. Diverging from previous efforts that have focused on raising awareness and promoting proper platform usage, there is a need to explore how companies can introduce friction into their platforms as a means of addressing different facets of misinformation. This includes enhancing users' capacity to identify and differentiate between fake and authentic content, as well as demonstrating the extent and ramifications of misinformation. The role of design in finding solutions to misinformation is already gaining political traction, as evidenced by the recommendation in the UK *Online Harms White Paper* for governments, civil society, and industry to collaborate on a "safety by design" framework (United Kingdom 2019).

Broadly speaking, "experience friction" refers to any element that impedes or slows down users in accomplishing their goals or completing tasks (Kollin 2018). Various platforms already incorporate friction models, such as requesting user confirmation for document deletion, signing out of accounts, and error anticipation. These friction models typically involve deliberate slowdowns in processes to ensure user awareness and supplementary information to enhance user understanding. Friction can also be incorporated into processes to challenge user perceptions and assumptions. For instance, Wells Fargo introduced an eye scan security feature in their mobile banking app. Although the technology processed scans rapidly, users felt that the process was too quick and doubted the reliability of the scan. To instill consumer trust, Wells Fargo increased the processing time for scans (Koren 2016). In jurisdictions like India, certain forms of friction have already been implemented on the WhatsApp platform, such as limiting the number of messages that can be forwarded to contacts (Singh 2019). However, methods to circumvent this limitation have already emerged, thereby necessitating further exploration.

Another example of introducing friction in online use is the incorporation of privacy nudges, which leverage heuristics, cognitive biases, and behavioral science to prompt users to make informed privacy decisions through various forms of nudges and notices (Wang et al. 2013). By infusing insights from behavioral science and societal norms related to misinformation into friction models, this system aims to combat the creation and dissemination of misinformation through social understandings, sanctions, and enforcement. Incorporating established societal values and norms that acknowledge the harm of misinformation within the platform can help shape how users evaluate, consume, generate, and share content.

WhatsApp is expected to continue training its algorithms to identify how information is shared rather than focusing on the content itself. Efforts may involve cracking down on "spam farms" and minimizing reliance on fact checkers (Rebelo 2018).

In India and other jurisdictions, as mentioned earlier, some forms of friction have already been brought into the WhatsApp platform by limiting the number of messages that can be forwarded to contacts. Though this may address, to an extent, the mass forwarding of messages, methods to circumvent this limitation have already emerged. While these solutions do not suffer from the same privacy challenges that traceability requirements have, they carry their own set of issues. Without any accountability mechanism, it is unlikely that WhatsApp will, on its own, implement any significant experience friction on the platform. The introduction of educational public service announcements makes the platform less attractive for users who are used to the seamless and fast design of messaging platforms.

A reasonable argument can be made that the same standards of private communication need not be applicable to a message that has spread beyond a certain number of people. In such cases, could there be a case for reduced privacy protection for such messages? The effect of such a step would be to render viral or highly forwarded messages, often in the form of multimedia posts, outside the scope of encrypted communication, and thus opening doors for increased friction on such messages if an assessment suggests that they qualify as misinformation or hate speech. However, arriving at any sliding scale that sets a threshold for when messages can be afforded less privacy protection is a tricky exercise wrought with many risks.

A safer approach may be to look at friction solutions that can be implemented on-device without tinkering with end-to-end encryption.

Some commentators have suggested solutions such as on-device context where a list of rumors (including image, audio, and video hashes) along with corresponding fact-checks can be regularly supplied to WhatsApp clients, so if someone receives a debunked rumor, WhatsApp can provide the relevant context or fact-check (similar to how WhatsApp currently flags suspicious links) (Ovadya 2021). It can be modeled on existing arrangements companies like Meta have with international consortiums such as the International Fact-Checking Network, which provide a trusted source of debunks and context around viral pieces of misinformation.

WhatsApp has also experimented with other forms of design solutions. For example, the platform has developed methods for recognizing spam-like forwarding actions using message and user metadata, without the need to access the actual message content. For instance, if an account primarily sends group messages rather than individual ones, WhatsApp can deduce potential involvement in prohibited mass messaging and take steps to deactivate the account. Another approach to identifying automated messaging behavior involves examining the presence of a "typing indicator" (the ellipsis bubble that appears when someone is composing a message in chat applications). As explained by Matt Jones, an engineer at WhatsApp, if a spammer's automated messaging script lacks a typing indicator before sending a message, the company will proceed to ban the account (USENIX Enigma Conference 2017).

Conclusion

The above analysis of the three approaches toward combating extreme speech on WhatsApp illuminates the complexities of the issue. The fact-checking approach is critical but suffers from severe limitations in a network where sender primacy, social contexts, and cognitive biases trump exposure to fact-checked information.

The traceability approach that the Indian government has relied upon is extremely misguided and emblematic of the tendency to shift responsibility of technology companies for social problems. We began this chapter with a series of examples of lynchings that was perpetrated through the use of WhatsApp to mobilize people. If we treat them merely as a technological misinformation problem, then we will miss the point. Arun (2019) uses examples to show that mob violence is often a result of how the media, local politicians, and representatives of the state instrumentalize what they perceive as offending behavior, such as cow slaughter. Often, it is the narrative color that these powerful actors give to an incident that leads to violence. Here, the role played by law enforcement and politicians in encouraging or condoning such acts of violence lends them legitimacy, or at least some measure of normalcy, whereas they should have been seen as the horrific acts they are. To pretend that such violence is brought about only by the advent of social media or messaging services and not by deep-rooted societal problems, often worsened by those tasked with making it better, would be a grave mistake. By framing a complex sociopolitical problem as merely technological, political actors are increasingly shifting the responsibility toward internet intermediaries. This will inevitably lead to more risk-averse behavior on the part of technology companies with adverse consequences for free speech.

In her study on Twitter, Tufekci (2017) argues that the nature and impact of censorship on social media are very different. Earlier, censorship was enacted by restricting speech. But now, it also works in the form of organized harassment campaigns, which use the qualities of viral outrage to impose a disproportionate cost on the very act of speaking out. Therefore, censorship plays out not merely in the form of the removal of speech but through disinformation and hate speech campaigns. In most cases, this censorship of content does not necessarily meet the threshold of hate speech, and free speech advocates have traditionally argued for counterspeech as the most effective response to such speech acts. However, the structural and organized nature of harassment and extreme speech often renders counterspeech ineffective. This ineffectual nature of counterspeech and failure of hate speech regulations to respond to online manipulation campaigns underscores the need to move beyond the "binary and normative divisions between acceptable and unacceptable speech" (Udupa and Pohjonen 2019).

Design solutions and other platform measures have the ability to address various aspects of extreme speech by engaging with how consumers respond to it. However, manipulation by polarizing speech is often a symptom of a deeper underlying social problem. To focus our energies simply on the specific message that leads to unreasonable actions, whether in the form of voting against one's own interests or engaging in mob violence, would be highly limiting. The ideals of the public's social existence, and the ability to form associations, offer useful guides on how best to navigate extreme speech on WhatsApp. It is necessary to create bottom-up community-based approaches that are facilitated and complemented by design solutions that illuminate the presence of extreme speech to users and offer some friction to impede the uncontrolled spread of misinformation and extreme speech on WhatsApp. Rather than focusing solely on legal solutions that remain narrow in their definitions of acceptable and unacceptable speech, or technical solutions like decryptions, which will inevitably compromise not only the rights to privacy but several other rights dependent on it, a sociotechnical approach that relies on an understanding of extreme speech and employs design interventions to address behavior appears more promising.

PART IV

Method

13

Methodological Challenges in Researching Disinformation on WhatsApp in Turkey

ERKAN SAKA

I had the opportunity to conduct research on EU-related disinformation in Turkey, thanks to an EU Delegation in Turkey grant in 2021 (Saka, 2021b). Most disinformation actors in Turkey currently invest their work in targets other than the European Union as their central focus. I could thus handle the relevant information flow easier than prevalent studies due to the relative disinterest in the EU, and I could slowly develop a multimodal research track. However, conspiracy theories and disinformation have always been part of the EU membership negotiations in Turkey, and some overarching themes through the decades could be found.

A broader context is due here as there have been a series of academic studies and think tank reports published in recent years on the topic of disinformation, misinformation, and conspiracy theories. This will help to contextualize where WhatsApp is situated in Turkey. Concerned with the circulation of COVID-19–related disinformation circulation in Turkey, Kirdemir's (2020) study emphasizes widespread belief in conspiracy theories. This work includes multiple platforms, but most examples are from YouTube content analysis, which makes it different from other works on the topic. He lists three basic types of misinformation: importing global conspiracy beliefs, references based on political polarization, and alternative realities promoted by viral YouTube videos.

Many recent studies on misinformation are unsurprisingly related to the pandemic. In the project "Understanding Disinformation Ecosystem in Global Politics," Parlar Dal and Erdoğan (2021) elaborate on disinformation as a security threat. According to the project report, disinformation such as that related to COVID-19 leads to increased uncertainty, which in turn leads to rising populism and authoritarianism and loss of confidence in democracy. In another related study, Erdoğan et al.

(2022) focused on the circulation of misinformation related to the pandemic. The research included face-to-face survey interviews, in-depth interviews, and data collection from Twitter. Another pandemic-related research focuses more on private messaging services: Koçer et al. (2022) did not particularly mention WhatsApp, but interviews implied new sources of misinformation that included WhatsApp.

Many studies focus more on the fact checkers than the platforms themselves. The well-known fact-checking organization Teyit.org released a report on false information that could be found on the internet that was related to the Nagorno-Karabakh conflict between Armenia and Azerbaijan (Arabacı, Mammadova, and Türkkan 2020). The data collected was based on media and internet monitoring among Turkish and Azeri users.

Erkan and Ayhan (2018) examine social media platforms as political disinformation tools and the role of fact-checking organizations after the early election decision in Turkey on June 24, 2018. They emphasize that, especially when social sensitivity is intense and accurate information becomes vital, the need for verification platforms becomes more evident. Aydın (2020) also analyzes investigations made by the fact-checking organization Teyit.org about the claims that circulated on social media during the COVID-19 pandemic. In another study, Çömlekçi (2019) compares Teyit.org and the U.S.-based fact-checking organization Snopes.com. The study focuses on the news verification methods and priorities of the platforms. In the research, the limitations of news verification platforms such as financial resources and political pressures are discussed, and it is underlined that it is important to raise public awareness and increase digital media literacy levels. Ünal and Çiçeklioğlu (2019) analyze the structure and functioning of fact-checking organizations in the context of preventing the propagation of fake news and improving digital literacy. The research involves content analysis of the verification activities of Teyit.org and in-depth interviews with the verification team. The study finds that fake content spreading on the internet predominantly consists of political issues. Karadağ and Ayten (2020) conducted a comparative study of Teyit.org and Doğruluk Payı, also a prominent fact-checking organization in Turkey. The scope of the research is a comparison of the structures and working manners of these organizations. Doğruluk Payı and Teyit.org have similarities in human resources, financing, and organization; however, they exhibit differences in the scope and process of verification/fact-checking and assessment.

Vanlıoğlu (2018) shifts the focus to another issue: political trolls. Emphasizing crowdsourcing as a source of propaganda and disinformation operations in cyberspace, he argues that there is a problem with arrangements where qualified people are permanently employed in large numbers in troll units. He notes that crowdsourcing allows for enhancing the performance and cost-efficiency of troll units. About political trolls, Unver (2019)'s study is rare since it focuses on external sources of disinformation. The research collected data from Twitter and focused on the impact and relevance of pro-Russian information operations in Turkey. Twitter-based data was gathered on specific dates, such as coup attempt days or the downing of a Russian jet by Turkish Air Forces. Unver (2020) later published the most comprehensive report on fact-checking organizations in Turkey. These organizations play a growing role in countering disinformation. Unver's work on Russian propaganda is further supported by Furman, Gürel, and Sivaslıoğlu (2023). The latter also relied on Twitter data and explicitly demonstrated how Sputnik's Turkish site circulated pandemic-related disinformation to weaken trust in Western establishments.

Dirini and Özsu (2020), on the other hand, collected data from specified hashtags on Twitter, Instagram, and YouTube to analyze hate speech that targeted specific groups such as Chinese citizens, LGBTQI+ individuals, and older citizens during the pandemic. Bozkanat (2021) investigated the features of fake news in Turkey and argued that fake news was circulating more on Facebook than on Twitter and Instagram.

WhatsApp has not been absent, but in the Turkish case, as the studies above reveal, it was not seen as a strong source of disinformation as it was in other countries in the Global South, such as the widespread disinformation campaigns and dissemination of conspiracy theories during the elections in Brazil in 2018 and India in 2019 (Resende et al, 2019b). My colleagues in this volume cite the global cases extensively, so instead of repeating them, I will focus on the Turkish case here.

As the cases above demonstrate, Turkish citizens are subject to disinformation at a high level through other platforms. Turkey has been identified as a country highly exposed to fake news and misinformation (Boyacı Yıldırım 2023). The volume of COVID-19–related misinformation has also overwhelmed fact checkers in Turkey (Kolluri, Liu, and Murthy 2022). For example, the withdrawal of Turkey from the Istanbul Convention, a human rights treaty, was accompanied by disinformation campaigns that targeted the convention and weaponized homophobia (Elmas, Overdorf, and Aberer 2021). Overall, the politically polarized environment in Turkey further influences users' trust in the news and their perceptions of misinformation on social media (Bozdağ and Koçer 2022). This situation is worsened by the fact that Turkish media is under ever-increasing political and economic pressure, and citizens do not have reliable media sources as the diversity in the quality of journalism has waned (Bas, Ogan, and Varol 2022; Bulut and Ertuna 2022).

In such a media context, WhatsApp usage as a source of disinformation needs further research. WhatsApp is a popular messaging app in Turkey, with a penetration rate of roughly 89 percent as of the third quarter of 2022 (Dierks 2023). Thus, this chapter focuses on WhatsApp (and other encrypted private messaging platforms [EMP]) and the methodological challenges of examining the specific role of WhatsApp in the propagation of disinformation.

This chapter stems from broader research on EU-related disinformation in relation to a rich media and political context in which disinformation flourishes in Turkey. The study was initially based on a specific politically charged topic, the European Union, and in turn also examined a series of issues connected to this, such as the "refugee crisis" and identity politics. However, two major events in Turkish history in 2023 provided further ethnographic insights and helped me to fine-tune my original arguments. After outlining some insights from this study, I will build on the insights and experiences gained in the study to emphasize the need for qualitative research on WhatsApp and highlight methodological challenges, specifically access to private groups, focused searches, shifting platform features, and subsequent changes in usage patterns as well as opportunities for multimodal and mixed methods.

Ethnographic Research on WhatsApp and Disinformation in Turkey

Tracking disinformation in a broad media ecosystem requires a multimodal perspective. An ethnographic overview helps provide a holistic view. Actors and narratives may quickly change according to diverse political contexts, and mere quantitative work such as an analysis of Twitter messages would not provide an explicit viewpoint in these dynamic and nonlinear conditions. My previous research thus took place in several layers: traditional media in national and local outlets, public social media, public intellectual outputs, and finally, in encrypted private messaging groups with different methodologies. However, one of the most exciting aspects of this research was the opportunity to focus on encrypted private messaging groups that included WhatsApp.

Researching encrypted private messaging (EPM) platforms is a new area of study within disinformation studies, and more substantive methods are yet to be found (Gursky and Woolley 2021). In a platform where data capture is relatively complex compared to public social media platforms, ethnographic research allows researchers to better understand the behaviors and interactions of the members, as well as analyze the context of the conversations to see what types of disinformation are shared for what purposes. If the research is conducted properly, studying WhatsApp triggers a series of new discussions on the "context collapse" debate as the content flow on this platform is more personal and private (Velasquez, Quenette, and Rojas 2021).

As part of the ethnographic approach, I interviewed members and collected additional screenshots from the group, as well as from other internet sources. WhatsApp is gated, but the content is ultimately digital, and one can observe a level of intertextuality in receiving disinformation. Call it intertextuality or cross-platform dynamics (Lukito et al. 2020); WhatsApp-focused research is inevitably located in the wider media ecology. Gursky et al. (2022) describe this as "cascade logic": information is moved upstream and downstream in chat app ecologies using cascade logic, which allows for the possibility of distortion along the path. Upstream information is moved from private conversations into the public sphere. Chat applications enable people to gradually withdraw and segregate into layered spaces of privacy and obscurity or to emerge from these spaces, according to cascade logic.

As insiders to political trolling circles, my research assistants and I got in touch with various groups, and we got access to some private groups. Some bigger private Telegram groups or channels were broadcasting the addresses of such groups on other public platforms, and we found them and joined them. For more private and smaller groups, the seemingly more traditional ethnographic practice was at work. Our connections in the field let us into some groups, and sometimes it was accidental incidents. For instance, my father's friend added me to a WhatsApp group for a conservative small-town community. In another, I could have a peek at a Quran-reading women's group, thanks to a close relative.

Within such groups, Telegram offers more opportunities for researchers in terms of focused searches. For instance, I could search predetermined keywords, such as "EU funding, Syrians, etc.," to focus on fragments of disinformation. Unlike Telegram, WhatsApp's group archive begins when one joins the group. While I could search the whole archive in a Telegram group, I could only start the search from the moment I joined a WhatsApp group. The subject matter determines what one can find. If one observes disinformation against the main opposition party, there is a steady stream of information. However, the EU focus is sporadic at best, and one may need long-term engagement to get a better grasp of disinformation that is related to the EU. It should also be noted that platform affordances change constantly. In the later months of our research, WhatsApp introduced more opportunities for group activities. This led to more group activity on the platform, and Telegram lost some of its appeal.

Ethnography, overall, may shape multimethod approaches, but wherever possible, there could be more quantitative methods. In a noteworthy piece of study, Maros, Almeida, and Vasconcelos (2021) examined more than forty thousand audio messages distributed across 364 publicly accessible groups in Brazil over six months of the nation's massive social mobilization. Researchers found that misinformation-containing audios frequently used the future tense, spoke directly to the listener, and had a greater presence of negative emotions. Additionally, misinformation-filled audio clips tended to go viral quickly and endure much longer in the network. However, in many instances where researchers face difficulty in getting automated (and sometimes anonymized) data, ethnographic approaches will have the upper hand. These instances can be practical or topical. For instance, Staudacher and Kaiser-Grolimund (2016) showed that WhatsApp could be used as a valuable tool in ethnographic research in three important fields of interaction and communication: first, between researchers and informants simultaneously in different places; second, as a tool to exchange information with field assistants; and third, to exchange information between researchers. Bueno-Roldan and Röder (2022) also found that WhatsApp is a flexible tool for conducting qualitative research with specific advantages over other messaging apps and voice-over-internet protocols. De Gruchy et al. (2021) demonstrated that WhatsApp can be used as a tool for data collection with migrant and mobile populations. Within these practical parameters, WhatsApp provides the potential to focus on the process of migration and the intersections it provides with access to health care and gender, suggesting the potential of WhatsApp as a research tool concerning migration and health (De Gruchy et al. 2021). Manji et al. (2021) found that WhatsApp can provide new and affordable opportunities for health research across time and place, potentially addressing the challenges of maintaining contact and participation involved in research with migrant and mobile populations. As Soares et al. (2021) suggest, it is also essential to study the role of political groups on WhatsApp in promoting nonadherence to COVID-19 guidelines and regulations. These groups may have played a central role in spreading disinformation due to their prior involvement in political propaganda (Soares et al. 2021).

However, my line of work can contribute to another track: there is still a methodological gap in the ongoing research on WhatsApp's role in civic and political engagement (Pang and Woo 2020). This position may also be supported with the findings of Soares et al. (2021). In comparing discursive strategies used to spread and legitimate disinformation on Twitter and WhatsApp during the 2018 Brazilian presidential election, their study found that tweets often framed disinformation as a "rational" explanation, while WhatsApp messages frequently relied on authorities and shared conspiracy theories in creating negative emotional framings.

In my further research on the earthquake and election instances in 2023, we found that political persecution and public shaming led users to be more active in WhatsApp groups. In the European Union case, nationalists expressed their hostility toward refugees. In later instances, more sections of users used WhatsApp. When the government was hostile against any criticism of how the Turkish state handled the earthquake survival efforts, misinformation and conspiracy theories abounded in the groups. The election process brought oppositional content producers to WhatsApp (along with Telegram and Discord). I do not claim that these were dominated by disinformation; however, election rigging claims could easily be found here but could not be verified by independent observers. Most of the claims would later be refuted. Since Turkey's disinformation law can imprison users, disinformation actors tend to prefer WhatsApp to make such claims.

Unlike our initial expectations, we could not find a pattern in which content was cooked here and later circulated on public platforms. There is more real-time engagement in the public spaces, and the content creators prefer to start there directly. There can, of course, be small groups that organize the public content at the outset, but we could not access these if they exist. In most cases, users followed prominent disinformation actors in public social media and acted accordingly. One significant finding is that the groups we zeroed in on mainly address secular nationalist young groups. Among these users, affiliation with İyi Party, a major nationalist opposition party, is frequent but not always present. These groups and their affiliations are in the political opposition, and their power comes from the rising anxieties about refugee-related demographic changes. This contrasts with the fact that most political troll studies are related to pro-government circles (Saka 2021a).

My further research and observations did not challenge this initial finding, but in other topical interests, WhatsApp began to play a more critical role, and it has the potential to reshape the flow of disinformation. For instance, a recent survey by the TRT, Turkey's public broadcaster, on the circulation of pandemic-related misinformation states that the top three channels through which information is shared are WhatsApp, face-to-face, and over the phone (TRT Akademi 2020). The survey examined the role of digital media in the infodemic, user experience in terms of information acquisition habits, and reliable sources of information. Five thousand and ten citizens over the age of eighteen in Turkey were contacted. The top three information sources during the pandemic were television news, internet news websites, and the Ministry of Health in Turkey.

Methodological Challenges of Qualitative Research on WhatsApp

Researching WhatsApp presents a series of challenges. One is related to data collection. Some researchers have used WhatsApp as a data collection tool to conduct focus group discussions (Anderson et al. 2021; Singer et al. 2023). Unlike public platforms, it is harder to get data from WhatsApp. At the time of publishing this chapter, there will probably be more software to gather data, but for an ethnographer with limited coding skills, the choices are limited at the moment. One must revert to manual practices such as saving screenshots or copy-pasting text. However, some groups may have changed settings to temporary appearances, and one needs to save content quicker than usual. In more serious cases, technical compatibility with forensic tools and mobile technologies when investigating digital crime evidence (Umar, Riadi, and Zamroni 2018) becomes more vital.

Feng et al. (2022) found that conventional content moderation techniques used by open platforms such as Twitter and Facebook are unfit to tackle misinformation on WhatsApp. However, this may not always be a bad sign. WhatsApp groups are relatively small compared to Telegram, and members are mostly vetted to become members. In order to maintain the group's survival, group administrators are quick to intervene. This may go both ways. Those challenging disinformation may be expelled, or those agents of disinformation may be the ones to go. During my observations, I witnessed the agitators being expelled, but among the more extremist groups, the opposite might be happening more if it ever happens. Because of the relative intimacy of the group, there are heated debates, but there are also conciliatory moves. In fact, more than expelling one, after a heated debate, some members prefer to leave the group.

A researcher in a WhatsApp group is not different from any other site in terms of investing time and emotions in the field. The flow of conversations may sometimes hide relevant topics and keywords. Group dynamics and jargon may also cloud some of the issues the researcher is interested in. Some researchers believe that an appropriate research methodology must be developed to detect disinformation topics accurately (Staender and Humprecht 2021). It is also challenging to find the sources and distributors of false information (Yustitia and Asharianto 2020), but in my case, my concern was to track the existence and flow of disinformation instead of finding actual sources. Another study (Nguyễn et al. 2022) points out that language barriers, private and closed information networks, and cultural and historical factors are other challenges that researchers experience. That, again, depends on the subject matter. It is a bigger challenge when one studies refugees. In my case, it was mostly capturing the political jargon or, in the earthquake, technical terms in describing the earthquake-related topics.

One of the ways to study disinformation on WhatsApp is to use a mixed methods approach that combines critical discourse analysis and quantitative analysis (Recuero, Soares, and Vinhas 2021). However, it is vital to observe how users make sense of a social phenomenon in groups. This is a process more than a mere moment, and it may not be bound to predetermined methods.

A final but equally important challenge is the ethical perspectives to develop during the research. Barbosa and Milan (2019) suggest that digital ethnography inside WhatsApp groups requires up-to-date, innovative, ethical guidelines, and researchers should take infrastructure seriously, embrace transparency, and guarantee full anonymization to research subjects. Staudacher and Kaiser-Grolimund (2016) also argue that WhatsApp constitutes a valuable tool in ethnographic research, but researchers need to consider the ethical implications of using the platform. Other researchers (De Gruchy et al. 2021) have warned of the ethical challenges that may arise when dealing with migrant and mobile populations. In authoritarian countries, studying political groups is an ethically serious task, and it is part of the ethnographer's duty not to be a risk to the informants. I preferred to use images and names when they became public on other public platforms. Otherwise, I switched to anonymized, descriptive, and predominantly textual narrations. Infrastructural changes are to be observed as WhatsApp's parent company Meta is notorious for making changes that affect users' privacy (Newcomb 2018). During my research, there was no explicit change, but this is a permanent issue to be aware of. In some critical cases, one should even be thinking about how to save research data, for instance, screenshots. These can give clues about informants' personal data, and they should at least be protected with encrypted solutions.

At another level of findings, national media environments seem to be influential in understanding WhatsApp's role. While the platform played a significant role in Brazil, it was not as effective in Turkey. In the Turkish case, the content was distributed among various relevant platforms. Here, Facebook groups and Telegram stole some of WhatsApp's role that could be seen in Brazil. An ideal situation of transmedia can be seen here as the speech is distributed across the whole media ecosystem.

End-to-end encryption, a WhatsApp feature that makes it easier for people to spread false information, enables writers, recipients, and sharers to remain anonymous. There is no information on the origin of the shared material, and its source cannot be determined (Rossini et al. 2021). Thus, it is a valuable source for disinformation agents. However, most trolls and other agents were not keen on hiding themselves in my research, so WhatsApp did not become a priority in this sense. The appeal of the platform is not privacy or security but immediate communication with family members, close friends, and a relatively close circle of activists. It is practical to communicate, considering its rate of penetration in society within a perceived level of social connectedness.

Infrastructural changes and regulative interventions may also change WhatsApp usage. Encrypted software may be criminalized, as correspondents of Vice had experienced on the Turkey-Syria border (Pizzi 2015). The authorities and the news items never stated the name of the software, but at the time of the arrests, I was told by my informants in the field that the software was probably one of the easily found but strongly encrypted softwares such as PGP or Tor Browser. WhatsApp stays in between, and the Turkish government has not initiated a WhatsApp-based crackdown or a regulative process toward the software. The recent surge in its use may change government attitudes. Amid the lack of public regulations, intraplatform measures are the only ones, such as restricting the number of forwarded messages. It will be a fine line to regulate these platforms without direct censorship. Wafa Ben-Hassine, a human rights lawyer and principal at Omidyar Network, believes the suboptimal design choices of the platforms themselves are the best solutions at the moment (Omidyar Network 2022).

Conclusion

In conclusion, researching and contextualizing WhatsApp as part of the disinformation media ecosystem presents several main research challenges. This work is based on a particular topic, EU-related disinformation, and further ethnographic observations of two major incidents in

Turkey. As outlined below, geographical and temporal parameters are decisive and further research is needed.

Obtaining data from WhatsApp presents challenges due to the platform's encryption and limited accessibility. Researchers need to develop appropriate methodologies for data collection, such as manual practices like saving screenshots or copy-pasting text. Technical compatibility with forensic tools and mobile technologies is crucial for investigating digital crime evidence.

Unlike other platforms, WhatsApp does not exhibit a clear pattern of content creation in private groups before being circulated publicly. Understanding how disinformation spreads within WhatsApp groups and transitions to public platforms requires a comprehensive approach that combines ethnographic research and quantitative analysis.

The nature of WhatsApp groups and the specific jargon used within them can challenge researchers in understanding and analyzing the content. Researchers need to navigate the dynamics of group interactions, including heated debates and potential exclusions or expulsions of members.

Conducting research on WhatsApp requires ethical guidelines to address issues of privacy, transparency, and anonymization. Researchers should consider the potential risks to informants, especially in authoritarian countries, and take steps to protect their identities and personal data. The ever-changing infrastructural and privacy policies of WhatsApp also require continuous ethical evaluation. WhatsApp is assumed to be more social and familial than other EPM outlets. This is directly related to some affordances, such as connection to a phone number and real identity and group limitations or mobility. However, this gives the researcher more responsibilities in protecting personal data.

WhatsApp's role in disseminating disinformation may vary depending on the national media environment. Understanding how WhatsApp interacts with other platforms and media channels is crucial for comprehending its impact and influence in the larger media ecosystem. The subject matter of disinformation also matters. The EU is not an issue of contention in Turkish daily lives, and it appears to circulate in "less social" outlets such as Telegram or Facebook pages or groups. This immediately changes when the topic moves to elections or health issues.

The lack of public regulations specific to WhatsApp poses challenges in addressing disinformation on the platform. Balancing measures to combat disinformation without resorting to direct censorship is a complex task, and researchers need to consider the potential impact of future regulative interventions.

Overall, researching WhatsApp in the context of disinformation requires innovative methodologies, a deep understanding of group dynamics, and consideration of ethical implications. Addressing these challenges is crucial for gaining insights into the role of WhatsApp in spreading disinformation and developing effective strategies to combat it.

14

Researching Political Communication on WhatsApp

Reflections on Method

TANJA BOSCH

Digital and social media are increasingly becoming primary spaces where people receive and circulate information about politics. This includes popular platforms like Facebook and Twitter as well as messaging apps like Telegram and WhatsApp. The use of WhatsApp for entertainment, e.g., sharing jokes and memes, has been "associated with a high perceived risk of exposure to misinformation" (Wasserman and Madrid-Morales 2022, 213). This chapter reflects on various methodological approaches that could be used to research WhatsApp, its use by citizens for political communication, and the challenges and opportunities thereof within a decolonial framework.

Messaging apps are one of the "primary sites for media activism to emerge and unfold . . . (but) have long remained under the radar of scholars of social movements and digital activism, largely because of the limited permeability of chat groups" (Barbosa and Milan 2019).

This chapter discusses existing, emerging, and suggested methodologies for researching WhatsApp, reflecting on the possibilities and challenges of researching WhatsApp as a platform for political communication, with a particular focus on the African context but with relevance for other contexts, too, particularly in the Global South or majority world. Countries of the majority world tend to share experiences and conditions that impact how disinformation is circulated and received (Wasserman and Madrid-Morales 2022).

While there is emergent literature on a range of topics related to WhatsApp, this chapter focuses primarily on WhatsApp and political communication. Researching WhatsApp and reflecting on suitable methodologies is important as social media (WhatsApp in particular) plays an important role in shaping public perceptions of politics and crises, and much of modern life is played out within social media platforms such as WhatsApp. The key argument made by this chapter is that multimethod approaches to researching WhatsApp are a key mechanism for a decolonial approach that prioritizes the qualitative and different ways of knowing. By focusing on users, WhatsApp researchers can highlight local context and experiences, and in doing so, they can illuminate the diverse and nuanced ways in which people use the app. Political communication on WhatsApp can vary greatly depending on cultural, social, and political contexts. Understanding the real-life experiences of users can inform more effective and culturally sensitive policymaking and practice. For example, insights into how political information spreads on WhatsApp can help in designing better communication strategies or interventions to combat misinformation. Qualitative research allows for a deeper understanding of these local contexts, which quantitative methods might overlook. This approach helps to challenge dominant narratives, fosters a more inclusive understanding of digital communication, and ensures that the voices and perspectives of marginalized communities are heard and valued in the research process.

WhatsApp in the Global South

Firstly, a brief note regarding the term "Global South" is in order. The term has been used to describe a range of developing countries and emerging economies, but the Global South is not a homogenous geographic space and encompasses diversity across culture, history, and political systems. Some countries in the Global South (e.g., Singapore and South Korea) have become high-income economies with strong political systems, while others are still marked by economic and political instability. "Global South" replaced the term "Third World" alongside shifts in the global political and economic landscape. The term emerged in the 1980s "as a shorthand to geographically indicate gaps in terms of resources and development between colonizing countries and their former colonies (not including the settler colonies of the United States, Canada, Australia, and New Zealand)" (Udupa and Dattatreyan 2023,

189). It has thus been used to refer to countries below the equator that are economically or politically marginalized in the global context, and tend to have a shared status as former colonies.

While the term "Global South" is potentially useful as an analytical tool to examine power structures and inequalities in the global system, the term is problematic as it could reinforce a sense of "otherness," reinforcing a binary opposition between Global North and South, while perpetuating a European perspective that centers the Global North as the norm and fails to acknowledge the diversity of the so-called South. In addition, it is not homogenous and the term leaves no room to encapsulate the diversity of the so-called Global South. As Santos (2012, 51) points out, the term is not a geographical concept because "south" also exists in the "north" "in the form of excluded, silenced, and marginalized populations, such as undocumented immigrants, the unemployed, ethnic or religious minorities and victims of sexism, homophobia and racism." Milan and Treré (2017, 321) have similarly argued that "the South is, however, not merely a geographical or geopolitical marker (as in 'Global South') but a plural entity subsuming *also* the different, the underprivileged, the alternative, the resistant, the invisible, and the subversive."

This chapter uses the term as a shorthand, acknowledging these critiques, and the idea that "the label itself is inherently slippery, inchoate, unfixed" (Comaroff and Comaroff 2015, 126). Importantly, social media in these contexts have developed distinct characteristics. "Social histories, community networks, and communication practices characteristic of these regions also impact the likelihood of media users in these regions to consume and share disinformation for reasons such as creating social awareness, contributing to humorous narratives, and acting out of social responsibility" (Wasserman and Madrid-Morales 2022, 211).

Reflection on Methods for Researching WhatsApp

The majority of research on WhatsApp and political activism originates from the Global South. Brazil and India feature widely, but there is a dearth of scholarly literature on the uses of WhatsApp in Africa. A range of methods has been used to research WhatsApp, ranging from qualitative approaches to virtual ethnography, interviews, and surveys of users.

Current methodological approaches to WhatsApp range from starting with notions of the internet and social media spaces as place or text, with an emphasis on the latter.

Researching WhatsApp Textually

One method to analyze WhatsApp discourse would be to consider messages or posts as texts and to analyze these in much the same way other texts might be researched. WhatsApp threads can be seen to be complex and dynamic texts that reflect and shape the social, cultural, and political contexts in which they are located. The method of qualitative content analysis could be one method of analyzing text conversations and text qualitatively. Qualitative content analysis is a research method used to analyze the content of texts such as written or historical documents, social media posts, interview transcripts, etc., to identify patterns and themes. Qualitative content analysis allows for systematic and in-depth analysis of content (Krippendorff 2018). This type of methodology could easily be applied to a corpus of WhatsApp messages, coding them to explore emergent themes in the conversation. Coding schemes allow for the systematic identification and categorization of content. Content analysis of WhatsApp message threads might thus be useful to "provide insights into the meanings and interpretations of the text, as well as to develop hypotheses or theories about the phenomena being studied" (Schreier 2012, 11).

Similarly, discourse analysis might also be a useful method to employ. Discourse analysis is a research method used to analyze the language, meaning, and social context of texts. This method can be used to study the discourses and power relations shaping online communication by exploring the language and rhetoric used in WhatsApp conversations. Exploring the discursive strategies used by political actors and citizens would allow researchers to uncover the power relations, ideologies, and social structures reflected and reproduced in language (Fairclough, 2013). Another related textual approach is conversation analysis, which focuses on a micro-level analysis of interaction, aiming to understand the sequential organization of talk, the ways in which participants coordinate their responses, and the ways in which social categories are constructed through conversation or talk (Heritage and Clayman 2011).

Text mining and computational approaches to text are also a possibility. WhatsApp chats can be text mined using Python libraries such as NLTK or spaCy, and sentiment analysis methodology could be applied. Sentiment analysis, also known as opinion mining, allows researchers to explore the emotional tone or sentiment of a text, i.e., whether a text is positive, negative, or neutral, which could be useful for researching political communication. Sentiment analysis is a natural language processing technique that involves the use of machine language algorithms that are trained on a large corpus of text data to identify patterns and relationships between words and sentiments.

In researching the text and content of WhatsApp messages, we should also consider expanding the datasets to include conversations that are in languages other than English.

Researching WhatsApp as "Place"

An alternate approach to the textual methods proposed above would be to think about WhatsApp as a space for the creation of online communities. Instead of only analyzing conversations and the textual output of the messaging app, this approach would be concerned with relationships between people and mapping relationships in an attempt to identify influential members of online networks. A starting point might be to think about the architectural structure and affordances of the app, using methods like the walkthrough method (Light, Burgess, and Duguay 2018), which was developed as a way to trace the technological mechanisms of an app to understand how this guides users. This approach brackets users and assumes individual access to mobile devices, and therefore, a more multidimensional approach to app studies might be more useful (Duguay and Gold-Apel 2023).

Exploring WhatsApp as a space could thus also include network analysis and online ethnographic approaches to researching the communities created by groups. This could include participant observation and autoethnography, with researchers seeking permission to join group chats and observing online interactions among participants. However, as Barbosa and Milan (2019) point out, the high frequency of exchange in groups poses a challenge for ethnographers. One example of this type of research is Barbosa's exploration of a group in Brazil in which he announced his research periodically, became an active member of a group, and promised to remain accountable with regard to public presentations of the research (Barbosa and Milan 2019). This digital ethnography allowed him to observe interactions and the dynamic realities of group engagement, as well as the formation of group identity. Interactions on messaging apps are "embedded, embodied, and everyday" (Barbosa and Milan 2019), and ethnography is thus a method well suited to studying WhatsApp.

In this instance, the data source would be the text of messages along with field notes analyzing the conversations and interactions between people, and this could be triangulated with group member interviews. This approach also necessitates keeping in mind the context in which messages on WhatsApp are circulated. The framework provided by Udupa (2023) suggests that to explore the contours of digital hate cultures, as we might see on WhatsApp, a close contextualization should be accompanied by deep contextualization, looking at both everyday practice and deeper histories, to account for historical continuities. "Disinformation, whether its production, reception, or responses to it, can only be properly understood within the social, political, economic, and historical contexts where it is consumed and spread" (Wasserman and Madrid-Morales 2022, 210). As Udupa (2023, 242) has further argued, "longer historical processes should be examined in relation to proximate contemporary contexts of digital circulation and practice—a kind of dual analysis that might be described as decolonial thinking."

Further audience research of the platform could involve exploring networks of connectivity and the role of network members in the flow of information. Social network analysis (SNA) would involve a relational and quantitative methodological approach to WhatsApp, exploring how members of groups are connected and researching who the influential members of the network are and how they control the flow of information. The methodology of SNA examines relationships between individuals (or groups) within a social network to explore how these relationships impact behavior, based on the premise that social relationships are not just incidental but have a significant impact on how people behave and make decisions. For example, Bursztyn and Birnbaum (2019) conducted an SNA of right- and left-wing users of WhatsApp in the 2018 Brazilian elections. To conduct a SNA of

WhatsApp, a Python script could be used to read chat data, construct a network graph, calculate centrality measures, and visualize the network to identify key users (influencers, connectors) and communication patterns. Other existing tools such as Gephi, NodeXL, or manual analysis with R could also be used.

Interviews and Ethnographic Approaches

As Schoon et al. (2020) emphasize, qualitative methodologies are especially important to develop decolonial approaches. As they note, "Qualitative methodologies are essential for developing decolonial scholarship that facilitates different ways of knowing that allow for the delinking from imperial mindsets and relationships" (Schoon et al. 2020). The qualitative-quantitative binary has long been debunked as unhelpful as it reinforces false assumptions and limits the potential for interdisciplinary research, but the prevailing epistemological dominance of positivism often prevails in scholarly research, particularly in the field of digital media studies. Other methodological approaches could thus also include phenomenological qualitative approaches, including focus groups and interviews. Although users could also be surveyed, the response rate might be low and impact the reliability and validity of the research. Qualitative methods, such as interviews and ethnography, can capture the subtleties of how political messages are created, shared, and interpreted by users. This nuance is crucial for understanding the complexities of political communication.

However, researching audiences by way of surveys or interviews might be one way to overcome ethical challenges concerning private chats (instead of analyzing the textual content of WhatsApp groups). Interviews with users are particularly helpful in allowing researchers to consider users' specific sociocultural experiences and contexts, keeping in mind that "context is paramount in the study of disinformation in the South" (Wasserman and Madrid-Morales 2022). Qualitative interviews or focus groups would provide rich, detailed data that can reveal how individuals use WhatsApp for political communication, including the motivations, meanings, and impacts of their interactions. Such interviews could be conducted in particular ways that allow for a consequent interaction with and reflection on the data produced by users. Alongside this, the scrollback method (Robards and Lincoln 2017) would be a useful approach, whereby an initial conversation with a participant is followed up with an interview where a user "scrolls back" through their social media feed and explains in detail what is shown, in this case reflecting on the messages they sent and received. This would allow researchers to gain a deeper understanding of how people engage with online messaging and the thought processes involved in choosing to either delete or forward messages that might be interpreted as misinformation. The affective turn in digital media studies is an important framework, as these messages and the choices made by users around them are most likely linked to affect or emotion. The "affective turn" refers to the recognition that emotions play a significant role in shaping people's interactions with digital media and that affective experiences can have important social and political implications. This method "brings back to life the digital trace, capturing the specific context(s) and contours within which . . . participants . . . [use WhatsApp] . . . to make disclosures that we could not intuit without them present" (Robards and Lincoln 2017, 720). Qualitative methods can thus integrate various aspects of users' lives, such as their social networks, cultural practices, and personal experiences, leading to a more holistic understanding of political communication.

Drawing on previous research conducted by this author (Bosch 2022; Schoon et al. 2020), this chapter advocates for a multimethod approach to studying WhatsApp, particularly in the context of the Global South. "Around the world, users are subject to the algorithms imposed by platforms; and Global South users of these largely Western designed and controlled platforms are subject to platform power and the infrastructural limitations of these platforms, which can impact and limit democratic culture, resulting in neo-colonial media culture" (Bosch 2022). Mixing methods in a qualitatively driven way, or even prioritizing qualitative methods, is important to avoid homogenizing social experiences on entire continents, as is often the case in scholarship on Africa. Qualitative methods such as interviews and ethnography, alongside content analysis, could thus be very useful. As Udupa (2023) argues, ethnographic exploration of media practices is a critical component of researching extreme speech. Extreme speech, conceptually, calls for us to place such practices into a broader context of contestations over power and a "grounded, historically aware analysis" (Udupa 2023, 233).

Ethical Considerations When Researching WhatsApp

As WhatsApp groups are closed, i.e., membership is managed by gate-keepers (group admins), this also raises various ethical issues for researchers. When WhatsApp groups are created, they can be set up either to be private or public. What differentiates these is the way in which they are managed and the level of access that users have to a group's content. Private groups are usually created by an individual user, who has control over who is invited to join the group. These groups are only accessible to people who are added or invited by a group's admin, and the content shared is only visible to members. Private groups are more secure and private, as they are not accessible to the general public and can only be joined by invitation. On the other hand, public WhatsApp groups are open to anyone with access to the group link or invitation. Anyone can join the group by clicking on a joining link. Public groups are usually managed by several admins who can add or remove members and control the content shared within the group. These groups are less secure, and unknown or unwelcome members could easily join. However, closed, private WhatsApp groups are more likely to be used for political communication and the spread of misinformation, raising various ethical issues for researchers as these are the types of groups we would be most interested in. Private groups allow for a more closed and exclusive environment, making it easier to share sensitive or controversial information without fear of public scrutiny. Private, closed groups can be difficult to monitor, making it easier for misinformation to spread.

The top sources of misinformation spread on the platform tend to be from a small number of users—in Nobre et al.'s (2022) study, ten users were responsible for up to 26 percent of all misinformation shared in a week. Moreover, the nature of the app lends itself to covert observation, and "deceiving participants is thus a concrete risk and a tempting possibility" (Barbosa and Milan 2019).

Ethical social media research is a challenge for researchers regardless of which platform they study, but it is particularly salient in the context of WhatsApp due to the bounded and often sensitive nature of WhatsApp groups. Navigating access to groups is tricky, but at the same time, announcing one's presence as a researcher could also impact the nature of subsequent conversations. Group members might be less likely to speak freely knowing that they are being observed. How do researchers gain informed consent, and how might their presence in groups impact conversations; what are the ethical dilemmas involved in "lurking" in such groups and engaging in participant observation? There is no single answer to this quandary—these decisions will have to be made in the field, on a case by case basis.

The public-private debate in social media research refers to the ways in which researchers balance the right to privacy with the potential benefits of studying social media data. Researchers could obtain informed consent from participants in closed groups and also de-identify the data to ensure that the research does not cause harm to participants. Aggregating the data to protect participants' identities and prevent identification of individual participants could be an additional solution. Informed consent, the standard practice for ethical research, is difficult to follow when members enter and exit a group constantly and authorization from the group admin would not constitute individualized consent (Piaia et al. 2022). One approach has been to obtain data via collaborators who voluntarily forward messages to the researcher, who does not become a member of the group (Piaia et al. 2022). Barbosa and Milan (2019) advocate for an approach that involves "moving past the consent form as the sole and merely regulatory moment of the researcher-subject relationship."

The Association of Internet Researchers (AoIR) has developed a set of ethical guidelines for conducting social media research, though specific guidelines do not yet exist for WhatsApp. The AoIR's *Internet Research: Ethical Guidelines 3.0* document (Franzke et al. 2019) calls for researchers to be transparent about research methods and data sources and to anonymize or de-identify data before analysis. This could be a useful starting point for WhatsApp researchers, but this would have to be on a case-by-case basis. For example, what happens if an admin removes a researcher from a group after data has already been collected, i.e., the chat history has been archived? What if the admin of a private group refuses to add a researcher, but the researcher is able to access the chat history from another user? What about instances in which the researcher encounters hate speech or inflammatory messages in the WhatsApp groups they are studying—would they be complicit if they didn't report hate speech that led to offline violence? These are ethical quandaries that require further consideration. As with other forms of research, researchers have to engage in constant reflection and discussion about these issues. In many countries outside of the Global North, institutions often do not require ethical clearance for this type of research. When actual WhatsApp data is collected, it should be anonymized so users' names and telephone numbers are not included in the dataset. Navigating the ethical challenges of researching WhatsApp requires a careful and thoughtful approach. By prioritizing privacy, obtaining informed consent, ensuring data security, maintaining transparency, and seeking ethical oversight, researchers can conduct their studies responsibly while respecting the rights and well-being of participants.

Decolonizing WhatsApp Research

Regardless of the specific method or approach selected, the broader context of decoloniality should be foregrounded. A decolonial methodology is an approach to research methods and knowledge production that challenges dominant Western epistemologies and ways of thinking and "knowing." It is focused on the fact that many existing knowledge fields have been shaped by colonialism, imperialism, and a Eurocentric bias. Decolonial methods focus on decentering Western perspectives and voices and recognizing and valuing the complexity of non-Western knowledge systems. This type of research highlights the epistemologies of the South to adequately account for the realities of the Global South. Decoloniality is a long-standing political and epistemological movement that has assumed various forms to deconstruct colonial matrices of power (Ndlovu-Gatsheni 2015).

A postcolonial critique drew attention to the legacy of colonialism and exploring the ways in which the West has constructed the "other." More recently, Connell's (2020) concept of southern theory focuses on highlighting texts and scholars from the Global South.

The growth of the mobile internet alongside the rise of digital platforms in the Global South has raised various issues regarding how researchers approach this so-called digital turn. People who are not "perpetually connected" use technology in ways that draw meaning from their specific social contexts (Arora 2016). In addition, we need to take into account the particularities of studying realities that are marked by longstanding inequalities in media access and use (Wasserman and Madrid-Morales 2022, 220). Udupa and Dattatreyan (2023) further highlight the digital as a historically constituted field of power and offer methodological opportunities to disrupt and "unsettle" the existing canon.

Importantly, Duguay and Gold-Apel (2023) warn that we should take care not to re-create the extractive principles we often attribute to and critique in relation to platforms. Self-reflexivity and an ethics of care are required for researching online spaces "that heeds whether the research is welcome in a community's digital environment and among its data, elevates the preservation of privacy, and identifies the researcher's personal responsibility protecting those implicated in research" (Duguay and Gold-Apel 2023, 7). This requires a degree of decolonial self-reflexivity, where we, as researchers, interrogate our own positionalities and scholarship in relation to decoloniality: "Our own positionalities are intricately interwoven with digital discourses" (Udupa 2023, 240). Moreover, an "iterative process of reflexivity" would compel researchers to "not only examine and reexamine the outputs of qualitative and quantitative analyses but also the very categories they build on, as researchers work through their own positionalities in the iterative process of refining the analytical categories they deploy" (Udupa and Dattatreyan 2023, 191). Qualitative research often involves closer interaction with participants, fostering ethical reflexivity. Researchers can be more attuned to the ethical implications of their work, ensuring that the research process respects the rights and dignity of participants.

Conclusion

To conclude, WhatsApp provides many opportunities for social media researchers to explore how the messaging app is used, particularly in the spread of political news and misinformation. The rise of WhatsApp in the Global South presents a particular research opportunity for scholars interested in the ways in which technology shapes everyday life. However, researchers working in this area should note the particularities of digital research in the Global South and embrace multimethod qualitative approaches, which take into account the local contexts out of which data emerges. The users of messaging apps are often those at the bottom of the data pyramid, and accordingly, ethical research involves traditional approaches to research ethics, such as the protection of users' privacy, but also necessitates a critical, decolonial approach, in which we conduct multisited, multilingual, qualitatively context-grounded research. This could include actively engaging with local communities to cocreate research agendas, ensuring that the research benefits the participants and not just the researchers, and acknowledging and addressing power imbalances. Additionally, it could involve giving participants control over their data, using participatory methods to involve them in the research process, and striving for transparency and accountability in how data is collected, used, and shared. Ethical research must also be flexible and adaptive to the cultural and social contexts of the participants, recognizing their agency and perspectives throughout the research process.

15

Collecting WhatsApp Data for Social Science Research

Challenges and a Proposed Solution

SIMON CHAUCHARD AND KIRAN GARIMELLA

Over the past few years, alarming press reports assigning blame to WhatsApp usage for a variety of events have proliferated. In countries like Brazil and India, analyses have repeatedly suggested that group-based interactions on WhatsApp distort beliefs among the electorate (e.g., Bengani 2019; Perrigo 2019; Tardáguila, Benevenuto, and Ortellado 2018), and beyond, that they impact various outcomes, including (but not limited to) individuals' propensity to engage in hostile, radical, or violent behaviors (Chopra 2019; Magenta, Gragnani, and Souza 2018; Ozawa et al. 2023).

While academic research has recently started to examine these dramatic narratives, much research admittedly remains to be done to quantitatively evaluate and disentangle the mechanisms through which WhatsApp may or may not be associated to these outcomes. Social scientists interested in hate speech or misinformation in the Global South will accordingly need access to WhatsApp data, potentially on a large scale, in years to come. Specifically, in order to ascertain the platform's role in the dissemination of problematic content, as well as the consequences of this dissemination, researchers still need to better understand (1) the type and style of hateful content and/or disinformation that circulates on the platform, (2) the overall volumes of such content, (3) their degree of virality, (4) the networks through which such content is most likely to circulate, and (5) the political, social, and contextual factors in which this content emerges and has real-world consequences. What is more, researchers will need to gain access to such data in a way that is practical, legal, and respectful of users' privacy.

Yet, access to WhatsApp data for research remains difficult and rare. While high-quality evidence about Facebook and Twitter users' "information diets" has existed for some time (Barberá et al. 2015; Guess, Nagler, and Tucker 2019), no comparable systematic evidence so far exists regarding WhatsApp, despite researchers' longstanding awareness that the platform is used to disseminate this type of content in much of the Global South (Tucker et al. 2018). Besides, when researchers do have access to *some* WhatsApp data, they most likely access samples of data that are too limited in scope to answer all the aforementioned questions, or they obtain such data in ways that may be suboptimal from an ethical standpoint.

How can researchers thus collect sufficiently interesting data in a way that minimizes these ethical concerns? To answer this question, this chapter presents a possible procedure (and the adjoining tool) to collect vast amounts of WhatsApp data. The *data donation* strategy we introduce minimizes the practical aspects of WhatsApp data collection, while conforming to dominant norms about privacy. We detail our general strategy and propose a protocol in section 3. In sections 4 and 5, we discuss the pros and cons of that strategy. To set the stage for this, the next section starts by reviewing the challenges associated with WhatsApp data collection.

Challenges of WhatsApp Data Collection

Researchers eager to engage in WhatsApp data collection may face technical, legal, privacy-related, and practical challenges.

Some of these challenges, however, strike us as being harder to overcome than others and, hence, worthy of more attention. Technical challenges are, for one, relatively limited: extracting data from private WhatsApp threads is *technically* easy once a thread participant (whether or not they are an admin) consents to extract it. Contrary to other platforms, WhatsApp makes it *very* easy for its users to archive the content of the conversations they are part of.[1] In our view, the most serious challenge is equally unlikely to be legal. *How* the platform will react if research we outline in the rest of this chapter becomes common remains

---

1. Concretely, in a matter of seconds, any user can go into a thread and press "export chat" and save the exported data in a variety of formats.

to be seen. Nonetheless, in our experience, and based on our admittedly limited discussions with WhatsApp representatives, the platform may be sympathetic to research that allows for the detection of problematic behaviors or for research on the causes of such behaviors. This is especially likely since WhatsApp's commercial promotion of its own encryption would make it difficult for the company to simultaneously research content circulating on the platform. In that sense, we hope that the delegation of this task to external researchers may not only be legally unproblematic but encouraged and supported by the platform.

These legal and technical hurdles notwithstanding, it remains that WhatsApp data collection on a large scale presents serious *privacy-related* and *practical* challenges. We detail these in the subsections that follow.

Privacy-Related Challenges

Since users may easily export data from the threads they are on, and since researchers cannot access private WhatsApp threads without going through a thread's participant, any WhatsApp data collection effort needs to be a data *donation* effort. That is, one or several users need to *consent* to give away some data from the threads included in their account, and to engage in a series of actions to export these.

Donating data from one's own account may be problematic from a privacy-protection point of view insofar as it may contradict guidelines, norms, or laws protecting individuals' privacy or limiting the processing of individuals' personal data. The European Union's General Data Protections Regulation (GDPR) currently constitutes the main example of such regulation, though privacy laws around the world—such as India's Personal Data Protection Bill, Brazil's General Data Protection Law (LGDP) or Canada's Personal Information Protection and Electronic Documents Act (PIPEDA)—echo most of the principles at the heart of the ruling when they exist. Besides, with regard to the handling of users' personal data, principles of user consent, anonymization, and limitations on the use of such data will likely deserve a discussion, whether a local equivalent to the GDPR exists or not.

So why and in what ways might donating data from one's own WhatsApp account violate the privacy of users, as defined in these norms? Though we acknowledge that norms and regulations will differ depending on the case chosen by researchers, we generally see five potential issues with a *WhatsApp* data donation program that relies on donations by what we will hereafter refer to as a consenting "gateway user":

There is, first, the issue of consent (or the lack thereof) of third-party group participants. These participants are the WhatsApp users who are *not* our gateway into a group, but whose phone numbers, profile pics, and messages, among other data, nonetheless feature on the threads whose data we harvest. While gateway users consent to give away their data, they cannot speak for these other users whose data will enter the dataset through their own data donation. This suggests that consent should be obtained from these third-party group participants, which may be impossible or undesirable from a methodological standpoint, or that the data of third-party users should be credibly anonymized to make their data unidentifiable.

A second issue concerns *how much* data a research project should be allowed to collect. Within the GDPR framework, this would be aligned with the "data minimization" principle—that is, the idea that personal data collected should be limited in time and scope to what is directly relevant and necessary. But more generally, in the context of WhatsApp, this may indirectly raise the question of the type of threads (one-on-one versus group, private versus public, etc.) that may be collected for research.

Once researchers have obtained access to data, the third and even greater problem will be that of anonymization—that is, the strategy used to credibly anonymize stored data and minimize the potential for reidentification, whether this is regarding gateway users or third-party users. This is because researchers may want or need to conceal the identity of discussion participants to protect their privacy. While protecting privacy is in and of itself important, this issue is especially likely to become an important issue if the data are later made available to others, as will tend to be the case under increasingly frequent open science agreements.

While anonymization will be of paramount importance to any WhatsApp data collection effort, how the data are handled *before* researchers are able to credibly anonymize or at least pseudonymize the data should also matter. Researchers will have to use safe, reliable, and credible strategies to transfer and store data in its pre-anonymization form, and relatedly, to ensure that these data cannot be accessed by un-authorized actors or lost at that stage, *before* they are anonymized.

Fifth and finally, any WhatsApp data donation effort will have to wrangle with the potential issue of "unexpected findings" and findings that are subject to legal disclosure obligations under international or local law. Concretely, this raises the question of what protocols research-ers will have in place if/when they stumble on data that are subject to legal disclosure obligations.

Practical Challenges

We add to these ethical and privacy-related concerns several practical challenges that researchers would face to make such a WhatsApp data donation program sufficiently useful from a research standpoint, in line with the balancing principle enumerated by Ohme and Araujo (2022).

Firstly, there are various technical issues to ensure the smooth, rapid, and private donation of such data. Practically, any effort to obtain data from gateway users is likely to fail (low participation rates, for instance) if this is a tedious, expensive, and/or time-consuming process. Similarly, it is also more likely to fail if the donation protocol requires that an enu-merator or another associate of the research team present during the data collection scrolls through the data or accesses it in any way in its pre-anonymization form.

A second and related challenge relates to the ability of researchers to convince a broad (and ideally representative) sample of gateway users to donate some of their data. To put it simply, the ambitious research goals enumerated above would require researchers to obtain data from a suf-ficiently diverse cross-section of the population in order to reach any scientifically valid conclusion about the populations targeted. This may require costly efforts by the research team to ensure that the sample of donors is sufficiently interesting.

A Possible Strategy

How can researchers overcome these many challenges?

As mentioned above, given end-to-end encryption, any ethical WhatsApp data collection effort by design needs to be a data *donation* effort. Considering the challenges we listed, this will additionally need to be a data donation strategy that facilitates a relatively effortless donation, in a privacy-preserving manner, and in a way that inspires confidence among a diverse group of donors.

As part of an ongoing research project requiring the collection of large amounts of private WhatsApp threads (the ERC POLARCHATS project[2]), we have spent much time developing such a solution over the past few years. As part of this process, we have developed a dedicated Web interface called *WhatsApp Explorer*. While further tests will be needed to validate our strategy, these efforts will arguably allow us to minimize all these challenges while allowing us to amass data that is interesting enough for research.

We detail the broad principles of this strategy in this section, before getting into the details of the data protocol in the next section. Our general strategy is to contact users and ask them to donate *some* of their WhatsApp data for social science research. The key technical innovation of the project is to make this donation process relatively seamless for consenting users so they may donate the data they wish to with minimal effort, with the aid of a research associate who will assist in the donation but never access the data.

To reward users for their time and contribution to research, we provide them with small amounts in phone credits. We also provide them with extensive guarantees regarding privacy and anonymization and highlight that their (anonymized-at-the-source) data will at no point be shared beyond the main members of the research team. Importantly, in the design we outline below, no field staff (enumerators and local partners) will have access to the data collected. That is, while field staff make the data collection possible, the data never transit through their own devices (they are instead instantly encrypted and uploaded to a secured server that only we, as principal investigators, have access to).[3]

---

2. The ERC-funded POLARCHATS project (2022–2027) documents the extent of the misinformation crisis in India and Brazil and explores the causes and consequences of exposure to this misinformation. The project relies on qualitative insights, quantitative descriptions, and experimental methods to achieve these objectives.

3. Nor do field staff subsequently have access to the server on which the data are securely stored.

Importantly, we also refrain from asking users to donate one-on-one threads, and we concentrate on group threads to limit privacy concerns (details in the full protocol below).

To overcome privacy-related challenges, we have devised an extensive strategy to anonymize data as it is uploaded to our servers. We do not store any raw de-anonymized data.

With regards to text data, we anonymize any personally identifiable information like the names, phone numbers, and emails from the dataset. The anonymization is done through state-of-the-art privacy-preserving algorithms that are a well-established and widely used library provided by Google called the Cloud Data Loss Prevention API (DLP API).[4]

Regarding visual content, we proceed to irreversibly anonymize most images we store as we upload them, with the exception of images/videos that are shared by at least k groups/threads (say k = 5) in our data. This ensures that we do not access the vast majority of unanonymized visual content. Importantly, the viral content we keep and analyze is extremely unlikely to be personal or private content, as it will be, by definition, content that is shared in many online communities. To anonymize visual content, we proceed in several steps, which we detail below. We first use automated tools to systematically blur faces and a few additional identifying features of images/videos (for instance, car plates). We also implement a second, human-supervised stage of anonymization *before* analyzing the data, to strengthen an already thorough anonymization strategy.

Illustration: A Possible Data Protocol

Here are how these general principles may translate into a concrete data collection protocol.

While a fully online process may be possible,[5] we think an in-person process may be more adapted to most Global South contexts, in which most users will not own the hardware necessary (concretely, two

---

4. Full technical specifications are provided.
5. We are currently exploring this strategy in one of our study sites to complement the in-person strategy we outline here.

screen-equipped devices, whether these are phones, tablets, or laptop or desktop computers) to complete the process themselves online, or will not have the skills to do so. In addition, in the Global South or elsewhere, we note that an in-person process may be necessary to efficiently deliver guarantees about privacy and to generate trust among respondents.

For this reason, we focus on describing a protocol for in-person collection. How would this look? Concretely, a research associate (a trained enumerator from a partner survey firm) would visit randomly selected citizens face-to-face (at their residence) and ask them to participate in a research study on their social media activity, particularly with regards to discussion groups they are part of, and on the content that circulates on these groups.

Concretely, this is how we envision the data collection will look, step by step:

1. Field enumerators explain the goal of the study to the individuals contacted and ask for consent, using a standard consent procedure. At this time, they also provide respondents with a printed flyer explaining who the researchers are, what their goals are, and how they can be contacted. This includes logos of all partner organizations, a link to the relevant registry of data processing activities that details the research plans and the legal basis for it, a hotline phone number for asking questions, and extensive details on our anonymization strategy in relatively untechnical language. Finally, this document also contains clear technical instructions on how to end participation in the project.[6]

2. If they consent to participate, the enumerator requests that the respondent scan our generated QR code through the WhatsApp app on their own smartphone. Concretely, the research associate—using the Web interface we designed (https://whatsapp.whats-viral .me/)—generates a QR code and asks the respondent to scan it with their phone (this is easily done within WhatsApp through the

---

6. A login code is stored to obtain data for two months, after which it is automatically deleted. The users, however, have a chance to log out any time before that on their own phones, using the instructions detailed here: https://faq.whatsapp.com/539218963354346/?locale=en_US.

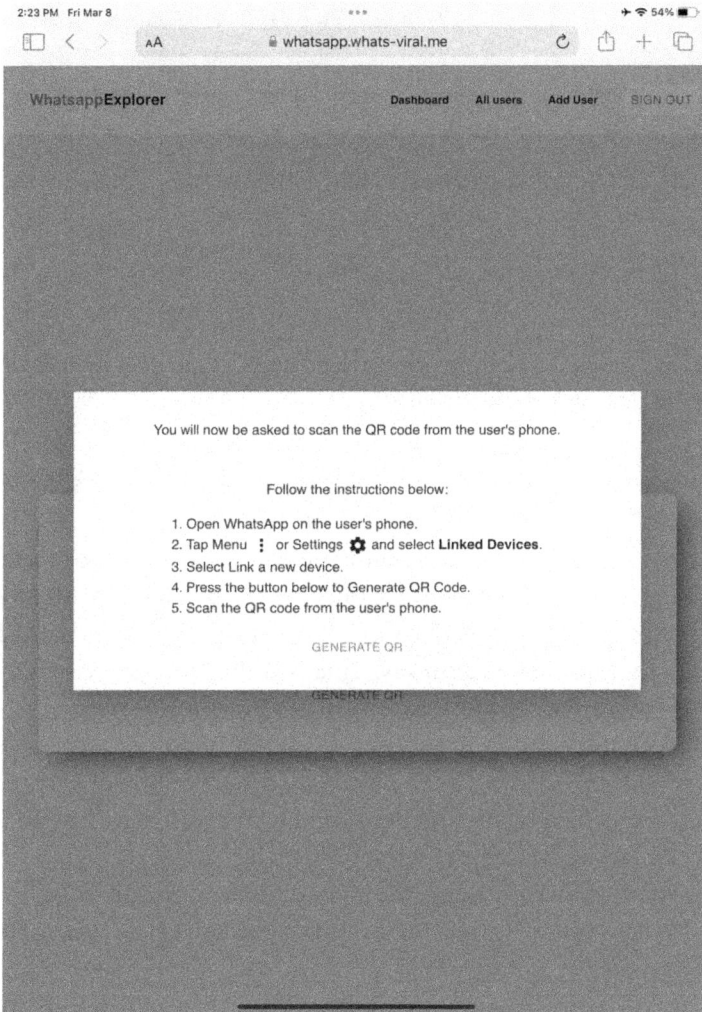

Figure 15.1: Screenshot of the front page of the WhatsApp Explorer interface.
Source: Authors

"linked device" function anyone can use to connect their WhatsApp, for instance, on a computer).

Importantly, throughout the process we describe here, the enumerators at no point need to handle the respondents' devices or see their content.

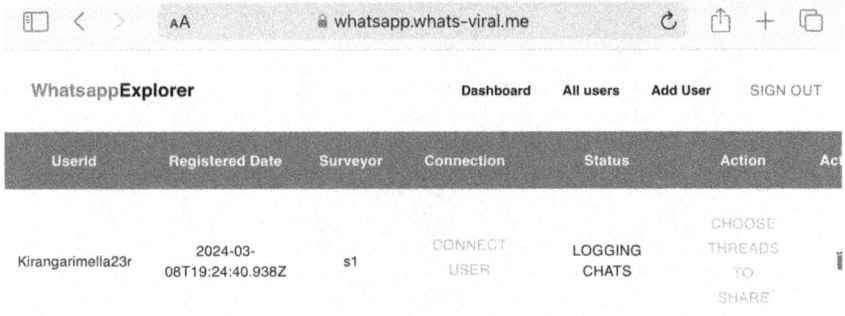

UserId	Registered Date	Surveyor	Connection	Status	Action	Act
Kirangarimella23r	2024-03-08T19:24:40.938Z	s1	CONNECT USER	LOGGING CHATS	CHOOSE THREADS TO SHARE	

Figure 15.2: Screenshot of the WhatsApp Explorer interface immediately after the QR code has been scanned. Source: Authors

Once this is done, the enumerator connects with the user's WhatsApp by pressing "connect user" on this screen.

Once this is done, the enumerator presses "choose threads to share":

At this point the enumerator can show the respondent the full list of WhatsApp group names that exist in the respondent's WhatsApp on the tablet they use for the survey. This is done without yet having access to their content or any other information besides metadata. Once more, throughout the process, the enumerators do not have access to the content of the threads, as these can be directly uploaded to a server that only the principal investigators will have access to at the end of the process, which we describe in what follows.

Importantly, at this stage we automatically exclude one-on-one threads to minimize the amount of data we manipulate, and to further protect privacy.[7]

We can order these group threads in three different ways: by date of the most recent post, by number of people in a group, and by total number of posts over the past two weeks.

On this screen, by default, we present to the participants all threads with six or more participants and with ten or more messages over the past two months.

---

7. Incidentally, we also believe this exclusion to constitute an important incentive for respondents to participate, as their most private data are likely to be in these one-on-one threads.

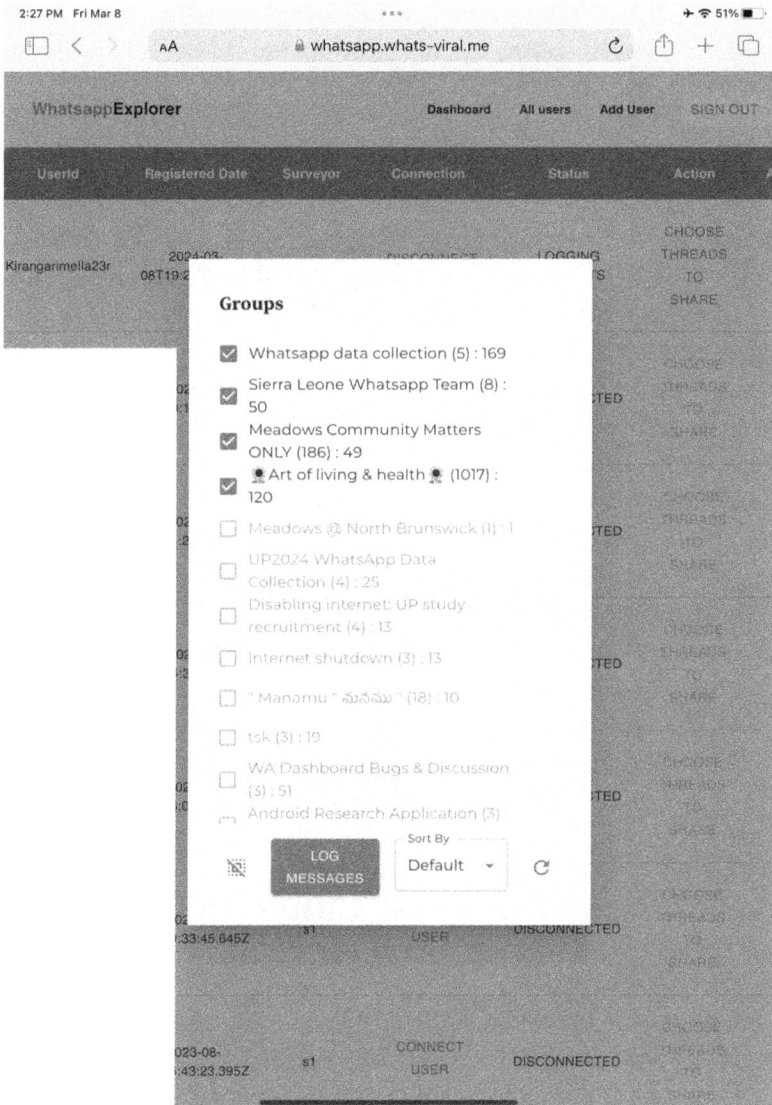

Figure 15.3: Screenshot of the WhatsApp Explorer interface during the last stage of the process, group selection. Source: Authors

However, this interface allows participating users to themselves select the groups they are willing to donate by ticking the corresponding box in the right panel. These steps are then followed:

1. At this point, we ask participants to share data from the subset of these groups that they selected for the two months before they entered the program (i.e., two months before the date of their interview) and for two months going forward, though we readily note that they may themselves choose to restrict their donation to either part of this (as shown below). That is, we ask that they provide us with *only* past or *only* future data, in addition to only sharing a subset of these groups.

2. As this inclusion/exclusion process takes place, our interviewer asks the type of data they are willing to give. Participants choose whether they want to give historical data (past two months), future data (in the next two months), or both. If the respondent indicates willingness to share data going forward, we explain what they ought to do to ensure the collection happens over the next two months, as well as what they can do to ensure it does not, should they change their mind. Importantly, they can take these steps in seconds, on their own phones should they want to disconnect. This is explained in detail in the draft flier we distribute when asking for users' consent. The enumerator may also demonstrate it in person.

3. Once this inclusion/exclusion process is completed, the enumerator presses the "log messages" button. Doing so leads to an upload of the past two months of content of the selected threads onto our secure server (details below), in an anonymized manner (details on anonymization strategy below). That is, we never upload or store non-anonymized content. The content is encrypted as it is transferred from the field staff's devices to our cloud server.[8] This guarantees that neither the principal investigator nor research staff ever has access to such content, either during the time of data

---

8. Concretely, we rely on the following procedure: (1) The unencrypted data is exported and temporarily saved in a location that the research team cannot temporarily access as its access is temporarily protected. (2) An automated procedure encrypts it. (3) The encrypted copy is stored safely in the research team's server. (4) As soon as step 3 is completed, the unencrypted data that was exported and temporarily saved is permanently deleted.

collection or in the future. This also generates an anonymized mirror copy of the selected threads, allowing us to collect data on these selected threads going forward. We see it as important to set such a limit for data going forward, after which the process is automatically disabled, with the user being disconnected and their contact information deleted from our database. Importantly, we will not have access to the data until the data collection period is concluded and/or the user disconnects. At that time, the temporarily saved unencrypted data that is saved in a location we cannot access is encrypted and sent to our server, where it is stored (and the unencrypted data is permanently deleted, as described above).

4. At this stage, once the messages/threads are logged, respondents answer a brief series of questions about up to twenty of the threads whose data were harvested, and about their own demographic characteristics.

5. Within a few weeks, they receive a reward (for instance, phone credits) directly on their phones. When they do, they once again receive contact information for the "hotline" and a link with information about the data donation program.

Advantages

Why do we believe this strategy is the right compromise for WhatsApp data collection, considering the ethical and practical challenges we listed above, and the need to strike the right balance between privacy and the need for research, as enumerated by Ohme and Araujo (2022)?

After several pretests, we are convinced that our strategy technically works, and that it is practical and rapid (hence solving the first practical challenge).

More importantly, we are confident that using WhatsApp Explorer (and its adjoining protocol) strikes the right balance between the need to collect WhatsApp data and respect for privacy and ethical standards.[9]

9. We reached this conclusion after a year-long ethical review process that involved the ethical review experts of the European Research Council (ERC), the University Carlos 3 Madrid's data protection officer (José Furones, whom we are especially grateful for, for his dedication to the project), the University Carlos 3 Madrid's Ethical Review Committee, and the Ethical Review Committee of the Fundação Getulio Vargas in Brazil. Importantly, all three institutions have

This is because of (1) the extensive, multistage anonymization we put in place, (2) the development of clear procedures to handle "unexpected findings," and (3) the procedure's respect for the data minimization principle. We detail each of these points in the following subsections.

Strong Anonymization and Privacy Protection

To understand why and how our anonymization strategy protects the privacy of users, it may first be useful to fully detail what data are collected through this strategy, as well as how and when it is anonymized.

As part of this process, we collect information on who users are connected to (i.e., users' address book), who they chatted with, how many groups they are part of, the respective size of these groups, and most importantly, the content from chats the users consented to share. The information we get from the content of the chats includes messages, images, and videos exchanged in the chat and the time stamps of when and by whom they were sent.

These data, however, go through several stages of anonymization. A first, automated anonymization stage happens before it is stored on our servers.[10] At that stage, we rely on automated procedures to anonymize any personally identifiable information like the names, phone numbers, and emails from the dataset. Each bit of sensitive information is encoded and replaced with a unique identifier. While it can technically be used for reidentification, we delete the original key to make this impossible as soon as we have verified that the data is safely stored. We do not store group icon pictures (i.e., profile pictures) or audio messages.

In addition, for pictures/videos included within the threads we collect, we create the following pipeline. We start by storing the pictures and videos securely on our servers but also create hashes of them. We then use the aforementioned hashes to identify if the same image/video was in data shared by multiple users. We proceed to irreversibly anonymize

---

now officially given us a green light on the design and protocol, hopefully creating a useful precedent within the GDPR (or the LGDP, Brazil's equivalent of GDPR) frameworks for research of this type.

10. On servers: the current plan is to store these data on UC3M servers, for which the necessary guarantees in terms of privacy and data transfer are already in place.

most images we store, with the exception of images/videos that are shared by at least five groups/threads in our data and that do not contain personal data (e.g., a nude that was forwarded around). Since this can eventually be limited to a small number of items, the principal investigator checks each of these images individually to decide whether this is the case. Overall, this procedure ensures that we do not access the vast majority of unanonymized visual content. Importantly, the viral content we keep and analyze is extremely unlikely to be personal or private content.

To anonymize visual contents, we rely on tools such as Brighter AI to blur out faces.[11] Such tools provide an automated (and hence convenient) procedure (the data never leaves our servers during this process) to blur faces and a few additional identifying features of images/videos (e.g., car plates). We, at that point, replace the unanonymized images and videos stored on our servers with these anonymized images and permanently delete the unanonymized originals.

Even with this extensive protocol, we however must acknowledge that perfect, foolproof anonymization is never possible, nor that it can be left to automated procedures alone. Hence, we also implement a second stage of human-driven anonymization. We implement this systematic anonymization audit (SAA) before analyzing the data, in order to potentially strengthen—and hopefully perfect—an already thorough anonymization strategy.

Concretely, the principal investigator and close associates with knowledge of the context systematically review all the text and visual content already anonymized using automated methods and evaluate the potential for reidentification of personal data.

Our strategy is to start by further anonymizing the text content. We systematically remove any mention of a location (neighborhood, city, region, etc.), identification numbers (e.g., a passport number or another ID number, though we will add to the list), or mentions of an individual physical, physiological, mental, economic, cultural, or social attribute.

We also conditionally redact information relating to individuals' possessions, if it is likely to enable identification (because it is rare and hence distinctive) and information relating to individuals' company or

11. See https://brighter.ai/.

social network, if it is likely to enable identification (again, because it is rare and hence distinctive).

Once this is done, we proceed to further anonymize the visual content, where we see a need to do so. While almost all identifying content has been removed from the textual part of the WhatsApp thread, this will be a necessary additional step in some cases.

Concretely, we make sure to blur street signs or store or office fronts indicating or providing hints or location; potentially distinctive or atypical landmarks in the background; dress of individuals if it is distinctive or identifying; and distinctive body marks (scars, tattoos, etc. if not on the face, since the face will be already blurred). Additionally, we pledge to add to this list as the need occurs.

Provision for Unexpected Findings

How do we deal with unexpected findings?

While we see this probability as low ex ante, we also recognize that either this data collection project or the analysis of data might lead to some "unexpected findings," that is, findings that fall outside of the scope of the principal research objectives but necessitate action on the part of the researchers, e.g., disclosure of information to appropriate or designated authorities.

In the context of a project on social media and violence, these ethics issues may be considered as "serious and/or complex" if the research yields unexpected or incidental findings that may require interventions to safeguard the well-being of research participants (e.g., signs of physical abuse, self-harm, or drug dependency or neglect in minors) or, alternately, findings that are subject to positive disclosure obligations under the national laws of countries in which the research takes place, requiring researchers to breach the confidence of research participants. Examples include criminal conduct such as crimes, child sexual exploitation, human trafficking, or terrorism.

If such content is detected, the suggested protocol is to consult the ethics advisory boards of respective projects. The approach is to refrain from establishing a blanket policy ex ante by considering the rapidly changing political landscapes in countries where research is carried out and the potential for political bias in the judicial systems.

Restraint in Amount of Data Collected

As explained above, we ask participants to share up to four months of data (two months prior, two months after) on a specific subset of their threads (i.e., threads with six or more participants that count ten or more messages over the past two months—all other threads, including one-on-one threads, are altogether excluded from the data collection), though we also provide them with an easy way to share only a subset of this subset.

That is, they may share either historical data or data going forward; they may also exclude any thread on the list we initially present them with, and as noted above, as many threads on that list as they wish. We also note that they may quit the program at any time after they consented.

This means, conversely, that we do not remain indefinitely connected and that we expressly restrain from collecting data from certain types of threads. We believe these parameters to constitute the right compromise between (1) the data minimization principle, (2) the feasibility of our anonymization-intensive strategy, and (3) our ability to conduct meaningful scientific enquiry—and especially statistical analyses—in the public interest.

Remaining Challenges and Conclusion

In sum, the strategy we detail here should—once the tool is fully implemented—provide researchers with an efficient, privacy-protecting, and secure methodology to collect WhatsApp data to answer a variety of research questions.

We acknowledge that this strategy has many severe limitations. First, we recognize that the multiple, extensive stages of anonymization we implement eventually fall short of eliminating 100 percent of the possible risks of identification of the individuals involved. We, however, believe it comes extremely close to doing that, in practice, and note that researchers willing to undertake WhatsApp research must, in one way or another, be willing to deviate from the strictest guidelines about privacy protection in order to make research in the public interest possible, or must be willing to interpret these guidelines creatively.

284 | SIMON CHAUCHARD, KIRAN GARIMELLA

Second, we still lack sufficient data to speak to the representativeness of the data we will eventually manage to extract. Until a larger study is run, we will remain unclear as to whether the strategy will function among some demographics, and the extent to which respondents will be selective in terms of the groups they choose to donate.[12] There is, in addition, little doubt that researchers focusing on populations by nature difficult to investigate (for instance, members of a rebel army or of a vigilante group, as several authors in this volume explore) will continue to struggle to obtain data to study the influence that WhatsApp networks may have in these processes. Our technology may not entirely change the reticence that many users may have when approached and asked to donate their smartphones' content.

Third and relatedly, our strategy is costly in labor, infrastructures, and resources, especially if researchers are going to provide rewards or incentives to potential donors. This implies that many researchers relying on it will not be able to collect large and/or representative datasets in the selected cases.

In spite of these important limitations, we believe the technology we present in this chapter, and which we will keep improving over the next few years, will dramatically improve current research opportunities and practice. Our early experiments in the field in India and Brazil on our own project (ERC POLARCHATS) suggest that we will be able to obtain large datasets from a diverse, if not representative, group of users. This is, in and of itself, an improvement over the status quo, and one that should allow us to answer some important research questions and monitor the virality of problematic content on the app.

Further, while we acknowledge that most researchers will not be able to collect as much data as we plan to due to the rather costly nature of our strategy, we also hope it will help set the standards for *how* to collect WhatsApp data, regardless of the amount of data collected by specific researchers. Important discussions about consent, privacy, and anonymization are at stake and need to be balanced with the need to access this data to document and analyze some pressing dangers. Even

---

12. Note that this may be one further argument for an automated online self-administered procedure, which would likely allow us to access different demographics.

if researchers assemble datasets more limited in scope than the ones we are planning to assemble, we believe their strategy should equally go through this balancing exercise and provide clear safeguards to users. In that sense, we hope this chapter will push researchers to reflect on what "fair" WhatsApp data collection should look like—a thorny issue we have tried to solve—in addition to assisting their practical needs.

16

Automating Data Collection from Public WhatsApp Groups

Challenges and Solutions

NICHOLAS MICALLEF, MUSTAQUE AHAMAD,
NASIR MEMON, AND SAMEER PATIL

The WhatsApp messaging service is one of the most popular mediums for broadening the reach of information dissemination (Srivastava and Singh 2021). Unlike other messaging platforms, the collection of real-world messaging data from WhatsApp is challenging and complicated because of end-to-end encryption, closed source code, and lack of publicly accessible application programming interfaces (APIs). Researchers have used two main ways to circumvent these issues and collect information propagated via WhatsApp: (1) setting up a dedicated WhatsApp number to which people can forward information and (2) joining public WhatsApp groups connected to information about topics of interest (e.g., health, elections, etc.). The latter of the two approaches has been the most popular technique reported in prior work involving WhatsApp data (Garimella and Tyson 2018; Melo et al. 2019; Reis et al. 2020; Resende et al. 2019b). However, such an approach requires considerable manual labor to curate the data collection (Melo et al. 2019; Reis et al. 2020), limiting the scale of the data collection efforts. In our research, we addressed the challenge of scale by investigating the barriers to automating data collection from public WhatsApp groups.

To achieve our research goals, we began with an exploratory investigation with one mobile device that the lead researcher used to manually discover, join, and observe the activities of several public WhatsApp groups. During this initial exploration, we uncovered various challenges that we classified into five broad categories: group discovery, group membership, group maintenance, inappropriate content or behavior, and other issues. Most of these challenges have received limited or no coverage in prior work, perhaps because of the continuously evolving nature of messaging and social media platforms (Gundecha and Liu 2012). Subsequently, we proposed and designed solutions to address the challenges we encountered. Next, we implemented most of the solutions on four mobile devices to study whether these achieved our goal of scaling WhatsApp data collection.

In the following sections, we first situate our research within the WhatsApp research context. Subsequently, we provide a detailed description of our device setup, the process we used to conduct the exploratory investigation, and ethical considerations. We then present the findings of our investigation and discuss their implications for automating future data collection on WhatsApp.

Related Work

WhatsApp emerged as a popular messaging platform that allows users to exchange messages privately (one-to-one) or within small groups (up to 256 members maximum) using an end-to-end encryption framework. In recent years, the platform has been increasingly exploited to spread misinformation (Coughlan 2020; Satish 2018). Researchers have analyzed the characteristics of the misinformation content shared on the platform (Moreno, Garrison, and Bhat 2017; Recuero, Soares, and Vinhas 2021; Reis et al. 2020; Vasconcelos et al. 2020) along with the underlying properties of the dissemination networks (Nobre, Ferreira, and Almeida 2020; Nobre, Ferreira, and Almeida 2022; Resende et al. 2019b). WhatsApp groups are often made available for public access by sharing an invitation link on the Web. Researchers have used these invitation links to join public WhatsApp groups for studying the content and network dynamics (Nobre, Ferreira, and Almeida 2022). In this section, we present prior work on the characteristics of the content shared on WhatsApp and its propagation within public WhatsApp groups. Afterward, we describe the WhatsApp data collection tools developed by the research community and discuss how researchers have addressed the challenges of automating data collection from public WhatsApp groups.

Characteristics and Spread of Misinformation on WhatsApp

Garimella and Tyson (2018) were the first to study WhatsApp content by examining the sharing practices of political groups. Bursztyn and Birnbaum (2019) extended the work by analyzing the characteristics of the content shared within politically affiliated WhatsApp groups on a larger scale. Multiple researchers studied the WhatsApp content shared during the Brazilian 2018 presidential elections (Recuero, Soares, and Vinhas 2021; Resende et al. 2019a). For instance, Resende et al. (2019b) found that images with fake content reached a broader audience and spread faster when compared to other content. Researchers have reported similar findings for audio (Maros et al. 2020) and textual (Resende et al. 2019a) content. Caetano et al. (2019) have reported that misinformation in political groups lasts longer and reaches more users than in nonpolitical groups. Follow-up work by Reis et al. (2020) suggests that explicitly flagging fake content to users could reduce the volume of misinformation on WhatsApp.

A few research efforts have examined the connections among WhatsApp users across several public groups. Melo et al. (2020) constructed a network of groups by clustering the public WhatsApp groups with users in common. Their goal was to study whether the spread of misinformation could be reduced by restricting message forwarding (Melo et al. 2020). Surprisingly, they found that misinformation would spread widely even if message forwarding was restricted; however, it happened at a slower pace. Concurrently, Resende et al. (2019b) studied the properties of the networks that contain users who share the same WhatsApp groups. They observed a network structure that allows content to go viral, similar to other social media platforms (Resende et al. 2019b). Research on WhatsApp users who share the same content in several WhatsApp groups revealed that strongly connected communities help spread content across groups (Nobre, Ferreira, and Almeida 2020). In addition, WhatsApp users who belong to the highest number of communities are the highest contributors to the broad reach of the content within those communities (Nobre, Ferreira, and Almeida 2020). More recently, Nobre, Ferreira, and Almeida (2022) studied the network characteristics of public WhatsApp groups by building three different levels of user aggregations that reflect the organizational components within the WhatsApp platform. The approach uncovered that the backbones of the communities are well established, which could signify coordinated efforts to broaden content spread (Nobre, Ferreira, and Almeida 2022). WhatsApp users who spread misinformation are those who are central to the communities.

The work described above sheds light on the characteristics of the content spread on the WhatsApp platform and the networks that connect WhatsApp users within public WhatsApp groups. However, the scale of the data collection and analyses is small, potentially limiting the insight that can be generated. Our work attempts to overcome this limitation by demonstrating approaches to automated larger-scale collection of WhatsApp data.

Tools for WhatsApp Data Collection and Automation

Researchers have developed tools to aid the collection of content shared in public WhatsApp groups (Bursztyn and Birnbaum 2019; Garimella and Tyson 2018; Melo et al. 2019). Several research efforts have used these tools and scripts to collect the WhatsApp data needed to address their specific research questions. Specifically, these tools extract the WhatsApp SQLite database from the phone and decrypt it to access the stored content (Bursztyn and Birnbaum 2019; Garimella and Tyson 2018; Melo et al. 2019). This task is essential for obtaining the information required to conduct the content and network analyses described earlier. Supporting scripts that accompany the tools enable searching for public WhatsApp group invites connected to specific keywords and scraping the corresponding links (Bursztyn and Birnbaum 2019; De Sá et al. 2021). The list of public groups obtained through search is typically manually curated by researchers to fit their specific research goals. Following this, separate supporting scripts can automate joining the public groups in bulk (Bursztyn and Birnbaum 2019; Garimella and Tyson 2018; Melo et al. 2019).

From an automation perspective, such tools and scripts mainly address the challenges related to setting up and extracting data from WhatsApp databases (Bursztyn and Birnbaum 2019; Melo et al. 2020; Reis et al. 2020). In fact, we used these tools and scripts as our starting point. However, we could not use the scripts to automate the joining of public

WhatsApp groups because WhatsApp has since implemented measures that prevent users from joining groups in bulk. Yet, there has been limited research on the challenges related to automating the discovery of relevant WhatsApp groups and maintaining membership in these groups. While prior work on WhatsApp mentions some of these challenges, the focus was on the specific goals of the respective studies, with limited attention to documenting the barriers to data collection. Moreover, the literature includes no concrete solutions proposed to address the barriers. Overcoming these barriers is essential for facilitating large-scale automated data collection for research on WhatsApp and similar platforms. In our work, we provide greater detail on the challenges encountered when collecting research data on the WhatsApp platform, with the overarching goal of facilitating the availability of large-scale WhatsApp datasets for the research community.

Method

In this section, we describe how we set up the devices used in our investigation, provide detail on the initial exploratory investigation, and discuss the ethical considerations involved in collecting WhatsApp data for research.

Device Setup

We procured four mobile devices (Xiaomi Redmi 6 [GSM Arena n.d.]) and four SIM cards. For transparency, we registered the SIM cards under the lead researcher's name. We obtained a unique phone number for each SIM card and registered WhatsApp accounts using these numbers so a WhatsApp confirmation message could be received to register the corresponding devices. Subsequently, we configured each device for data collection using existing research tools and scripts mentioned in the previous section. The scripts periodically extracted the WhatsApp SQLite databases from the devices (Bursztyn and Birnbaum 2019; Garimella and Tyson 2018) and transferred them to a standalone computer for processing and analysis.

As an alternative, we considered writing our own Python scripts using Beautiful Soup (Richardson 2023) to extract the data from each device using the WhatsApp Web tool.[1] We decided against such an approach for several reasons: (1) it would have returned less information than that stored in the WhatsApp SQLite database, (2) it would have been time-consuming to develop the required scripts, and (3) no prior work available at the time (i.e., mid-2019) used such an approach. However, we recommend that future researchers consider following such an approach and sharing the scripts with other researchers (see the Findings section for a more in-depth discussion). The scripts we developed for this chapter can be accessed at https://tinyurl.com/3atuhnmd.

Initial Exploratory Investigation

We began the research with an initial exploratory investigation with only one device. During the first two months of the exploration (i.e., August 2019 to September 2019), we manually conducted each of the following tasks on the device: (1) discovering WhatsApp groups through manual Web searches of specific topics (e.g., health) and manually extracting the invitation links from the search results, (2) joining the manually discovered WhatsApp groups, (3) maintaining membership in the joined WhatsApp groups, and (4) monitoring the activity in the joined WhatsApp groups by observing who was sharing content, what type of content was shared, and how often. In the subsequent two months (i.e., October 2019 to November 2019), we designed and developed solutions to address most of the challenges we encountered during the first two months of the exploration. Later, we implemented some of these solutions on all four mobile devices to confirm that they were successful in scaling the data collection.

Ethical Considerations

Collecting data from public WhatsApp groups may be considered human-subjects research since it cannot be determined in advance whether a piece of content is generated and shared by humans or bots. Our observations of the collected data suggest that large amounts of information shared in public WhatsApp groups is posted by bots.

1. See https://web.whatsapp.com/.

However, we do not have the evidence to make concrete claims about these observations.

We obtained a waiver of informed consent for conducting the research as it was deemed the most appropriate approach. Obtaining consent from the humans who were sharing information in the public WhatsApp groups included in our research would have been impracticable and would have affected the integrity of our findings (Hudson and Bruckman 2004). Moreover, our initial inspection revealed that there was little personally identifiable content in the messages, with most messages (greater than 70 percent) sharing nontext content, such as images, videos, links, etc. Therefore, simple approaches for anonymizing the data after collection were enough to protect the privacy of any human members of the WhatsApp groups (Melo et al. 2019). Specifically, we anonymized every identifiable attribute (e.g., mobile phone number) in the data before every analysis. We further protected the privacy of members by storing the decrypted WhatsApp SQLite databases on a single standalone computer that was encrypted and password protected. In addition, we limited data access by researchers other than the lead researcher, even within the project team.

Findings

We classified the challenges uncovered in the initial exploration into five broad categories: group discovery, group membership, group maintenance, inappropriate content or behavior, and other issues. We describe the solutions we devised to address the issues. We then present the results of deploying most of the devised solutions for large-scale data collection of activity from public WhatsApp groups over a period of eighteen months.

Group Discovery

Before collecting data from public WhatsApp groups, it is necessary to determine which groups to join. A simple query on an online search engine for WhatsApp groups yields numerous sites with links to public WhatsApp groups on various topics.[2] Researchers must then visit the

2. WhatsApp Group Links, https://whatsgrouplink.com/.

sites, manually choose the groups relevant to their research, and join the selected groups. Such a process cannot scale when the data collection activities involve hundreds or thousands of groups covering many topics. Another reason group discovery is challenging to automate is that there is no indication whether the group invites in the search results are related to the topics of interest to the researchers.

To address these challenges, we automated the discovery of public WhatsApp groups by developing a process that scrapes group invites from public Web searches periodically (i.e., every night) and subsequently checks the group description to filter out irrelevant groups. Although the keywords used in the group descriptions were sometimes not accurate, we found that our solution still managed to filter out a large proportion of irrelevant groups.

Group Membership

Once the list of relevant WhatsApp groups is identified, researchers need to join these groups to start collecting the content shared within them. Since the number of groups of interest in a large-scale data collection effort is quite large, it is impractical to join every group manually. Therefore, we developed a script to automate the joining of public WhatsApp groups. When developing the script, we encountered several barriers to automation. First, we found that the WhatsApp platform mechanisms temporarily ban a WhatsApp account if it attempts to join too many groups within a short interval. Reinstating the account requires manually contacting the WhatsApp support team. Second, we could not join various groups because at the time (i.e., in 2019), each group was limited to a maximum 256 members (Bursztyn and Birnbaum 2019). The upper limit on membership made it highly likely that a public WhatsApp group would already be full before we attempted to join it.

To overcome these two barriers, we developed a semiautomated method. Our approach spread out the automatic joining attempts over time intervals long enough to avoid getting banned by WhatsApp. In addition, we kept track of the groups that could not be joined because they were full and periodically repeated the join attempts until we succeeded in joining. We implemented the semiautomated approach as an Android app that takes a list of WhatsApp groups as input and prompts the researcher at specific times to join a small subset of the groups in the list. The app flags the groups that could not be joined because of the upper limit on membership and prompts the researcher later to attempt again to join these groups.

Group Maintenance

After joining the desired public WhatsApp groups and starting to collect data, we faced additional challenges that have received limited attention in the literature. We immediately noticed that the moderators of some groups tend to remove those who do not interact, most likely due to the upper limit on the number of people that can be in a WhatsApp group (i.e., 256 at the time of research, which has now increased to 1,024). Since research purposes typically involve passive collection of the activities within the WhatsApp group without interacting with the group members, the WhatsApp account used for data collection purposes may get removed by the group moderators. In contrast, moderators of some WhatsApp groups sometimes add members in their groups to other public groups that might not be relevant to the goals of data collection. Being added to irrelevant groups can affect data collection because the number of groups a device can handle has an upper limit that varies by device specification (in our case, it was four hundred). Being added to too many groups can overload device resources, causing the device to malfunction (e.g., switch off on its own) and compromising the integrity of the data collection process. In addition, we observed that some public WhatsApp groups have no activity or the activity within them is unrelated to the research goals. The inclusion of such groups in the data collection efforts can impact the integrity and validity of the data.

To address the above issues, we developed scripts to detect when other parties removed the WhatsApp account used for data collection from the joined groups or added it to new groups. Any groups from which the account was removed by others were added to the list of groups input to the Android app described above so these could be joined again at the next available opportunity. In future work, we plan to enhance the scripts for periodically checking whether the activity within each joined WhatsApp group is relevant to the goals of the data collection. The functionality could be further enhanced with the addition of the capability to check whether the groups from which the research account is removed are worth rejoining and the groups to which it is added are worth keeping.

Inappropriate Content or Behavior

Some of the challenges we encountered might not affect the automation or scaling of the data collection but are still important to consider because of the potential to cause harm. More specifically, we experienced various instances in which WhatsApp members in the joined groups sent direct messages or tried to call the devices involved in the data collection. It is important that researchers are aware that such situations might occur and cannot be avoided. More importantly, we encountered a large amount of inappropriate (e.g., X-rated) content being shared on some of the public WhatsApp groups we joined. Frequent exposure to such content could have a harmful effect on the recipients as reported in research on the effect of exposure to such content on moderators (Steiger et al. 2021). Since researchers will likely need to engage in at least some manual examination of the collected data, exposure to such content might be unavoidable. To minimize exposure to potentially problematic content, we implemented filters to exclude WhatsApp groups with X-rated keywords in their descriptions. We suggest further enhancing the functionality to exclude any joined WhatsApp groups that contain large amounts of inappropriate content.

Other Issues

Besides the challenges described in the previous subsections, we uncovered additional issues during the exploratory investigation that could not be classified in any of the above categories. For instance, we found that media files (i.e., images, videos, apps, documents, etc.) shared in public WhatsApp groups were not automatically downloaded to the device and needed to be downloaded manually by the person managing the device despite the auto-download setting being enabled. This is a major barrier to automating the data collection process because no media files could be collected without manual action for each media file. Considering that the data we collected over eighteen months involves over 1.1 million media files from 1,200 WhatsApp groups, it is obvious that the manual approach cannot scale. However, media files cannot simply be excluded from data collection because they are typically a critical aspect of research investigations (Resende et al. 2019b).

The above challenges could be addressed in two alternative ways. One approach is to employ scripts that go through all available WhatsApp groups and initiate the download of all media files using the WhatsApp Web tool. Alternatively, an app on the device involved in the data collection can continuously interact with the WhatsApp application on the device to navigate through all groups and trigger the download of all media files shared in the group. However, security restrictions of the device operating system might make it infeasible to implement such an app, even on rooted devices. As a precautionary note, we found that the media files shared in public WhatsApp groups may contain malware. Therefore, the devices used to collect WhatsApp data need to be protected with appropriate antimalware measures that inspect the downloaded media files.

Large-Scale Longitudinal Deployment

We developed and deployed most of the solutions described above on four mobile devices to automate the data collection from public WhatsApp groups over a period of eighteen months from December 2019 to May 2021. Using the solutions, we collected more than six million messages from around 1,200 public WhatsApp groups. By analyzing the collected data, we have been able to shed light on several aspects of misinformation distribution on WhatsApp. For instance, we found that a large amount of misinformation in public WhatsApp groups is embedded as text in images, requiring a multistage pipeline that combines multiple tools for extracting the text from these images before it can be checked for misinformation.

We encountered additional challenges over the eighteen months of large-scale data collection. Specifically, we found that the WhatsApp SQLite databases on all four devices became so large over time that they filled up the internal device storage. Running low on available storage led some of the devices to malfunction (e.g., frequently rebooting to clear space). To address this issue, data from the WhatsApp SQLite databases should be flushed at regular intervals to external storage, such as a standalone computer.

Discussion and Implications

Our findings highlight several important points regarding automated, large-scale collection of research data on the WhatsApp platform. We discuss each in turn.

Continuously Evolving Platform Is a Barrier to Automation

When analyzing the challenges uncovered during our exploratory investigation, we found that most of the challenges that pose a barrier to automation have received limited or no coverage in the literature (Bursztyn and Birnbaum 2019; Garimella and Tyson 2018; Melo et al. 2019). The lack of coverage in the literature may be attributable to these issues being relatively new. For example, when Garimella and Tyson (2018) and Bursztyn and Birnbaum (2019) developed tools for automatically joining public groups in 2017 and 2018, WhatsApp had not deployed sophisticated measures to prevent bulk joining of public groups. Similarly, in 2017 and 2018, media files were downloaded automatically on the device without requiring the user to take any action or enable a setting. WhatsApp changed this operation in one of the updates released after 2018. It should be evident from these examples that the continuous evolution of the WhatsApp platform is one of the main barriers to automated data collection. There is no easy solution to the barrier because messaging and social media platforms keep evolving to restrict automation for various reasons, including: (1) securing the platform from malicious actors who use bots and (2) protecting the business interests of the platform. This implies that researchers need to be proactive in identifying potential new barriers to automation by frequently examining whether their data collection processes need to be adjusted, especially after major platform updates.

Manual Intervention Is Inevitable

In our exploratory investigation, we found that certain tasks during data collection inevitably require manual intervention. For instance, joining automatically discovered WhatsApp groups required us to use a semi-automated application to join public WhatsApp groups at specific time intervals. Moreover, we needed to leave certain public groups manually to reduce excessive exposure to inappropriate content. Future work needs to investigate whether automated approaches could be developed to address these challenges more effectively. Nonetheless, our experience indicates that some manual intervention would be inevitable during automated data collection because researchers need to verify periodically that the data collection process is progressing as expected. The optimal frequency of such manual checks could change over time as it depends on several factors, such as the scale of the data collection, the number of researchers in the team, etc. For example, manual verification could be conducted more frequently at the beginning when the new public WhatsApp groups are being joined and decrease progressively as the list of targeted groups stabilizes and the researchers gain confidence about the reliability of the automated data collection.

Mechanisms Are Needed to Access Public Information

Our initial exploratory investigation helped us automate the subsequent larger-scale data collection from public WhatsApp groups over a duration of eighteen months. Although we designed solutions to address most of the barriers that surfaced during the initial exploration, we continued to encounter additional challenges during the longitudinal data collection. Apart from the evolution of the WhatsApp platform as described above, the additional issues were related to the complexity of data collection task at hand, i.e., automating the data collection from a platform that does not provide an API. It is important to note that WhatsApp provides end-to-end encryption to protect user privacy. Making private messages exchanged via the platform available to others would defeat the purpose of end-to-end encryption. However, the presence of publicly available WhatsApp groups indicates that not all communication on WhatsApp is meant to be protected from public disclosure. It would be useful if the platform could explore mechanisms to facilitate automated collection of such public content by trusted parties such as researchers.

Conclusion

In an exploratory investigation, we manually discovered, joined, and monitored the activity of publicly available WhatsApp groups for four months. The exploration resulted in uncovering several barriers to discovering, joining, and staying in public WhatsApp groups, along with challenges related to inappropriate content and user behavior within the groups. By developing solutions to address most of these challenges, we subsequently managed to automate the large-scale collection of data from public WhatsApp groups over eighteen months. We found that platform evolution and security measures create additional challenges during longitudinal automation of public data collection from a closed platform. We addressed these challenges via manual intervention and adjustments to the solutions we developed.

Our solutions provide valuable guidelines for collecting data from public WhatsApp groups at scale, serving a variety of stakeholders such as researchers, journalists, fact checkers, and media organizations. For instance, such data can help earlier identification of disinformation campaigns that originate on public WhatsApp groups and subsequently spread to other platforms. We call for an exploration of additional mechanisms that can help trusted parties obtain automated access to public data on closed platforms.

Acknowledgments

The research described in this chapter was partially supported by funds from the Center for Cyber Security at New York University Abu Dhabi.

PART V

Reflections on Policy

17

Fact-Checking on WhatsApp in Africa

Challenges and Opportunities

CAYLEY CLIFFORD

Africa Check has been verifying the accuracy of information in the public domain since 2012. We fact-check claims made by public figures, organizations, and the media, as well as viral information shared on social media.

Our work is based on the understanding that people need access to accurate information in order to make informed decisions. While fact-checking alone will not solve the problem of "fake news" and online hate, it is an attempt to cut through the noise and provide the public with information they can trust.

This is a relatively straightforward exercise when fact-checking claims made in the public domain, for example, in an online news article. Africa Check's researchers can identify the claimant, contact them, and ask for the evidence to support the claim before moving on to other steps in the fact-checking process.

Fact-checking claims circulating on closed end-to-end encrypted messaging platforms like WhatsApp is more difficult. Unless readers forward Africa Check messages they've received on WhatsApp, we cannot know with certainty what kinds of false information are spreading on the platform. When false information is not surfaced, fact checkers are unable to counter it.

A second challenge with fact-checking on WhatsApp is that users tend to place more trust in the information that is shared with them (Kuru et al. 2022, 2). As opposed to being exposed to information on a public platform, users receive direct messages from people they likely know and trust. This could include family, friends, and colleagues.

At Africa Check, we understand that our regular mechanisms for surfacing misinformation on social media may not suffice on WhatsApp. Encouraged by the opportunity to take a more creative approach, in 2019 we launched *What's Crap on WhatsApp?*, a voice-note show that debunks the worst misinformation on WhatsApp every two weeks. The podcast show is jointly run and managed by Africa Check and podcasting company Volume.

What's Crap on WhatsApp?

Fact-checking on WhatsApp via a podcast required the establishment of a dedicated WhatsApp tip-off line that subscribers are asked to save to their mobile phones. When subscribers receive a message on WhatsApp they suspect is false, they are able to forward it directly to Africa Check. Every two weeks, Africa Check rounds up three or four of the most popular and harmful claims received and fact-checks them as per our regular fact-checking process. The exception is that the fact-checks are not simply published on our website. Rather, the content is contained on the WhatsApp platform. The fact-checks are used as the basis for a script that is recorded as a voice note and sent to subscribers.

There are several advantages to fact-checking misinformation on WhatsApp in this way. An age-old concern for fact checkers, supported by research, is that fact-checked information may not be reaching those who need it most (Vosoughi, Roy, and Aral 2018). By keeping this service within WhatsApp, we are responding directly to users' concerns about particular information they've received. Efforts are also made to ensure that the voice note remains between five and seven minutes long. In this way, it provides subscribers with the information they need without becoming arduous. The short length of the voice note means it is low cost to produce and share. It does not require much data to download and is in a forwardable message format, enabling subscribers to send the voice note to other contacts on WhatsApp.

Subscribers to *What's Crap on WhatsApp?* are managed through broadcast lists. While there is currently no limit to the number of broadcast lists one can create, each list has a limit of 256 contacts (WhatsApp Help Center n.d.). With over five thousand subscribers and counting, the logistics of sending and receiving messages can be complicated and time-consuming. While this challenge may be easily solved by making use of WhatsApp's application programming interface (API) for business, we felt strongly that the process should not be automated. Rather, the WhatsApp line is manned by Africa Check staff who manually interact with subscribers. This has been an important method through which to build rapport with the audience, making it more likely that they interact with the service.

Trends and Patterns of Misinformation on WhatsApp

Messages sent to the tip-off line not only surface misinformation for Africa Check to counter but also provide insight into trends and patterns on WhatsApp. By making note of every message and categorizing it by theme or topic, Africa Check has been able to build a database that facilitates understanding of how and why false information spreads on WhatsApp. This, in turn, allows Africa Check to better respond to people's concerns and fears with information they can trust.

For example, of the 274 messages sent to the *What's Crap on WhatsApp?* tip line in 2022, 144 were labeled as hoaxes and fabrications. This refers to new content that contains fabricated claims about nonexistent entities, events, people, or simply incorrect information. Examples include messages circulating on WhatsApp in January 2022 that claimed "massive" roadblocks were being set up around the country to round up "illegal" Zimbabweans. In March 2022, similar messages claimed all fuel stations across the country were to be closed. Hoaxes and fabrications, left unchallenged, can result in different types of harm ranging from the creation of panic and civil unrest to the discrimination of minority groups.

A further fifty-seven messages were categorized as conspiracy theories and predictions. This refers to information that claims a covert but influential or powerful organization is responsible for unexplained events. Examples include messages circulating on WhatsApp in August 2022 that claimed U.S. industrialist and philanthropist John D. Rockefeller "paid scientists to call oil a 'fossil fuel' to induce the idea of scarcity, in order to set a 'world price for oil.'" Also in August 2022, the WhatsApp line received a number of messages claiming the South African government would soon introduce health legislation that would make it illegal to refuse medical examination. Conspiracy theories and predictions, like hoaxes and fabrications, can lead to mistrust in authority.

For evidence that these types of false information can lead to real-world harm, one need not look further than COVID-19, when a flood of both correct and incorrect information about the virus made it difficult for people to make informed decisions, in a phenomenon the World Health Organization termed an "infodemic" (World Health Organization n.d.).

While there is evidence of lives being directly impacted by misinformation, it is also true that not all those who come across misinformation will act on it. A 2020 study that analyzed user behavior and misinformation shared with Africa Check on WhatsApp showed that African users frequently shared COVID-19 misinformation on the platform, either to individual contacts or to one or more WhatsApp groups. Although some users could and did assess the validity of information before sharing it, more still appeared to delete messages, ignore them, or share them in any event. Some users worryingly acted on misinformation (Africa Check and the Africa Centre for Evidence 2020, 9).

However, rather than behavior being a deliberate attempt to deceive, survey respondents were motivated by a desire to raise awareness about the pandemic and provide helpful information to those they care about. The study found that social media users were strongly influenced by their social circles, which appears specifically relevant to WhatsApp, where the moral obligation to share helpful information with family and friends is strong (Africa Check and the Africa Centre for Evidence 2020, 10).

Exploring the evidence base to understand what motivates users to respond to misinformation in particular ways, the study found that the type of content, who had shared it with them, the emotions it triggered, their trust in social media, and their tendency toward conformity all shaped their behavior (Africa Check and the Africa Centre for Evidence 2020, 31). Of particular interest is that survey respondents rated their trust in messages from "legitimate news sources," like established media outlets, to be higher than their trust in government sources, for example, a Ministry of Health (Africa Check and the Africa Centre for Evidence 2020, 40). It is likely that contradictory messages from government sources on regulations to curb the spread of COVID-19 contributed to this phenomenon.

Lessons Learned

Africa Check's experience of fact-checking on WhatsApp has provided a number of key takeaways.

First, proactive key messages and positive reinforcement around WhatsApp users' need to be helpful in a time of crisis can be useful. When fact-checkers simply correct false information, there is a risk that it reinforces someone's beliefs. This is sometimes referred to as the "backfire effect" (Swire-Thompson, DeGutis, and Lazer 2020).

However, if fact-checkers are able to tap into social media users' desire to be helpful, evidence suggests they may better internalize correct information. This was the finding of a 2021 study that evaluated whether sustained exposure to fact-checks reduces citizens' susceptibility to misinformation and, in turn, promotes accurate beliefs that guide informed behaviors. In one of the treatment arms, regular Africa Check fact-checks were augmented with empathetic language emphasizing the narrator's understanding of how fear and concern about family and friends might lead individuals to be fooled by misinformation. The study showed that interventions are most effective when accompanied by emotive appeals that increase the resonance of corrective information with consumers (Bowles et al. 2023).

Second, it may be useful to leverage users' social circles to champion evidence-based information. Africa Check experimented with this approach in 2021–2022 through a "Fact Ambassadors" project that aimed to increase the reach of accurate information through credible, trusted champions. Across four countries—South Africa, Nigeria, Senegal, and Kenya—fact ambassadors distributed content to their peers on their social media accounts, including WhatsApp (Africa Check 2021). Lessons learned are being incorporated into a separate study that seeks to understand how social media influencers can be harnessed to counter misinformation (Social Science Research Council n.d.). Such undertakings will provide further insight into how social circles can be successfully leveraged and absorbed into the work of fact-checkers.

Third, however, little of this work will be of much use unless fact-checking organizations are also able to extend the fact-checking "circle of trust" by building partnerships with trustworthy media, government bodies, civil society partners, religious leaders, and big tech companies.

This includes advocating for clearer communication, particularly during times of crisis, as conflicting information creates fertile ground for the spread of disinformation, particularly on WhatsApp. For example, Africa Check is a third-party fact-checking partner with Meta, which allows us to fact-check claims that are flagged as potentially false on Facebook. We have also worked with Twitter and TikTok around events such as elections. During this time, we see a spike in disinformation on social media, and partnering with platforms so potentially harmful posts are flagged as soon as possible has proven to be an effective strategy.

Lastly, fact-checking organizations need to continue empowering individuals to take control of misinformation through media literacy and social media campaigns. The sheer number of messages circulating on WhatsApp means fact checkers cannot address each one. A more sustainable approach includes equipping WhatsApp users with the skills they need to counter misinformation on their own. To this end, Africa Check has produced a number of guides, in text and video format, which detail steps for verifying voice notes, links, videos, images, and text on WhatsApp.

18

Challenges of Fact-Checking WhatsApp Messages in India

JENCY JACOB

In April to May 2017, BOOM had just revamped operations, transitioning from a broad digital outfit covering stories of human interest to India's first full-fledged fact-checking newsroom. One of our early encounters with the dangers of misinformation/disinformation came in the form of viral messages spreading on the popular messaging platform WhatsApp. These messages consisted of an image of bodies of little children and a video showing a child being kidnapped from a street with a text message warning parents to take care of their children.

Back then, fact-checking of images and videos were at a nascent stage, but we soon found that neither of them had anything to do with the viral claim. The video was an edited clip, part of a public service campaign from Pakistan (Rebelo 2018b) to spread awareness about missing children in Karachi. Similarly, the image of the bodies of little children was originally shot in Syria in 2013. But in the hands of bad actors, they had become a potent weapon to spread hate against unsuspecting strangers, as we would soon discover in a matter of months.

While our fact-check (Rebelo 2017) went largely unnoticed, the image and video kept circulating on social media, especially WhatsApp groups. Between the months of April to July 2018, twenty-four people lost their lives (IndiaSpend 2018) due to mob attacks provoked by child kidnapping rumors. BOOM did several ground reports after the mob lynching incidents and found that in all the cases, several messages had gone viral on community groups warning them about child kidnapping gangs and an organ harvesting racket. And in several of these cases, the public service messaging video from Pakistan and image from Syria were used as a tool to create an atmosphere of fear in the minds of citizens.

We never found out if these messages were made viral as part of an orchestrated campaign or plain coincidence. But the pressure on WhatsApp was rising as both the state governments and central government identified the platform as being one of the major sources for such unverified information. The platform was warned to take corrective action. The platform did take steps: messages could no longer be mass forwarded to more than five individuals or groups in a bid to force friction and change user behavior, forwarded messages were labeled (BOOM Fact Check Team 2018) to give better information to the users about the origin of the message, and close monitoring of those accounts flagged by users for sending mass spam messages was implemented. Coupled with mass public service campaigns by the state and the center and some active policing measures, mob lynching cases due to child kidnapping rumors did come to an end.

While the deaths due to child kidnapping rumors were the most potent dangers visible due to rumors spreading on WhatsApp, the subsequent years have shown that the platform does have a problem that it has not been able to fully address. One of the major reasons for its inability to nail down bad actors has been the encrypted nature of the platform and the assurance made by the platform to its users that the platform cannot read the messages. End-to-end encryption is a double-edged sword. Weakening encryption will result in a flight of users to safer platforms while protecting encryption exposes them to regulatory agencies who have had better success dealing with other platforms where identities cannot be easily hidden.

Since BOOM's fact-checking operations began in 2017, our newsroom has actively run two helplines inviting users to send us viral messages that look suspect to them. From elections to neighborhood conflicts with China and Pakistan, terror attacks, natural disasters, and in recent years the dreaded COVID-19 pandemic, the spread of disinformation has come in the form of waves with periods of calm in between. While messages around politics, communal fear mongering, and health-related misinformation are the most popular, recent years have also seen a rise in false claims related to corporates, brands, and financial institutions.

To study the impact of misinformation, BOOM ran a pilot project in the months of October to December 2022 during a closely tracked state election in Western India. We invited citizens to join us as "truth warriors" and become part of communities formed as groups divided along city and regional boundaries. We were keen to closely monitor the spread of misinformation in real time and also study the impact on voting patterns.

At the end of the three-month project, we understood that the communities of citizen groups helped us better prioritize fact-checks that were in tune with their needs. We were able to reduce the time to identify misinformation pieces before they went viral. Conversely, with limited resources we found the whole exercise overwhelming and tiring as our users expected quick responses to claims sent by them. Also, we found it difficult to deal with spam messages, especially crypto spammers who would enter our groups, mass forward links and text messages, and then exit the group. Blocking their numbers didn't prove to be very useful as they always found new ways and unidentified numbers to use to reenter the groups after a few days.

There is no doubt that WhatsApp has struggled to respond to misuse of its platform by ideologically aligned political and nonpolitical groups who have created mass hysteria and hatred against minorities and specific religious communities. The challenge lies in identifying users who are using methods to create and distribute hate messages on an industrial level thereby defeating the original purpose of creating the platform—a medium used by friends and family to share photos and videos in a safe and secure manner.

The Indian government has made further amendments to its IT Rules that mandate platforms take down posts that have been fact-checked as false by a fact-check unit designated by them (Hakim 2023). The details of the fact-check unit are yet to be finalized as the move got challenged in the courts by individuals and media associations. While fact checkers are already working with WhatsApp through tip lines and encouraging users to surface questionable claims, we have been demanding an in-platform mechanism to fact-check high-volume forwarded messages without compromising encryption. This could be done by giving users the right to report a message to WhatsApp, which can then use it as a signal based on virality to send such claims to all the verified tip lines of fact-checking newsrooms. The platform can also use its marketing muscle to popularize our tip lines to its massive user base.

WhatsApp has also launched a new feature called Channels where it allows influencers and organizations to attract their followers and engage with them (Meta 2023). Channels are currently outside the purview of fact-checking newsrooms, and it would help if WhatsApp allowed its users to report the content published there by the admins. Fact checkers have already expressed their concerns to the platform that political party operatives will use the Channels feature to spread disinformation ahead of elections, and such operatives cannot be allowed to go unchallenged.

19

The Policy Problems of Coordinated Harm on WhatsApp in Africa

From Calculation to Observation

SCOTT TIMCKE

Following the growth of people using the internet in sub-Saharan Africa over the past decade, telecommunication policy researchers have been on the lookout for the "digital dividends" that would herald the arrival of economic transformation (World Bank 2016).[1] While there are massive changes occurring in African telecommunication infrastructure construction and market development—indeed Southwood (2022) argues that the label "revolution" is entirely appropriate—there is value in studying how people themselves are making and making sense of these changes. One way to do so is through understanding how people use platforms like WhatsApp. Knowledge of these everyday practices can ensure that there is a thorough understanding of all dimensions of a particular issue, and that these dimensions are duly considered when designing programs to achieve social equality through policy interventions. And as much as WhatsApp mediates productive social engagements, the platform can be used to coordinate harm. The study of the latter component from a policy perspective is the subject of this essay.

Problematically Inaccurate Information on WhatsApp

There are documented cases of coordinated harm on WhatsApp in Africa (Hlomani et al. 2023; Timcke, Orembo, and Hlomani 2023). For example, in 2020, 3,300 women were surveyed in Addis Ababa, Nairobi,

---

1. The percentage of people using the internet in sub-Saharan Africa grew from 6 percent in 2010 to 36 percent in 2021 (World Bank 2023).

Kampala, Dakar, and Johannesburg with nearly 40 percent expressing that they were somewhat or very concerned with their online safety; many had become more concerned with their online safety in recent years primarily because of sustained instructive harassment by men on platforms (Iyer, Nyamwire, and Nabulega 2020). In countries like South Africa that have high levels of institutional democratization and wide support for a human rights culture, the most vulnerable undocumented migrants nevertheless face considerable hardships, like xenophobia and its online counterparts. As Dratwa's (2023) early findings show, "the Put South Africans First movement was born on Twitter in April 2020, evolving out of the identically named hashtag. From there it moved to Facebook, to WhatsApp, and then to the streets."[2]

Typically, telecommunication policy researchers prescribe that platforms adopt more stringent content modernization practices (e.g., Forum on Information and Democracy 2023; UNESCO 2023) that may include redesigning recommender algorithms to deprioritize specific content, the banning of accounts that undertake or encourage coordinated harm, encouraging verification of account holders, or even altering the economic incentives of platforms and users.

Given that their main businesses are data arbitrage and commodifying audiences, platforms tend to undertake voluntary content moderation provided it satisfies the bulk of active users, thereby satisfying the desires of advertisers. On occasion, platforms invite third parties to provide limited fact-checking services on viral content. Depending on the type of third parties, these organizations may or may not work for the public internet. Concurrently, content producers adapt to these changes, including those engaged in coordinated harm. This series of constant adaptations means that automated content moderation is currently unlikely to keep problematically inaccurate information in check (Caplan, Hanson, and Donovan 2018). These issues are further compounded by platforms typically not catering to languages that are not in their major markets (Udupa, Maronikolakis, and Wisiorek 2023; UNESCO 2023).

---

2. In South Africa, coordinated xenophobia attacks are experienced most acutely by Black African immigrants and are perpetuated mostly—but not exclusively—by poorer Black South Africans in major urban areas like Johannesburg and Durban. Events like these exemplify the xenophobic attitudes in geographies where targeted overt and covert violence coordinated via platforms is common.

At the same time, it would be an error to scapegoat platforms for root social troubles and weaknesses in state social protection policy. Certainly, anger is directed at the most vulnerable undocumented migrants on platforms, for example, thereby jeopardizing the integrity and image of platform companies. But it is noticeable how states tend to downplay social conditions and their history of governing those conditions. Put differently, platforms can be co-opted into projects that circulate narratives of hate. Nevertheless, highlighting what prompts these narratives to be shared and how users bolster their prominence within a network through deploying extreme speech is foundational to the study of content circulating on platforms like WhatsApp.

Selected Factors Shaping Current Policy Research on WhatsApp

There are good case studies of how WhatsApp is used to coordinate harm, stoke hate, and spread misinformation (e.g., El-Masri, Riedl, and Woolley 2022; Kazemi et al. 2022; Vasudeva and Barkdull 2020); still, there are hurdles surrounding the design of quantitative studies to comprehend the scope of these issues. These include the difficulties around tracking the dissemination of content in detail, or determining the magnitude of effects through proxy measurements. Consider how platforms like Facebook or Twitter have built-in metrics that allow researchers to use data scraping techniques to trace "the social life of data" (Beer 2016, 78) or undertake rudimentary sentiment analysis. As WhatsApp does not have these kinds of public metrics, it is difficult to know the reach of information, how it has traveled, or to quantify the audience that has seen this information on the platform.

A related matter is that tracking coordinated harm requires a fixed definition. This can prove difficult as both coordination and harm turn on the intent of content producers, their grammar with a primary audience, and the intent of people who then share content to other audiences. And then there are matters around a climate of stochastic violence and definitions. Concepts like disinformation, misinformation, and malinformation seek to strike a balance around intent, primary audience, and harm, although not in ways that satisfy all researchers in the field (Lenoir and Anderson 2023; Timcke et al. 2023). What this means for policy formation is that content moderation can be challenging.

Indeed, insisting on content moderation is controversial for platforms like WhatsApp, which offer end-to-end encryption for its messages. Without question the privacy of users always outweighs the curiosity of researchers. And so, policy researchers need to be less "platform centric" and instead lean into prevailing sociological models to form intuitions about what social-technical processes explain acts of coordinated harm that occur on platforms.

An Ethnographic Turn in Telecommunication Policy Research

Telecommunication policy research on WhatsApp can absolutely benefit from social scientific ethnographic field observation for data collection over the medium term. This could partly be accomplished by drawing upon the relatively new techniques of "comparative ethnography" (Simmon and Smith 2019; chapters in this volume). By looking at continuities and differences between meaning making, decisions, and actions across field sites, this technique is especially well suited to hypothesis testing about how problematically inaccurate information circulates. This technique can cater to an examination of the conditions and considerations that feed into brewing incitement. As digital media anthropologists well understand, platforms can dramatize and "spectacularize" social life, and as contemporary everyday cultures are so thoroughly interwoven with digital platforms, ethnographic methods are well suited to adequately grasp the multiple, multisited, and mobile character of sociotechnical phenomena.

By ethnographically identifying the causal pathways that drive incitement, telecommunication policy researchers can study how platforms could play a role in curbing xenophobia in South Africa; for instance, whether through "on platform" interventions or through avenues to create model "off platform" projects. This kind of policy intervention can help advance the cause of minimizing the negative experiences people encounter on platforms. Furthermore, methods like these can encourage studies of what kinds of capacities need to be fostered to effectively manage a social climate where poverty and stratification as well as an increasing sense of frustration fuel political violence and hinder peace. To better grasp coordinated harm, telecommunication policymakers must better understand the sociology of social conflict.

A comparative ethnography of WhatsApp can focus on matters related to scale, mobility, and agency to develop explanations predicated upon local actors' own articulations, which is useful when xenophobic events can be unexpected and fluid. Conducting basic and applied research of this sort can reveal incongruencies between policy assumptions and everyday digital practices, thereby bridging the gap between knowledge and policy formation. That said, recent academic research on measurement and corrective interventions to counter misinformation shows mixed results (Jerit and Zhao 2020), including for example whether "backfire effects" exist when interventions are undertaken (e.g., Haglin 2017; Nyhan and Reifler 2010). Given this type of uncertainty, there is value in using ethnographic fieldwork to explore how emic worldviews make sense of specific situations, which can help to understand the internal causal pathway about how people's attitudes and behaviors are formed. Fulfilling this goal makes observational research into sociocultural phenomena necessary, especially when examining the link between online and offline social life to trace the meaning making around incitement.

An Intervention into Policy Research

As it stands, UNESCO's (2023) guidelines for multistakeholder regulation of digital platforms is the highwater mark for current thinking around policy interventions on platforms like WhatsApp. These guidelines seek to synthesize practical recommendations from a decade of academic study from across the world. For example, the guidelines point to obligations for states to provide meaningful, universal access to the internet for their citizens, safeguarded by the judiciary and independent information regulators that have investitive powers and can levy financial penalties (Shilongo, Sey, and Hlomani 2022). Similarly, UNESCO's guidelines ask platforms to provide consistent services across all regions and for all users, like reporting mechanisms for abuse in a user's primary language. Doing so, UNESCO argues, is the first step in promoting a transparent human rights culture in the business world, which would help bring equality between the Global North and Global South, Africa included (see Timcke et. al 2023).

The main weakness is that UNESCO's recommendations are innocent of capitalism and imperialism; they stress volition while being silent on how commodification and profit maximization shapes platforms' business operations, and they offer no plan for the decommodification of global platforms. I raise these issues because unless there are substantive enforcement mechanisms that can and do jail executives and state officials for the violation of human rights, then little else matters. While the Rome Statute exists for state officials, a similar treaty is needed that permits and facilitates extradition of the executives that oversee commercial crimes. These issues ought to be the priority in global policy agenda in the next decade.

In the meantime, I have suggested a few avenues for policy researchers to alter their approaches, methods, and techniques to study coordinated harm on WhatsApp. Too often policy researchers are "platform centric," focusing too much on the "easily accessible" and ignoring the ways in which these platforms are used and experienced in different parts of the world. And they recoil from more sustained consideration of wider social forces. One way to address this problem is to adopt methods for studying digital technologies that are sensitive to local contexts. Undertaking comparative ethnographic studies of digital media can help prompt telecommunication policy researchers to better understand the grounded sociocultural and political dimensions of platforms. Methods like comparative digital ethnography can greatly help develop more effective ways to pursue policy objectives. Ultimately, there is merit in moving from calculation to observation.

ACKNOWLEDGMENTS

This book flows from a workshop jointly organized by LMU Munich and Stellenbosch University at the Stellenbosch Institute for Advanced Study (STIAS) in 2023. Support from the Millennium Trust, European Research Council (grant agreement number 714285), and LMU-Centre for Advanced Studies to host the conference and subsequent workshops is gratefully acknowledged. We thank Furqan Sayeed and Eric C. Zinner at NYUP for their meticulous editorial support and anonymous readers for their valuable feedback.

BIBLIOGRAPHY

Adams, Bouddih. 2023. "Human Rights Activist, Tambe Tiku, Crusades against Security/Defence Forces Searching Phones." *The Post*, January 16.

Adekoya, Clement Ola. 2021. "Information and Misinformation during the #EndSARS Protest in Nigeria: An Assessment of the Role of Social Media." *Covenant Journal of Library and Information Science* 4 (1): 1–11.

Africa Center for Strategic Studies. 2024. "Mapping a Surge of Disinformation in Africa." https://africacenter.org.

Africa Check. 2021. "Africa Check Launches Fact Ambassador Programme." https://africacheck.org.

Africa Check, and the Africa Centre for Evidence. 2020. *Tackling Misinformation on WhatsApp in Kenya, Nigeria, Senegal, and South Africa: Effective Strategies in a Time of Covid-19*. Johannesburg: Africa Check.

Aguirre, Adalberto, Edgar Rodriguez, and Jennifer K. Simmers. 2011. "The Cultural Production of Mexican Identity in the United States: An Examination of the Mexican Threat Narrative." *Social Identities* 17 (5): 695–707. doi:10.1080/13504630.2011.595209.

Agur, Colin, and Nicholas Frisch. 2019. "Digital Disobedience and the Limits of Persuasion: Social Media Activism in Hong Kong's 2014 Umbrella Movement." *Social Media + Society* 5 (1). doi:10.1177/2056305119827002.

Ahmed, Saifuddin, Dani Madrid-Morales, and Melissa Tully. 2023. "Online Political Engagement, Cognitive Skills, and Engagement with Misinformation: Evidence from Sub-Saharan Africa and the United States." *Online Information Review* 47 (5): 989–1008. doi:10.1108/OIR-11-2021-0634.

Ajakaiye, Olu, and Mthuli Ncube. 2010. "Infrastructure and Economic Development in Africa: An Overview." *Journal of African Economies* 19 (suppl_1): i3–12. doi:10.1093/jae/ejq003.

Akindipe, Dayo. 2023. "The 2023 Presidential Elections and the Emerging Risks to Cybersecurity in Nigeria." *Redeemer's University Law Journals* 3 (1). dx.doi:10.2139/ssrn.4393079.

Allan, Stuart, Prasun Sonwalkar, and Cynthia Carter. 2007. "Bearing Witness: Citizen Journalism and Human Rights Issues." *Globalisation, Societies and Education* 5 (3): 373–89.

Allcott, Hunt, and Matthew Gentzkow. 2017. "Social Media and Fake News in the 2016 Election." *Journal of Economic Perspectives* 31 (2): 211–36.

Allison, Simon. 2022. "Reinventing the Newspaper for the WhatsApp Age." In *WhatsApp and Everyday Life in West Africa: Beyond Fake News*, edited by Idayat Hassan, and Jamie Hitchen, 181–91. London: Zed Books.

Allred, Kristen. 2018. "The Causes and Effects of 'Filter Bubbles' and How to Break Free," April 13, https://medium.com.

Almond, Gabriel A., and Sidney Verba. 1963. *The Civic Culture: Political Attitudes and Democracy in Five Nations*. Princeton: Princeton University Press.

Altheide, David. 2001. "Ethnographic Content Analysis." *Ethnography* 4: 174–84.

Alzouma, Gado. 2005. "Myths of Digital Technology in Africa: Leapfrog-ging Development?" *Global Media and Communication* 1 (3): 339–56. doi:10.1177/1742766505058128.

Amnesty International. 2015. *Cameroon: Human Rights under Fire: Attacks and Viola-tions in Cameroon's Struggle with Boko Haram*. London: Amnesty International.

———. 2017. "Cameroon: Thousands Worldwide Demand Release of Students Jailed for Sharing Boko Haram Joke." www.amnesty.org.

———. 2022. "Cameroon: Amnesty International Urges Release of Abdul Karim Ali, a Peace Activist Detained without Charge for More than Four Months." www .amnesty.org.

Anastasiadou, Athina, Artem Volgin, and Douglas R. Leasure. 2023. "War and Migra-tion: Quantifying the Russian Exodus through Yandex Search Trends." *SocArXiv*. doi:10.31235/osf.io/92zam.

Anderson, Benedict. 2006. *Imagined Communities: Reflections on the Origin and Spread of Nationalism*, rev. ed. London: Verso Books.

Anderson, Elizabeth, Mary Koss, Ana Lucía Castro Luque, David Garcia, Elise Lopez, and Kacey Ernst. 2021. "WhatsApp-Based Focus Groups among Mexican-Origin Women in Zika Risk Area: Feasibility, Acceptability, and Data Quality." *JMIR For-mative Research* 5 (10). doi:10.2196/20970.

Appadurai, Arjun. 1990. "Technology and the Reproduction of Values in Rural West-ern India." In *Dominating Knowledge: Development, Culture, and Resistance*, edited by Frédérique Apffel Marglin, and Stephen A. Marglin, 185–216. Oxford: Oxford University Press.

Arabacı, Ali Osman, Sayyara Mammadova, and Seçil Türkkan. 2020. "Azerbaycan Ermenistan Çatışmaları." https://127.0.0.1:6567.

Ariely, Gal. 2017. "Global Identification, Xenophobia and Globalisation: A Cross-National Exploration." *International Journal of Psychology* 52 (S1): 87–96. doi:10.1002/ijop.12364.

Arif, Ahmer, John J. Robinson, Stephanie A. Stanek, Elodie S. Fichet, Paul Townsend, Zena Worku, and Kate Starbird. 2017. "A Closer Look at the Self-Correcting Crowd." In *Proceedings of the 2017 ACM Conference on Computer Supported Cooperative Work and Social Computing*, 155–68. Portland, OR: Association for Computing Machinery. doi:10.1145/2998181.2998294.

Arora, Payal. 2016. "Bottom of the Data Pyramid: Big Data and the Global South." *International Journal of Communication* 10 (19): 1681–99.

Arun, Chinmayi. 2019. "On WhatsApp, Rumours, and Lynchings." *Economic & Political Weekly* 54 (6): 30–35.

Ashukem, Jean-Claude N. 2021. "To Give a Dog a Bad Name to Kill It—Cameroon's Anti-Terrorism Law as a Strategic Framework for Human Rights' Violations." *Journal of Contemporary African Studies* 39 (1): 119–34. doi:10.1080/02589001.2020. 1839633.

Atkinson, Paul. 2007. *Ethnography: Principles in Practice*. New York: Routledge.

Atton, Chris. 2007. "Current Issues in Alternative Media Research." *Sociology Compass* 1 (1), 17–27. doi:10.1111/j.1751-9020.2007.00005.x.

Auerbach, David. 2012. "Anonymity as Culture: Treatise." www.canopycanopycanopy .com.

Auxier, Brooke, and Monica Anderson. 2021. *Social Media Use in 2021*. Washington, DC: Pew Research Center.

Avritzer, Leonardo, and Lúcio Rennó. 2023. "Populism, the Pandemic, and the Crisis of Bolsonarismo." In *Right-Wing Populism in Latin America and Beyond*, edited by Anthony W. Pereira. New York: Routledge.

Aydın, Ali Fikret. 2020. "Post-Truth Dönemde Sosyal Medyada Dezenformasyon: Covid-19 (Yeni Koronavirüs) Pandemi Süreci." *Asya Studies* 4 (12): 76–90.

Bachelet, Michelle. 2022. "Crisis and Fragility of Democracy in the World," speech delivered by the UN High Commissioner for Human Rights, Boston, August 3.

Bailur, Savita, and Emrys Schoemaker. 2016. "WhatsApp, Facebook, and Pakapaka: Digital Lives in Ghana, Kenya, and Uganda." *LSE* (blog), April 18. https://blogs.lse .ac.uk.

Banks, Nicola, Melanie Lombard, and Diana Mitlin. 2020. "Urban Informality as a Site of Critical Analysis." *Journal of Development Studies* 56 (2): 223–38. doi:10.1080/002 20388.2019.1577384.

Barberá, Pablo, John T. Jost, Jonathan Nagler, Joshua A. Tucker, and Richard Bonneau. 2015. "Tweeting from Left to Right: Is Online Political Communication More than an Echo Chamber?" *Psychological Science* 26 (10): 1531–42.

Barberá, Pablo, and Gonzalo Rivero. 2015. "Understanding the Political Representative-ness of Twitter Users." *Social Science Computer Review* 33 (6): 712–29.

Barbosa, Sérgio, and Stefania Milan. 2019. "Do Not Harm in Private Chat Apps: Ethical Issues for Research on and with WhatsApp." *Westminster Papers in Communication and Culture* 14 (1): 49–56. doi:10.16997/wpcc.313.

Bas, Ozen, Christine L. Ogan, and Onur Varol. 2022. "The Role of Legacy Media and Social Media in Increasing Public Engagement about Violence against Women in Turkey." *Social Media + Society* 8 (4): 1–13.

Bayer, Judit, and Petra Bárd. 2020. *Hate Speech and Hate Crime in the EU and the Evaluation of Online Content Regulation Approaches*. Brussels: Policy Department for Citizens' Rights and Constitutional Affairs, Directorate-General for Internal Policies, European Parliament.

BBC News. 2023. "Brazil Congress: Mass Arrests as Lula Condemns 'Terrorist' Riots." January 9. www.bbc.com.

Becker, Howard. 1949. "The Nature and Consequences of Black Propaganda." *American Sociological Review* 14 (2): 221–35. doi:10.2307/2086855.

Beek, Jan. 2018. How Not to Fall in Love: Mistrust in Online Romance Scams. In *Mistrust: Ethnographic Approximations*, edited by Florian Mühlfried, 49–69. Bielefeld: Transcript.

Beer, David. 2016. *Metric Power*. London: Palgrave Macmillan.

Belair-Gagnon, Valerie, Colin Agur, and Nicholas Frisch. 2017. "The Changing Physical and Social Environment of Newsgathering: A Case Study of Foreign Correspondents Using Chat Apps during Unrest." *Social Media + Society* 3 (1). doi:10.1177/2056305117701163.

Belinskaya, Yulia. 2023. "'Insider News' on Russian Telegram: Resembling Truth, Proximity and Objectivity." *Journal of Applied Journalism and Media Studies*. doi:10.1386/ajms_00108_1.

Bengani, Priyanjana. 2019. "India Had Its First 'WhatsApp Election.' We Have a Million Messages from It." *Columbia Journalism Review*, October 16. www.cjr.org.

Bentivegna, Sara. 2006. "Rethinking Politics in the World of ICTs." *European Journal of Communication* 21 (3): 331–43. doi:10.1177/0267323106066638.

Bernal, Victoria. 2014. *Nation as Network: Diaspora, Cyberspace, and Citizenship*. Chicago: University of Chicago Press.

Bhushan, Kul. 2019. "Here's How WhatsApp Is Dealing with the Menace of Spam, Automated Accounts." *Tech Hindustan Times*, February 7. https://tech.hindustantimes.com.

Binder, Larissa, Simone Ueberwasser, and Elisabeth Stark. 2020. "Gendered Hate Speech in Swiss WhatsApp Messages." In *Language, Gender and Hate Speech: A Multidisciplinary Approach*, edited by Giuliana Giusti and Gabriele Iannàccaro, 59–74. Venice: Fondazione Università Ca' Foscari.

Bode, Leticia, and Emily K. Vraga. 2018. "See Something, Say Something: Correction of Global Health Misinformation on Social Media." *Health Communication* 33 (9): 1131–40. doi:10.1080/10410236.2017.1331312.

Bohman, James. 2004. "Realizing Deliberative Democracy as a Mode of Inquiry: Pragmatism, Social Facts, and Normative Theory." *Journal of Speculative Philosophy* 18 (1): 23–43.

BOOM Fact Check Team. 2018. "5 Things to Know about WhatsApp's 'Forwarded Label' Feature to Fight Fake News." www.boomlive.in.

Borealis. 2024. "Cameroon: Database of Atrocities." https://borealisdata.ca.

Bornman, Jon. 2021. "Operation Dudula Pushes ahead with Hateful Politics." https://adf.org.za.

Bosch, Tanja. 2013. "Youth, Facebook and Politics in South Africa." *Journal of African Media Studies* 5 (2): 119–30. doi:10.1386/jams.5.2.119_1.

———. 2022. "Decolonizing Digital Methods." *Communication Theory* 32 (2): 298–302.

———. 2023. "Epistemic Decolonization of Digital Media (Research) in the Global South." Unpublished conference paper. *Southern Digitalities*. Doha: Northwestern University, Qatar.

Boulianne, Shelley, and Yannis Theocharis. 2020. "Young People, Digital Media, and Engagement: A Meta-Analysis of Research." *Social Science Computer Review* 38 (2): 111–27. doi:10.1177/0894439318814190.

Bowker, Geoffrey C., and Susan Leigh Star. 2000. *Sorting Things Out: Classification and Its Consequences*. Cambridge: The MIT Press.

Bowles, Jeremy, Kevin Croke, Horacio Larreguy, Shelley Liu, and John Marshall. 2023. *Sustaining Exposure to Fact-Checks Can Inoculate Citizens against Misinformation in the Global South*. Cambridge: Harvard University.

Boyacı Yıldırım, Merve. 2023. "COVID-19 Infodemic: A Study on the Fragile Five Countries." *Journal of Public Affairs* 23 (1). doi:10.1002/pa.2846.

Boyd-Barrett, Oliver. 2019. "Fake News and 'RussiaGate' Discourses: Propaganda in the Post-Truth Era." *Journalism* 20 (1): 87–91. doi:10.1177/1464884918806735.

Bozdağ, Çiğdem, and Suncem Koçer. 2022. "Skeptical Inertia in the Face of Polarization: News Consumption and Misinformation in Turkey." *Media and Communication* 10 (2): 169–79.

Bozkanat, Esra. 2021. "Detecting Fake News on Social Media: The Case of Turkey." In *Analyzing Global Social Media Consumption*, edited by Patrick Kanyi Wamuyu. Hershey: IGI Global.

Bradshaw, Samantha, and Philip N. Howard. 2018. "Challenging Truth and Trust: A Global Inventory of Organized Social Media Manipulation." *The Computational Propaganda Project* 1: 1–26.

Brant, João. 2023. *Social Media 4 Peace: Local Lessons for Global Practices*. Paris: UNESCO. https://unesdoc.unesco.org.

Brantner, Cornelia, Joan Ramon Rodríguez-Amat, and Yulia Belinskaya. 2021. "Structures of the Public Sphere: Contested Spaces as Assembled Interfaces." *Media and Communication* 9 (3): 16–27. doi:10.17645/mac.v9i3.3932.

Braun, Virginia, and Victoria Clarke. 2006. "Using Thematic Analysis in Psychology." *Qualitative Research in Psychology* 3 (2): 77–101.

Breese, Elizabeth Butler. 2011. "Mapping the Variety of Public Spheres." *Communication Theory* 21 (2): 130–49. doi:10.1111/j.1468–2885.2011.01379.x.

Brenes Peralta, Carlos M., Rolando Pérez Sánchez, and Ignacio Siles González. 2022. "Individual Evaluation vs Fact-Checking in the Recognition and Willingness to Share Fake News about COVID-19 via WhatsApp." *Journalism Studies* 23 (1): 1–24. doi:10.1080/1461670X.2021.1994446.

Bright, Jonathan, Antonella Perini, Anne Ploin, Reja Wyss, United Nations Office on Genocide Prevention and the Responsibility to Protect, and UNESCO. 2021. *Addressing Hate Speech on Social Media: Contemporary Challenges*. Paris: UNESCO. https://unesdoc.unesco.org.

Bruns, Axel. 2011. "Gatekeeping, Gatewatching, Real-Time Feedback: New Challenges for Journalism." *Brazilian Journalism Research* 7 (11): 117–36. doi:10.25200/BJR. v7n2.2011.355.

Bryman, Alan. 2016. *Social Research Methods*. Oxford: Oxford University Press.

Bueno-Roldan, Rocio, and Antje Röder. 2022. "WhatsApp? Opportunities and Challenges in the Use of a Messaging App as a Qualitative Research Tool." *The Qualitative Report* 27 (12): 2961–76.

Bulut, Ergin, and Can Ertuna. 2022. "The Pandemic Shock Doctrine in an Authoritarian Context: The Economic, Bodily, and Political Precarity of Turkey's Journalists during the Pandemic." *Media, Culture & Society* 44 (5): 1003–20.

Bundesministerium der Justiz. 2017. "Gesetz zur Verbesserung der Rechtsdurchsetzung in Sozialen Netzwerken (Netzwerkdurchsetzungsgesetz—NetzDG)." https://www.gesetze-im-internet.de.

Bureau of Democracy, Human Rights, and Labor. 2021. *2021 Country Reports on Human Rights Practices: Cameroon*. Washington, DC: U.S. Department of State.

Bursztyn, Victor S., and Larry Birnbaum. 2019. "Thousands of Small, Constant Rallies: A Large-Scale Analysis of Partisan WhatsApp Groups." In *Proceedings of the 2019 IEEE/ACM International Conference on Advances in Social Networks Analysis and Mining*, 484–88. New York: Association for Computing Machinery. doi:10.1145/3341161.3342905.

Business Insider Africa. 2022. "5 Reasons Africans Love Using WhatsApp Messenger." August 27. https://africa.businessinsider.com.

BusinessTech. 2022. "New WhatsApp Change Could Lead to Legal Trouble for Group Admins in South Africa." https://businesstech.co.za.

Caetano, Josemar Alves, Gabriel Magno, Marcos Gonçalves, Jussara Almeida, Humberto T. Marques-Neto, and Virgílio Almeida. 2019. "Characterizing Attention Cascades in WhatsApp Groups." In *Proceedings of the 10th ACM Conference on Web Science*, 27–36. New York: Association for Computing Machinery.

Cai, Welyl, and Ford Fessenden. 2020. "Immigrant Neighborhoods Shifted Red as the Country Chose Blue." *New York Times*, December 20.

Calderón, César, and Luis Servén. 2010. "Infrastructure and Economic Development in Sub-Saharan Africa." *Journal of African Economies* 19 (suppl_1): i13–87. doi:10.1093/jae/ejp022.

Cameroon News Agency. 2023. "This Has Become a Daily Routine in the University Community of Bambili, Mezam Division in the North West Region." Facebook post, January 15. www.facebook.com.

Caplan, Robyn, Lauren Hanson, and Joan Donovan. 2018. *Dead Reckoning: Navigating Content Moderation after "Fake News."* New York: Data & Society Research Institute.

Carlson, Matt. 2017. *Journalistic Authority: Legitimating News in the Digital Era*. New York: Columbia University Press.

———. 2020a. "Fake News as an Informational Moral Panic: The Symbolic Deviancy of Social Media during the 2016 U.S. Presidential Election." *Information, Communication & Society* 23 (3): 374–88. doi:10.1080/1369118X.2018.1505934.

———. 2020b. "Journalistic Epistemology and Digital News Circulation: Infrastructure, Circulation Practices, and Epistemic Contests." *New Media & Society* 22 (2): 230–46. doi:10.1177/1461444819856921.

Cave, Andrew. 2017. "Deal That Undid Bell Pottinger: Inside Story of the South Africa Scandal." *The Guardian*, September 5. www.theguardian.com.

Ceci, Laura. 2022. "WhatsApp—Statistics and Facts." www.statista.com.

Census of India. 2011. "Tiruvannamalai District—Population 2011–2024." https://www.census2011.co.in.

Cesarino, Letícia. 2019. "Identidade e Representação no Bolsonarismo: Corpo Digital do Rei, Bivalência Conservadorismo-neoliberalismo e Pessoa Fractal." *Revista de Antropologia* 62 (3). doi:10.11606/2179-0892.ra.2019.165232.

———. 2022. *O Mundo do Avesso: Verdade e Política na Era Digital.* São Paulo: UBU.

Chadwick, Andrew, Cristian Vaccari, and Ben O'Loughlin. 2018. "Do Tabloids Poison the Well of Social Media? Explaining Democratically Dysfunctional News Sharing." *New Media & Society* 20 (11): 4255–74. doi:10.1177/1461444818769689.

Chagas, Viktor. 2022. "WhatsApp and Digital Astroturfing: A Social Network Analysis of Brazilian Political Discussion Groups of Bolsonaro's Supporters." *International Journal of Communication* 16.

Chakrabarti, Santanu. 2018. *Duty, Identity, Credibility: "Fake News" and the Ordinary Citizen in India.* London: BBC.

Chakrabarti, Santanu, Claire Rooney, and Minnie Kewon. 2018. *Verification, Duty, Credibility: Fake News and Ordinary Citizens in Kenya and Nigeria.* London: BBC.

Cheeseman, Nic, Jonathan Fisher, Idayat Hassan, and Jamie Hitchen. 2020. "Social Media Disruption: Nigeria's WhatsApp Politics." *Journal of Democracy* 31 (January): 145–59. doi:10.1353/jod.2020.0037.

Cheruiyot, David, and Raul Ferrer-Conill. 2018. "'Fact-Checking Africa:' Epistemologies, Data and the Expansion of Journalistic Discourse." *Digital Journalism* 6 (8): 964–75. doi:10.1080/21670811.2018.1493940.

Choane, Mamokhosi, Lukong Stella Shulika, and Mandla Mthombeni. 2011. "An Analysis of the Causes, Effects and Ramifications of Xenophobia in South Africa." *Insight on Africa* 3 (2): 129–42. doi:10.1177/0975087814411138.

Chopra, Rohit. 2019. "In India, WhatsApp Is a Weapon of Antisocial Hatred." *The Conversation*, April 23.

CHRDA. 2022a. "The Government of Cameroon Should Urgently Open an Investigation into All Reported Acts of Torture by Chief Ewume John (Chifed Moja Moja) in the South West Region." https://www.chrda.org.

CHRDA. 2022b. "Recent Selected Incidents of Violence Committed by Elements of the Defence and Security Forces and Non-State Armed Groups." https://www.chrda.org.

Coddington, Mark Allen. 2019. *Aggregating the News: Secondhand Knowledge and the Erosion of Journalistic Authority.* New York: Columbia University Press.

Cole, Matthew A., Robert J. R. Elliott, Giovanni Occhiali, and Eric Strobl. 2018. "Power Outages and Firm Performance in Sub-Saharan Africa." *Journal of Development Economics* 134: 150–59. doi:10.1016/j.jdeveco.2018.05.003.

Collier, David. 2011. "Understanding Process Tracing." *PS: Political Science & Politics* 44 (4): 823–30.

Comaroff, Jean, and John L. Comaroff. 2015. *Theory from the South: Or, How Euro-America Is Evolving toward Africa.* New York: Routledge.

Çömlekçi, Mehmet Fatih. 2019. "Sosyal Medyada Dezenformasyon ve Haber Doğrulama Platformlarinin Pratikleri." *Gümüşhane Üniversitesi İletişim Fakültesi Elektronik Dergisi* 7 (3): 1549–63.

Communications Regulatory Authority of Namibia (CRAN). 2022. *Quarterly Statistics Newsletter: Q4 of 2022*. Windhoek: CRAN. www.cran.na.

Connell, Raewyn. 2020. *Southern Theory: The Global Dynamics of Knowledge in Social Science*. New York: Routledge.

Cottle, Simon. 2013. "Journalists Witnessing Disaster: From the Calculus of Death to the Injunction to Care." *Journalism Studies* 14 (2): 232–48. doi:10.1080/1461670X.2012.718556.

Coughlan, Sean. 2020. "WhatsApp Rumours Fear over BAME Covid Vaccine Take Up." *BBC News*, December 16. www.bbc.co.uk.

Creech, Brian. 2018. "Bearing the Cost to Witness: The Political Economy of Risk in Contemporary Conflict and War Reporting." *Media, Culture & Society* 40 (4): 567–83. doi:10.1177/0163443717715078.

Crozier, Michael J., Samuel P. Huntington, and Joji Watanuki. 2012. "The Crisis of Democracy. Report on the Governability of Democracies to the Trilateral Commission." *Sociología Histórica* 1.

Cruz, Edgar Gómez, and Ramaswami Harindranath. 2020. "WhatsApp as 'Technology of Life': Reframing Research Agendas." *First Monday* 25 (January). doi:10.5210/fm.v25i12.10405.

Curran, James, and Tamar Liebes. 2002. *Media, Ritual and Identity*, 2nd ed. New York: Routledge.

Dalton, Russell J. 2004. *Democratic Challenges, Democratic Choices: The Erosion of Political Support in Advanced Industrial Democracies*. Oxford: Oxford Academic Books.

Das, Veena. 2007. *Life and Words: Violence and the Descent into the Ordinary*. Berkeley: University of California Press.

———. 2020. *Textures of the Ordinary: Doing Anthropology after Wittgenstein*. New York: Fordham.

Das, Veena, and Deborah Poole. 2004. "State and Its Margins: Comparative Ethnographies." In *Anthropology in the Margins of the State*, edited by Veena Das and Deborah Poole, 3–34. Santa Fe, NM: School of American Research Press.

Da Silva, Issa Sikiti. 2020. "Covid-19 Reveals Digital Divide as Africa Struggles with Distance Learning." *TRT World*. https://www.trtworld.com/magazine/covid-19-reveals-digital-divide-as-africa-struggles-with-distance-learning-37299.

Data.ai. 2022. "Digital 2022 Chile." Slideshare, February 17. www.slideshare.net.

Datareportal. 2021. "Digital 2021: Cameroon." www.datareportal.com.

———. 2023. *Digital 2023: Nigeria*. https://datareportal.com/reports/digital-2023-nigeria.

Deeks, Ashley. 2016. *The International Legal Dynamics of Encryption*. Hoover Institution Series Paper No. 1609. www.hoover.org.

#DefyHateNow. 2020. "Common Digital Platforms in Cameroon and Their Usage." www.defyhatenow.org.

Degenhard, J. 2023. WhatsApp Users in India 2017–2025. www.statista.com.

De Gruchy, Thea, Jo Vearey, Calvin Opiti, Langelihle Mlotshwa, Karima Manji, and Johanna Hanefeld. 2021. "Research on the Move: Exploring WhatsApp as a Tool for Understanding the Intersections between Migration, Mobility, Health and Gender in South Africa." *Globalization and Health* 17 (1): 1–13.

De Meo, Pasquale, Emilio Ferrara, Giacomo Fiumara, and Alessandro Provetti. 2014. "On Facebook, Most Ties Are Weak." *Communications of the ACM* 57 (11): 78–84.

De Sá, Ivandro Claudino, José Maria Monteiro, José Wellington Franco da Silva, Leonardo Monteiro Medeiros, Pedro Jorge Chaves Mourao, and Lucas Cabral Carneiro da Cunha. 2021. "Digital Lighthouse: A Platform for Monitoring Public Groups in WhatsApp." In *Proceedings of the 23rd International Conference on Enterprise Information Systems* (1): 297–304.

Desai, Nachiketa. 2019. "Pratik Sinha of Alt News: Mischief Mongers Creating Havoc on Internet." *National Herald*, April 28. www.nationalheraldindia.com.

Diepeveen, Stephanie. 2021. *Searching for a New Kenya: Politics and Social Media on the Streets of Mombasa*. Cambridge: Cambridge University Press.

Dierks, Zeynep. 2023. "Turkey: Social Network Penetration 2022." www.statista.com.

Dirini, İlden, and Gökçe Özsu. 2020. *COVID-19 Pandemi Sürecinde: Sosyal Medyada Nefret Söylemi Raporu*. Ankara: Alternatif Bilişim.

Di Salvo, Philip. 2022. "Information Security and Journalism: Mapping a Nascent Research Field." *Sociology Compass* 16 (3). doi:10.1111/soc4.12961.

Dlamini, Lucky Brian, and Glenda Daniels. 2023. "Scrutinising South African Media Companies' Strategies for Generation Z's News Consumption." *Media, Culture & Society* 45 (4): 702–19. doi:10.1177/01634437221135979.

Domańska, Maria. 2023. *Russian Civil Society Actors in Exile: An Underestimated Agent of Change*. SWP Comment, 26/2023. Berlin: Stiftung Wissenschaft und Politik. doi:10.18449/2023C26.

Dosek, Tomas. 2021. "Snowball Sampling and Facebook: How Social Media Can Help Access Hard-to-Reach Populations." *PS: Political Science & Politics* 54 (4): 651–55. doi:10.1017/S104909652100041X.

Doshi, Vidhi. 2017. "India's Millions of New Internet Users Are Falling for Fake News—Sometimes with Deadly Consequences." *Washington Post*, October 1. www.washingtonpost.com.

Dos Santos, João Guilherme Bastos, Miguel Freitas, Alessandra Aldé, Karina Santos, and Vanessa Cristine Cardozo Cunha. 2019. "WhatsApp, Política Mobile e Desinformação: A Hidra nas Eleições Presidenciais de 2018." *Comunicação & Sociedade* 41 (2): 307–34.

Dourish, Paul, Connor Graham, and Dave Randall. 2010. "Theme Issue on Social Interaction and Mundane Technologies." *Personal and Ubiquitous Computing* 14: 171–80. https://doi.org/10.1007/s00779-010-0281-0.

Dowling, Melissa-Ellen. 2022. "Cyber Information Operations: Cambridge Analytica's Challenge to Democratic Legitimacy." *Journal of Cyber Policy* 7 (2): 230–48.

Downey, John, and Natalie Fenton. 2003. "New Media, Counter Publicity, and the Public Sphere." *New Media & Society* 5 (2): 185–202. doi:10.1177/1461444803005002003.

Dratwa, Bastien. 2023. "Digital Xenophobia Is on the Rise in South Africa," *LSE* (blog), March 7. https://blogs.lse.ac.uk.

Duffy, Andrew, Edson Tandoc, and Rich Ling. 2020. "Too Good to Be True, Too Good Not to Share: The Social Utility of Fake News." *Information, Communication & Society* 23 (13): 1965–979. doi:10.1080/1369118X.2019.1623904.

Duguay, Stefanie, and Hannah Gold-Apel. 2023. "Stumbling Blocks and Alternative Paths: Reconsidering the Walkthrough Method for Analyzing Apps." *Social Media + Society* 9 (1). doi:10.1177/20563051231158822.

Dukalskis, Alexander. 2021. *Making the World Safe for Dictatorship*. Oxford: Oxford University Press.

Dutta, Soumya, and Saswati Gangopadhyay. 2019. "Digital Journalism: Theorizing on Present Times." *Media Watch* 10 (3): 713–22.

Dvir-Gvirsman, Shira, Keren Tsuriel, Tamir Sheafer, Shaul Shenhav, Alon Zoizner, Liron Lavi, Michal Shamir, and Israel Waismel-Manor. 2022. "Mediated Representation in the Age of Social Media: How Connection with Politicians Contributes to Citizens' Feelings of Representation. Evidence from a Longitudinal Study." *Political Communication* 39 (6): 779–800.

Dwyer, Maggie, and Thomas Molony, eds. 2019. *Social Media and Politics in Africa: Democracy, Censorship and Security*. London: Zed Books.

Eley, Geoffrey. 1990. *Nations, Publics, and Political Cultures: Placing Habermas in the Nineteenth Century*. Center for Research on Social Organization Working Paper 417.

Ellis, Stephen. 1989. "Tuning In to Pavement Radio." *African Affairs* 88 (352): 321–30.

Ellison, Nicole, and Danah M. Boyd. 2013. "Sociality through Social Network Sites." In *The Oxford Handbook of Internet Studies*, edited by William H. Dutton, 510–78. Oxford: Oxford University Press.

Elmas, Tuğrulcan, Rebekah Overdorf, and Karl Aberer. 2021. "Tactical Reframing of Online Disinformation Campaigns against the Istanbul Convention." *ArXiv*. doi:10.36190/2021.42.

El-Masri, Azza, Martin J. Riedl, and Samuel Woolley. 2022. "Audio Misinformation on WhatsApp: A Case Study from Lebanon." *Harvard Kennedy School Misinformation Review* 3 (4). doi:10.37016/mr-2020–102.

Erdoğan, Emre, Pınar Uyan Semerci, Birnur Eyolcu Kafalı, and Şaban Çaytaş. 2022. *İnfodemi ve Bilgi Düzensizlikleri: Kavramlar, Nedenler ve Çözümler*. Istanbul: İstanbul Bilgi Üniversitesi Yayınları.

Erkan, Gülsüm, and Ahmet Ayhan. 2018. "Siyasal İletişimde Dezenformasyon ve Sosyal Medya: Bir Doğrulama Platformu Olarak Teyit. Org." *Akdeniz Üniversitesi İletişim Fakültesi Dergisi* 29: 202–23.

European Commission. 2017. "Counter-Narratives: How to Support Civil Society in Delivering Effective Positive Narratives against Hate Speech Online." https://ec.europa.eu.

Evangelista, Rafael, and Fernanda Bruno. 2019. "WhatsApp and Political Instability in Brazil: Targeted Messages and Political Radicalisation." *Internet Policy Review* 8 (4): 1–23. doi:10.14763/2019.4.1434.

Facchini, Regina, and Horacio Sívori. 2017. "Conservadorismo, Direitos, Moralidades e Violência: Situando um Conjunto de Reflexões a Partir da Antropologia." *Cadernos Pagu* 50.

Fairclough, Norman. 2013. *Critical Discourse Analysis: The Critical Study of Language*. New York: Longman.

Farias, Juliana. 2020. *Governo de Mortes: Uma Etnografia da Gestão de Populações de Favelas no Rio de Janeiro*. Rio de Janeiro: Papeis Selvagens.

Farooq, Gowhar. 2018. "Politics of Fake News: How WhatsApp Became a Potent Propaganda Tool in India." *Media Watch* 9 (1): 106–17. dx.doi:10.17613/45p0-vg35.

Fast Check CL. 2020. "Fraude Electoral: Como se Repite la Misma Narrativa de Desinformación en la Región." September 5. www.fastcheck.cl.

Feng, Kevin K. J., Kevin Song, Kejing Li, Oishee Chakrabarti, and Marshini Chetty. 2022. "Investigating How University Students in the United States Encounter and Deal with Misinformation in Private WhatsApp Chats during COVID-19." In *Proceedings of the Eighteenth Symposium on Usable Privacy and Security*, 427–46. Berkeley, CA: The Advanced Computing Systems Association.

Findlay, Kyle. 2021. "What Social Media Tells Us about Political Parties' Local Government Election Strategies—and Voter Manipulation." *Daily Maverick*, October 5. www.dailymaverick.co.za.

Fine, Gary Alan. 2010. "The Sociology of the Local: Action and Its Publics." *Sociological Theory* 28 (4): 355–76. doi:10.1111/j.1467-9558.2010.01380.x.

Flick, Uwe, ed. 2013. *The SAGE Handbook of Qualitative Data Analysis*. New York: Sage.

Forum on Information and Democracy. 2023. *Pluralism of News and Information in Curation and Indexing Algorithms*. https://informationdemocracy.org.

Franzke, Aline Shakti, Anja Bechmann, Michael Zimmer, and Charles M. Ess. 2019. *Internet Research: Ethical Guidelines 3.0*. Chicago: Association of Internet Researchers.

Fraser, Nancy. 1989. *Unruly Practices: Power, Discourse and Gender in Contemporary Social Theory*. Minneapolis: University of Minnesota Press.

———. 1992. "Rethinking the Public Sphere: A Contribution to the Critique of Actually Existing Democracy." In *Habermas and the Public Sphere*, edited by Craig Calhoun, 109–42. Cambridge: The MIT Press.

———. 2019. *The Old Is Dying and the New Cannot Be Born: From Progressive Neoliberalism to Trump and Beyond*. Brooklyn: Verso Books.

Freelon, Deen G. 2010. "ReCal: Intercoder Reliability Calculation as a Web Service." *International Journal of Internet Science* 5 (1): 20–33.

———. 2018. "Campaigns in Control: Analyzing Controlled Interactivity and Message Discipline on Facebook." *Journal of Information Technology & Politics* 14 (2): 168–81.

Fortes, Meyer. 2005. *Kinship and the Social Order: The Legacy of Lewis Henry Morgan*. New York: Routledge.

Fuchs, Christian. 2015. *Culture and Economy in the Age of Social Media*. New York: Routledge.

Furman, Ivo Ozan, Kurt Bilgin Gürel, and Fırat Berk Sivaslıoğlu. 2023. "'As Reliable as a Kalashnikov Rifle': How Sputnik News Promotes Russian Vaccine Technologies in the Turkish Twittersphere." *Social Media + Society* 9 (1): 1–14.

Gaber, Ivor. 2009. "Exploring the Paradox of Liberal Democracy: More Political Communications Equals Less Public Trust." *Political Quarterly* 80 (1): 84–91. doi:10.1111/j.1467–923X.2009.01961.x.

Gagliardone, Iginio, Stephanie Diepeveen, Kyle Findlay, Samuel Olaniran, Matti Pohjonen, and Edwin Tallam. 2021. "Demystifying the COVID-19 Infodemic: Conspiracies, Context, and the Agency of Users." *Social Media + Society* 7 (3). doi:10.1177/20563051211044233.

Galletta, Anne. 2013. *Mastering the Semi-Structured Interview and Beyond: From Research Design to Analysis and Publication*. New York: New York University Press.

Garimella, Kiran, and Dean Eckles. 2020. "Images and Misinformation in Political Groups: Evidence from WhatsApp in India." *Harvard Kennedy School Misinformation Review* 1 (5). doi:10.37016/mr-2020–030.

Garimella, Kiran, and Gareth Tyson. 2018. "WhatsApp Doc? A First Look at WhatsApp Public Group Data." In *Proceedings of the 12th International AAAI Conference on Web and Social Media*, vol. 12, 511–17. doi:10.1609/icwsm.v12i1.14989. Washington, DC: Association for the Advancement of Artificial Intelligence.

Garrett, R. Kelly. 2017. "The 'Echo Chamber' Distraction: Disinformation Campaigns Are the Problem, Not Audience Fragmentation." *Journal of Applied Research in Memory and Cognition* 6 (4): 370–76. doi:10.1016/j.jarmac.2017.09.011.

George, Cherian. 2016. *Hate Spin: The Manufacture of Religious Offense and Its Threat to Democracy*. Cambridge: The MIT Press.

Germany. 2018. The Network Enforcement Act (Netzwerkdurchsetzungsgesetz), 18/12356, 18/13013.

Giddens, Anthony. 1984. *The Constitution of Society: Outline of the Theory of Structuration*. Berkeley: University of California Press.

Gil de Zúñiga, Homero, Alberto Ardèvol-Abreu, and Andreu Casero-Ripollés. 2021. "WhatsApp Political Discussion, Conventional Participation and Activism: Exploring Direct, Indirect and Generational Effects." *Information, Communication & Society* 24 (2): 201–18. doi:10.1080/1369118X.2019.1642933.

Gil de Zúñiga, Homero, Trevor Diehl, Brigitte Huber, and James H. Liu. 2019. "The Citizen Communication Mediation Model across Countries: A Multilevel Mediation Model of News Use and Discussion on Political Participation." *Journal of Communication* 69 (2): 144–67. doi:10.1093/joc/jqz002.

Gilili, Chris. 2022. "Xenophobic Violence: Zimbabweans Live in Fear of Vigilantes in Diepsloot | News24." *News24*, April 11. www.news24.com.

Giusti, Giuliana, and Gabriele Iannàccaro, eds. 2020. *Language, Gender and Hate Speech: A Multidisciplinary Approach*. Venice: Venice University Press.

Global Witness. 2023. "'We Need to Kill Them': Xenophobic Hate Speech Approved by Facebook, TikTok and YouTube." www.globalwitness.org.

Goggin, Gerard. 2020. "Digital Journalism after Mobility." *Digital Journalism* 8 (1): 170–73. doi:10.1080/21670811.2019.1711434.

Goldman, Alvin I. 2001. "Experts: Which Ones Should You Trust?" *Philosophy and Phenomenological Research* 63 (1): 85–110.

Gordon, Steven. 2022. "Are Foreigners Welcome in South Africa? An Attitudinal Analysis of Anti-Immigrant Sentiment in South Africa during the 2003–2018 Period." In *Paradise Lost: Race and Racism in Post-Apartheid South Africa*, edited by Gregory Houston, Modimowabarwa Kanyane, and Yul Derek Davids, 245–66. Leiden: Brill.

Graan, Andrew, Adam Hodges, and Meg Stalcup. 2020. "Fake News and Anthropology: A Conversation on Technology, Trust, and Publics in the Age of Mass Disinformation Part I." https://polarjournal.org.

Granovetter, Mark S. 1973. "The Strength of Weak Ties." *American Journal of Sociology* 78 (6): 1360–80.

Graphika and Stanford Internet Observatory. 2022. *Unheard Voice: Evaluating Five Years of Pro-Western Covert Influence Operations*. Stanford: Stanford University.

Graves, Lucas. 2016. *Deciding What's True: The Rise of Political Fact-Checking in American Journalism*. New York: Columbia University Press.

Grisham, Kevin. 2021. "Far-Right Groups Move to Messaging Apps as Tech Companies Crack Down on Extremist Social Media." *The Conversation*, January 22. http://theconversation.com.

Grover. n.d. "The Electronics Price Index." www.grover.com.

Grover, Gurshabad, Tanaya Rajwade, and Divyank Katira. 2021. "The Ministry and the Trace: Subverting End-To-End Encryption." *NUJS Law Review* 2 (April–June): 223–53.

GSM Arena. n.d. "Xiaomi Redmi 6." www.gsmarena.com.

Guess, Andrew M., Pablo Barberá, Simon Munzert, and JungHwan Yang, J. 2021. "The Consequences of Online Partisan Media." *Proceedings of the National Academy of Sciences* 118 (14). doi:10.1073/pnas.2013464118.

Guess, Andrew, and Benjamin Lyons. 2020. "Misinformation, Disinformation, and Online Propaganda." In *Social Media and Democracy: The State of the Field, Prospects for Reform*, edited by Nathaniel Persily, and Joshua A. Tucker, 10–33. Cambridge: Cambridge University Press.

Guess, Andrew, Jonathan Nagler, and Joshua Tucker. 2019. "Less than You Think: Prevalence and Predictors of Fake News Dissemination on Facebook." *Science Advances* 5 (10). doi:10.1126/sciadv.aau4586.

Guest, Peter. 2023. "Britain Admits Defeat in Controversial Fight to Break Encryption." *Wired*. https://www.wired.com.

Gundecha, Pritam, and Huan Liu. 2012. "Mining Social Media: A Brief Introduction." *New Directions in Informatics, Optimization, Logistics, and Production* (September): 1–17.

Gupta, Manjul, Denis Dennehy, Carlos M. Parra, Matti Mäntymäki, and Yogesh K. Dwivedi. 2023. "Fake News Believability: The Effects of Political Beliefs and Espoused Cultural Values." *Information & Management* 60 (2). doi:10.1016/j.im.2022.103745.

Guribye, Frode, and Lars Nyre. 2017. "The Changing Ecology of Tools for Live News Reporting." *Journalism Practice* 11 (10): 1216–30. doi:10.1080/17512786.2016 .1259011.

Gursky, Jacob, Martin J. Riedl, Katie Joseff, and Samuel Woolley. 2022. "Chat Apps and Cascade Logic: A Multi-Platform Perspective on India, Mexico, and the United States." *Social Media + Society* 8 (2). doi:10.1177/20563051221094773.

Gursky, Jacob, Martin J. Riedl, and Samuel C. Woolley. 2021. "The Disinformation Threat to Diaspora Communities in Encrypted Chat Apps." www.brookings.edu.

Gursky, Jacob, and Samuel Woolley. 2021. *Countering Disinformation and Protecting Democratic Communication on Encrypted Messaging Applications*. Washington, DC: Brookings.

Gwagwa, Arthur E. 2018. "Juste à Temps. Bandwidth Throttling in Cameroon's 2018 Elections." Global South Initiative. https://www.academia.edu/37894469/Bandwidth _throttling_in_Cameroons_2018_Elections.

Habermas, Jürgen. 1975. *Legitimation Crisis*. Boston: Beacon Press.

Haglin, Kathryn. 2017. "The Limitations of the Backfire Effect." *Research & Politics* 4 (3): 1–5. doi:10.1177/2053168017716547.

Hakim, Sharmeen. 2023. "Bombay High Court to Pronounce Verdict in Challenge to IT Rules Amendment on December 1, Centre Won't Notify Fact Check Unit till Then." www.livelaw.in.

Hampton, Keith N., Oren Livio, and Lauren Sessions Goulet. 2010. "The Social Life of Wireless Urban Spaces: Internet Use, Social Networks, and the Public Realm." *Journal of Communication* 60 (4): 701–22. doi:10.1111/j.1460–2466.2010.01510.x.

Hartmann, Martin. 2011. *Die Praxis Des Vertrauens*. Berlin: Suhrkamp.

Hassan, Idayat, and Jamie Hitchen, eds. 2022. *WhatsApp and Everyday Life in West Africa: Beyond Fake News*. London: Zed Books.

Heritage, John, and Steven Clayman. 2011. *Talk in Action: Interactions, Identities, and Institutions*. Hoboken: Wiley-Blackwell.

Hetherington, Marc J. 2005. *Why Trust Matters: Declining Political Trust and the Demise of American Liberalism*. Princeton: Princeton University Press.

Hill, Kevin A., and John E. Hughes. 1998. *Cyberpolitics: Citizen Activism in the Age of the Internet*. Lanham: Rowman & Littlefield.

Hillyard, Paddy. 1993. *Suspect Community. People's Experience of the Prevention of Terrorism Acts in Britain*. London: Pluto Press.

Hindman, Matthew. 2008. *The Myth of Digital Democracy*. Princeton: Princeton University Press.

The Hindu Businessline. 2018. "Govt Sends Second Notice to WhatsApp to Control Fake News." July 19. www.thehindubusinessline.com.

Hine, Christine. 2015. *Ethnography for the Internet: Embedded, Embodied and Everyday*. New York: Bloomsbury.

Hiropoulos, Alexandra. 2017. "Analysis: Mapping, Understanding & Preventing Xenophobic Violence in SA." http://africacheck.org.

Hitchen, Jamie, Jonathan Fisher, Idayat Hassan, and Nic Cheeseman. 2019. *WhatsApp and Nigeria's 2019 Elections: Mobilising the People, Protecting the Vote*. Birmingham: University of Birmingham.

Hlomani, Hanani, Liz Orembo, Zara Schroeder, Theresa Schültken, and Scott Timcke. 2023. *Policy Reinforcements to Counter Information Disorders in the African Context*. Policy Brief 2/2023. Cape Town: Research ICT Africa.

Horner, Brett. 2020. "Racist Rant Attributed to Trump Is a Recycled Claim Also Tied to Other Leaders in the Past." *AFP*, July 29. https://factcheck.afp.com.

Horner, Christy Galletta, Dennis Galletta, Jennifer Crawford, and Abhijeet Shirsat. 2021. "Emotions: The Unexplored Fuel of Fake News on Social Media." *Journal of Management Information Systems* 38 (4): 1039–66. doi:10.1080/07421222.2021.1990610.

Howe, Jeff. 2008. *Crowdsourcing: How the Power of the Crowd Is Driving the Future of Business*. London: Random House Business.

Huckfeldt, Robert, Jeanette Morehouse Mendez, and Tracy Osborn. 2004. "Disagreement, Ambivalence, and Engagement: The Political Consequences of Heterogeneous Networks." *Political Psychology* 25 (1): 65–95. doi:10.1111/j.1467–9221.2004.00357.x.

Hudson, James M., and Amy Bruckman. 2004. "'Go Away': Participant Objections to Being Studied and the Ethics of Chatroom Research." *Information Society* 20 (2): 127–39.

Human Rights Watch. 2009. "South Africa: Events of 2008." www.hrw.org.

———. 2022a. "Activist in Cameroon Detained Again." www.hrw.org.

———. 2022b. "Cameroon: Army Killings, Disappearances, in North-West Region." www.hrw.org.

———. 2022c. "Cameroon: Separatist Abuses in Anglophone Regions." www.hrw.org.

Hutchison, Marc L., and Kristin Johnson. 2017. "Political Trust in Sub-Saharan Africa and the Arab Region." In *Handbook on Political Trust*, edited by Sonja Zmerli, and Tom W. G. van der Meer, 461–87. Northampton: Edward Elgar Publishing.

Ikegami, Eiko. 2000. "A Sociological Theory of Publics: Identity and Culture as Emergent Properties in Networks." *Social Research* 67 (4): 989–1029.

Ikenze, Adaora. 2023. "How Meta Is Preparing for Nigeria's 2023 General Elections." https://about.fb.com.

Independent Electoral Commission. 2021. "Electoral Commission on Multi-Stakeholder Partnership to Combat Disinformation in 2021 Municipal Elections," October 19, www.gov.za.

IndiaSpend. 2018. "Child-Lifting Rumours: 33 Killed in 69 Mob Attacks since Jan 2017. Before That Only 1 Attack in 2012." www.boomlive.in.

Indo Asian News Service. 2019a. "WhatsApp Sends Cease and Desist Letters to Bogus Companies in India." *mint*, May 16. www.livemint.com.

———. 2019b. "WhatsApp Tipline of No Use for 2019 Lok Sabha Polls." *Economic Times*, April 6. https://economictimes.indiatimes.com.

Internet World Stats. 2023. "Internet World Stats: Usage and Population Statistics." www.internetworldstats.com.

Ipsos, and UNESCO. 2023. "Survey on the Impact of Online Disinformation and Hate Speech." https://www.ipsos.com.

Isaacs, Lauren. 2022. "Sanef Calls for Removal of Fake 'Tembisa 10' Story from Global Awards Shortlist." *EWN*, April 7. https://ewn.co.za.

Iyer, Neema, Bonnita Nyamwire, and Sandra Nabulega. 2020. *Alternate Realities, Alternate Internets: African Feminist Research for a Feminist Internet.* Johannesburg: Association for Progressive Communications.

Jacomy, Mathieu, Tommaso Venturini, Sebastien Heymann, and Mathieu Bastian. 2014. "ForceAtlas2, a Continuous Graph Layout Algorithm for Handy Network Visualization Designed for the Gephi Software." *PLoS ONE* 9 (6). doi:10.1371/journal.pone.0098679.

Jakesch, Maurice, Kiran Garimella, Dean Eckles, and Mor Naaman. 2021. "#Trend Alert: How a Cross-Platform Organization Manipulated Twitter Trends in the Indian General Election." In *Proceedings of the ACM on Human-Computer Interaction* 5 (CSCW2): 1–19. New York: Association for Computing Machinery. doi:10.1145/3479523.

Jayarajan, Sreedevi. 2018. "TN Horror: Mob Mistakes 65-Year-Old Woman for Child Trafficker, Lynches Her to Death." *The News Minute*, May 10. www.thenewsminute.com.

Jerit, Jennifer, and Yangzi Zhao. 2020. "Political Misinformation." *Annual Review of Political Science* 23: 77–94. doi:10.1146/annurev-polisci-050718-032814.

Johns, Amelia, and Niki Cheong. 2021. "The Affective Pressures of WhatsApp: From Safe Spaces to Conspiratorial Publics." *Continuum: Journal of Media & Cultural Studies* 35 (5): 732–46. doi:10.1080/10304312.2021.1983256.

Johnson, Constance. 2014. "Cameroon: New Law on Repression of Terrorism Passed." www.loc.gov.

Joseff, Katie, Anastasia Goodwin, and Samuel Wooley. 2020. "Nanoinfluencers Are Slyly Barnstorming the 2020 Election." *Wired*, August 15. https://www.wired.com.

Jowett, Garth S., and Victoria O'Donnell. 2012. *Propaganda and Persuasion*, 5th ed. London: SAGE Publications.

Judis, John B. 2001. *The Paradox of American Democracy: Elites, Special Interests, and the Betrayal of Public Trust.* New York: Routledge.

Kamalov, Emil, Veronika Kostenko, Ivetta Sergeeva, and Margarita Zavadskaya. 2022. *Russia's 2022 Anti-War Exodus: The Attitudes and Expectations of Russian Migrants.* PONARS Eurasia Policy Memo. Washington, DC: Elliott School of International Affairs.

Karadağ, Gökmen Hakan, and Adem Ayten. 2020. "A Comparative Study of Verification/Fact-Checking Organizatons in Turkey: Dogrulukpayi.com and Teyit.org." *Motif Akademi Halkbilimi Dergisi* 13 (29): 483–501.

Kaseke, Nyasha, and Stephen G. Hosking. 2013. "Sub-Saharan Africa Electricity Supply Inadequacy: Implications." *Eastern Africa Social Science Research Review* 29 (2): 113–32. doi:10.1353/eas.2013.0009.

Kazemi, Ashkan, Kiran Garimella, Gautam Kishore Shahi, Devin Gaffney, and Scott A. Hale. 2022. "Research Note: Tiplines to Uncover Misinformation on Encrypted Platforms: A Case Study of the 2019 Indian General Election on WhatsApp." *Harvard Kennedy School Misinformation Review* 3 (1). doi:10.37016/mr-2020-91.

Kelly, John. 2009. *Red Kayaks and Hidden Gold: The Rise, Challenges and Value of Citizen Journalism.* Oxford: Reuters Institute for the Study of Journalism.

Kertysova, Katarina. 2018. "Artificial Intelligence and Disinformation: How AI Changes the Way Disinformation Is Produced, Disseminated, and Can Be Countered." *Security and Human Rights* 29 (1–4): 55–81. doi:10.1163/18750230-02901005.

Kindzeka, Moki Edwin. 2017. "Report: Cameroon Using Anti-Terror Law to 'Silence Critics.'" *VOA News*, September 20. www.voanews.com.

Kirchgaessner, Stephanie, Manisha Ganguly, David Pegg, Carole Cadwalladr, and Jason Burke. 2023. "Revealed: The Hacking and Misinformation Team Meddling in Elections." *The Guardian*, February 15. www.theguardian.com.

Kirdemir, Baris. 2020. *Exploring Turkey's Disinformation Ecosystem.* Instanbul: EDAM.

Kleinberg, Bennett, Isabelle van der Vegt, and Paul Gill. 2021. "The Temporal Evolution of a Far-Right Forum." *Journal of Computational Social Science* 4 (1): 1–23. doi:10.1007/s42001-020-00064-x.

Kligler-Vilenchik, Neta. 2021. "Collective Social Correction: Addressing Misinformation through Group Practices of Information Verification on WhatsApp." *Digital Journalism* 10 (2): 300–18. doi:10.1080/21670811.2021.1972020.

Kligler-Vilenchik, Neta, and Ori Tenenboim. 2020. "Sustained Journalist–Audience Reciprocity in a Meso News-Space: The Case of a Journalistic WhatsApp Group." *New Media & Society* 22 (2): 264–82. doi:10.1177/1461444819856917.

Knuttila, Lee. 2011. "User Unknown: 4chan, Anonymity and Contingency." *First Monday* 16 (10). doi:10.5210/fm.v16i10.3665.

Koçer, Suncem, Bahadır Öz, Gülten Okçuoğlu, and Fezal Tapramaz. 2022. "Folk Theories of False Information: A Mixed-Methods Study in the Context of Covid-19 in Turkey." *New Media & Society.* doi:10.1177/14614448221142310.

Kollin, Zoltan. 2018. "Designing Friction for a Better User Experience." *Smashing Magazine*, January 10. www.smashingmagazine.com.

Kolluri, Nikhil, Yunong Liu, and Dhiraj Murthy. 2022. "COVID-19 Misinformation Detection: Machine-Learned Solutions to the Infodemic." *JMIR Infodemiology* 2 (2). doi:10.2196/38756.

Konings, Piet, and Francis B. Nyamnjoh. 1997. "The Anglophone Problem in Cameroon." *Journal of Modern African Studies* 35 (2): 207–29. doi:10.1017/S0022278X97002401.

———. 2019. "Anglophone Secessionist Movements in Cameroon." In *Secessionism in African Politics: Aspiration, Grievance, Performance, Disenchantment*, edited by Lotje de Vries, Pierre Englebert, and Mareike Schomerus, 59–89. Cham: Springer International Publishing.

Koren, James Rufus. 2016. "Cutting Edge: Wells Fargo Looks to Eye-Scan Security." *Los Angeles Times*, March 6. www.latimes.com.

Kouagheu, Josiane. 2021. "Cameroun: La Mort d'une Fillette Fait Craindre une Nouvelle Poussée de Violences en Zone Anglophone." *Le Monde*, October 18.

Koutsokosta, Efi. 2023. "Why the Far-Right Is Increasingly Getting into Power across Europe." *Euronews*, June 19. www.euronews.com.

Krafft, P. M., and Joan Donovan. 2020. "Disinformation by Design: The Use of Evidence Collages and Platform Filtering in a Media Manipulation Campaign." *Political Communication* 37 (2): 194–214. doi:10.1080/10584609.2019.1686094.

Krippendorff, Klaus. 2018. *Content Analysis: An Introduction to Its Methodology*. Thousand Oaks: SAGE Publications.

K. S. Puttaswamy v. Union of India. 2017. 10 SCC 1.

Kumar, Srishti. 2022. "Traceability and End-To-End Encryption: An Analysis of India's Intermediary Rules Mandating Traceability." dx.doi:10.2139/ssrn.4344579.

Kuru, Ozan, Scott W. Campbell, Joseph B. Bayer, Lemi Baruh, and Richard Ling. 2022. "Encountering and Correcting Misinformation on WhatsApp." In *Disinformation in the Global South*, edited by Herman Wasserman, and Dani Madrid-Morales, 88–107. Hoboken, NJ: Wiley-Blackwell.

Ladini, Riccardo, Moreno Mancosu, and Cristiano Vezzoni. 2020. "Electoral Participation, Disagreement, and Diversity in Social Networks: A Matter of Intimacy?" *Communication Research* 47 (7): 1056–78. doi:10.1177/0093650218792794.

Larkin, Brian. 2013. "The Politics and Poetics of Infrastructure." *Annual Review of Anthropology* 42: 327–43. doi:10.1146/annurev-anthro-092412-155522.

Leatherby, Lauren, Arielle Ray, Anjali Singhvi, Christiaan Triebert, Derek Watkins, and Haley Willis. 2021. "How a Presidential Rally Turned into a Capitol Rampage." *New York Times*, January 12. www.nytimes.com.

Lee, Ronan. 2019. "Extreme Speech | Extreme Speech in Myanmar: The Role of State Media in the Rohingya Forced Migration Crisis." *International Journal of Communication* 13.

Lenoir, Théophile, and Chris Anderson. 2023. "Introduction Essay: What Comes after Disinformation Studies." *Center for Information, Technology & Public Life, University of North Carolina at Chapel Hill*. https://citap.pubpub.org.

Levenshtein, Vladimir I. 1966. "Binary Codes Capable of Correcting Deletions, Insertions and Reversals." *Soviet Physics Doklady* 10 (8): 707–10.

Lewandowsky, Stephan, Ullrich K.H. Ecker, and John Cook. 2017. "Beyond Misinformation: Understanding and Coping with the 'Post-Truth' Era." *Journal of Applied Research in Memory and Cognition* 6 (4): 353–69. doi:10.1016/j.jarmac.2017.07.008.

Lewin, Kurt. 1947. "Frontiers in Group Dynamics: II. Channels of Group Life; Social Planning and Action Research." *Human Relations* 1 (2):5–41. doi:10.1177/001872674700100201.

Light, Ben, Jean Burgess, and Stefanie Duguay. 2018. "The Walkthrough Method: An Approach to the Study of Apps." *New Media & Society* 20 (3): 881–900.

Lindlof, Thomas R., and Bryan C. Taylor. 2017. *Qualitative Communication Research Methods*, 4th ed. Thousand Oaks: SAGE Publications.

LiveMint. 2024. "Lok Sabha Elections 2024: From WhatsApp to Social Media Influencers: Here's How Parties Are Wooing Voters." https://www.livemint.com.

Lowenkron, Laura. 2015. *O Monstro Contemporâneo: A Construção da Pedofilia em Múltiplos Planos*. Rio de Janeiro: Eduerj.

Lukito, Josephine, Jiyoun Suk, Yini Zhang, Larissa Doroshenko, Sang Jung Kim, Min-Hsin Su, Yiping Xia, Deen Freelon, and Chris Wells. 2020. "The Wolves in Sheep's Clothing: How Russia's Internet Research Agency Tweets Appeared in U.S. News as Vox Populi." *International Journal of Press/Politics* 25 (2): 196–216.

Lünenborg, Margreth, and Christoph Raetzsch. 2018. "From Public Sphere to Performative Publics: Developing Media Practice as an Analytic Model." In *Media Practices, Social Movements, and Performativity: Transdisciplinary Approaches*, edited by Susanne Foellmer, Margreth Lünenborg, and Christoph Raetzsch. New York: Routledge.

Lunt, Peter, and Sonia Livingstone. 2013. "Media Studies' Fascination with the Concept of the Public Sphere: Critical Reflections and Emerging Debates." *Media, Culture & Society* 35 (1): 87–96. doi:10.1177/0163443712464562.

Machado, Caio, Beatriz Kira, Vidya Narayanan, Bence Kollanyi, and Philip Howard. 2019. "A Study of Misinformation in WhatsApp Groups with a Focus on the Brazilian Presidential Elections." In *Companion Proceedings of the 2019 World Wide Web Conference*, 1013–19. New York: Association for Computing Machinery. doi:10.1145/3308560.3316738.

Madianou, Mirca, and Daniel Miller. 2013. "Polymedia: Towards a New Theory of Digital Media in Interpersonal Communication." *International Journal of Cultural Studies* 16 (2): 169–87.

Madrid-Morales, Dani, Herman Wasserman, Gregory Gondwe, Khulekani Ndlovu, Etse Sikanku, Melissa Tully, Emeka Umejei, and Chikezie Uzuegbunam. 2021. "Motivations for Sharing Misinformation: A Comparative Study in Six Sub-Saharan African Countries." *International Journal of Communication* 15: 1200–19.

Madung, Odanga. 2022. *From Dance App to Political Mercenary: How Disinformation on TikTok Gaslights Political Tensions in Kenya*. San Francisco: Mozilla. https://foundation.mozilla.org.

Magenta, Matheus, Juliana Gragnani, and Felipe Souza. 2018. "How WhatsApp Is Being Abused in Brazil's Elections." *BBC News*, October 24. www.bbc.com.

Majeed, Bakare. 2022. "2023: Tinubu's Chicago Affair That Won't Go Away." *Premium Times*, November 13. https://www.premiumtimesng.com.

Makananise, Fulufhelo Oscar. 2022. "Youth Experiences with News Media Consumption: The Pursuit for Newsworthy Information in the Digital Age." *Journal of African Films & Diaspora Studies* 5 (2): 29–50.

Malik, Sajjad. 2008. *Media Literacy and its Importance*. Islamabad: Society for Alternative Media and Research.

Manji, Karima, Johanna Hanefeld, Jo Vearey, Helen Walls, and Thea De Gruchy. 2021. "Using WhatsApp Messenger for Health Systems Research: A Scoping Review of Available Literature." *Health Policy and Planning* 36 (5): 774–89.

Manovich, Lev. 2001. *The Language of New Media*. Cambridge: The MIT Press.

Mare, Admire. 2023. *Digital Spaces, Rights and Responsibilities: Towards a Duty of Care Model in Southern Africa*. Digital Rights and Access to Information Series 7. Windhoek: Friedrich Ebert Stiftung Media.

Mare, Admire, and Allen Munoriyarwa. 2022. "Guardians of Truth? Fact-Checking the 'Disinfodemic' in Southern Africa during the COVID-19 Pandemic." *Journal of Media Studies* 14: 63–79.

Markowitz, David M., Timothy R. Levine, Kim B. Serota, and Alivia D. Moore. 2023. "Cross-Checking Journalistic Fact-Checkers: The Role of Sampling and Scaling in Interpreting False and Misleading Statements." *PLoS One* 18 (7). doi:10.1371/journal.pone.0289004.

Maros, Alexandre, Jussara M. Almeida, Fabrício Benevenuto, and Marisa Vasconcelos. 2020. "Analyzing the Use of Audio Messages in WhatsApp Groups." In *Proceedings of the Web Conference 2020*, 3005–11. New York: Association for Computing Machinery. doi:10.1145/3366423.3380070.

Maros, Alexandre, Jussara M. Almeida, and Marisa Vasconcelos. 2021. "A Study of Misinformation in Audio Messages Shared in WhatsApp Groups." In *Disinformation in Open Online Media: Third Multidisciplinary International Symposium*, Vol. 3, 85–100. Cham: Springer International Publishing.

Marwick, Alice E. 2018. "Why Do People Share Fake News? A Sociotechnical Model of Media Effects." *Georgetown Law Technology Review* 2 (2): 474–512.

Marwick, Alice E., and Rebecca Lewis. 2017. *Media Manipulation and Disinformation Online*. New York: Data & Society Research Institute.

Masip, Pere, Jaume Suau, Carles Ruiz-Caballero, Pablo Capilla, and Klaus Zilles. 2021. "News Engagement on Closed Platforms. Human Factors and Technological Affordances Influencing Exposure to News on WhatsApp." *Digital Journalism* 9 (8): 1062–84.

Masnick, Mike. 2022. "EU Parliament Votes to Require Internet Sites to Delete 'Terrorist Content' in One Hour (by 3 Votes)." www.techdirt.com.

Matassi, Mora, Pablo J. Boczkowski, and Eugenia Mitchelstein. 2019. "Domesticating WhatsApp: Family, Friends, Work, and Study in Everyday Communication." *New Media & Society* 21 (10): 2183–200. doi:10.1177/1461444819841890.

Maximova, Svetlana, Oksana Noyanzina, Daria Omelchenko, Irina Molodikova, and Alla Kovaleva. 2019. "The Russian Diaspora: A Result of Transit Migrations or Part of Russia." *Opción* 34 (15): 1016–44.

Melo, Philipe, Carolina Coimbra Vieira, Kiran Garimella, Pedro OS Vaz de Melo, and Fabrício Benevenuto. 2020. "Can WhatsApp Counter Misinformation by Limiting Message Forwarding?" In *Complex Networks and Their Applications VIII*, vol. 2, edited by Hocine Cherifi, Sabrina Gaito, José Fernendo Mendes, Esteban Moro, and Luis Mateus Rocha, 372–84. New York: Springer International Publishing.

Melo, Philipe, Johnnatan Messias, Gustavo Resende, Kiran Garimella, Jussara Almeida, and Fabrício Benevenuto. 2019. "WhatsApp Monitor: A Fact-Checking System for WhatsApp." In *Proceedings of the International AAAI Conference on Web and Social*

Media, vol. 13, 676–77. doi:10.1609/icwsm.v13i01.3271. Washington, DC: Association for the Advancement of Artificial Intelligence.

Meta. 2023. "WhatsApp Channels: Here's Everything You Need to Know." https://about.fb.com.

Mhlambi, Thokozani. 2019. "African Pioneer: KE Masinga and the Zulu 'Radio Voice' in the 1940s." *Journal of Radio & Audio Media* 26 (2): 210–30. doi:10.1080/19376529.2018.1468445.

Michener, Greg. 2011. "FOI Laws Around the World." *Journal of Democracy* 22 (2): 145–59.

Milan, Stefania, and Emiliano Treré. 2017. "Big Data from the South: The Beginning of a Conversation We Must Have." SSRN. dx.doi.org/10.2139/ssrn.3056958.

Miller, Daniel. 2021. "The Anthropology of Social Media." In *Digital Anthropology*, 2nd ed., edited by Haidy Geismar and Hannah Knox, 85–100. New York: Bloomsbury Academic.

Miller, Daniel, Laila Abed Rabho, Patrick Awondo, Maya Vries, Marília Duque, Pauline Garvey, et al. 2021. *The Global Smartphone: Beyond a Youth Technology*. London: UCL Press.

Ministry of Electronics & Information Technology. 2023. "The Information Technology (Intermediary Guidelines and Digital Media Ethics Code) Rules, 2021." https://www.meity.gov.in.

Molina, Paula. 2022. "La 'Brutal' Desinformación Sobre la Nueva Constitución Propuesta para Chile (y Algunas de las Confusiones más Difundidas)." *BBC News*, July 21. www.bbc.com.

Moon, Ruth. 2022. "Moto-Taxis, Drivers, Weather, and WhatsApp: Contextualizing New Technology in Rwandan Newsrooms." *Digital Journalism* 10 (9): 1569–90.

———. 2023. *Authoritarian Journalism: Controlling the News in Post-Conflict Rwanda*. Oxford: Oxford University Press.

Moreno, Andrés, Philip Garrison, and Karthik Bhat. 2017. "WhatsApp for Monitoring and Response during Critical Events: Aggie in the Ghana 2016 Election." In *Proceedings of the 14th International Conference on Information Systems for Crisis Response and Management*, edited by Tina Comes, Frédérick Bénaben, Chihab Hanachi, and Matthieu Lauras.

Mouffe, Chantal. 2000. *The Democratic Paradox*. Brooklyn: Verso Books.

Mougoué, Jacqueline B. T. 2017. "The Coffin Revolution in Cameroon." www.africasacountry.com.

Muddiman, Ashley, and Natalie Jomini Stroud. 2017. "News Values, Cognitive Biases, and Partisan Incivility in Comment Sections." *Journal of Communication* 67 (4): 586–609. doi:10.1111/jcom.12312.

Mukhopadhyay, Debabrata, and Arun Kumar Mandal. 2019. "Smartphone, Internet and Digitalization in India: An Exploratory Analysis." *PRAGATI: Journal of Indian Economy* 6 (2): 22–43. doi:10.17492/pragati.v6i2.187355.

Munoriyarwa, Allen, and Sarah H. Chiumbu. 2019. "Big Brother Is Watching: Surveillance Regulation and Its Effects on Journalistic Practices in Zimbabwe." *African Journalism Studies* 40 (3): 26–41. doi:10.1080/23743670.2020.1729831.

N., Liliane. 2021. "Le Ministère de la Défense s'Explique sur la Mort d'un Civil Survenu à Bamenda 3 (Nord-Ouest)." *Agence Cameroun Presse*, July 7.

Ndlovu, Mphathisi, and Makhosi Nkanyiso Sibanda. 2022. "Digital Technologies and the Changing Journalism Cultures in Zimbabwe: Examining the Lived Experiences of Journalists Covering the COVID-19 Pandemic." *Digital Journalism* 10 (6): 1059–78.

Ndlovu-Gatsheni, Sabelo J. 2015. "Decoloniality as the Future of Africa." *History Compass* 13 (10): 485–96. doi:10.1111/hic3.12264.

Nemer, David. 2021. "Disentangling Brazil's Disinformation Insurgency." *NACLA Report on the Americas* 53 (4), 406–13. doi:10.1080/10714839.2021.2000769.

———. 2022. *Technology of the Oppressed: Inequity and the Digital Mundane in Favelas of Brazil*. Cambridge, MA: The MIT Press.

Newcomb, Alyssa. 2018. "A Timeline of Facebook's Privacy Issues—and Its Responses." *NBC News*, March 24. www.nbcnews.com.

Newman, Nic, Richard Fletcher, Kirsten Eddy, Craig T. Robertson, and Rasmus Kleis Nielse. 2023. *Reuters Institute Digital News Report 2023*. Oxford: Reuters Institute for the Study of Journalism. https://reutersinstitute.politics.ox.ac.uk.

Newman, Nic, Richard Fletcher, Craig T. Robertson, Kristen Eddy, and Rasmus Kleis Nielsen. 2022. *Reuters Institute Digital News Report 2022*. Oxford: Reuters Institute for the Study of Journalism.

Newman, Nic, Richard Fletcher, Anne Schulz, Simge Andı, Craig T. Robertson, and Rasmus Kleis Nielsen. 2021. *Reuters Institute Digital News Report 2021*. Oxford: Reuters Institute for the Study of Journalism.

Neyazi, Taberez Ahmed. 2020. "Digital Propaganda, Political Bots and Polarized Politics in India." *Asian Journal of Communication* 30 (1): 39–57. doi:10.1080/01292986.2019.1699938.

Ng, Sheryl Wei Ting, and Taberez Ahmed Neyazi. 2023. "Self- and Social Corrections on Instant Messaging Platforms." *International Journal of Communication* 17: 426–46.

Ngange, Kingsley Lyonga, and Moki Stephen Mokondo. 2019. "Understanding Social Media's Role in Propagating Falsehood in Conflict Situations: Case of the Cameroon Anglophone Crisis." *Studies in Media and Communication* 7 (2): 55–67.

Nguyễn, Sarah, Rachel Kuo, Madhavi Reddi, Lan Li, and Rachel E. Moran. 2022. "Studying Mis- and Disinformation in Asian Diasporic Communities: The Need for Critical Transnational Research beyond Anglocentrism." *Harvard Kennedy School Misinformation Review*, March 24. doi:10.37016/mr-2020-95.

Nizaruddin, Fathima. 2021. "Role of Public WhatsApp Groups within the Hindutva Ecosystem of Hate and Narratives of 'CoronaJihad.'" *International Journal of Communication* 15 (0): 1102–19.

Nobre, Gabriel Peres, Carlos Henrique Gomes Ferreira, and Jussara Marques Almeida. 2020. "Beyond Groups: Uncovering Dynamic Communities on the WhatsApp Network of Information Dissemination." In *Proceedings of Social Informatics 12th International Conference*, edited by Samin Aref, Kalina Bontcheva, Marco Braghieri,

Frank Dignum, Fosca Giannotti, Francesco Grisolia, and Dino Pedreschi, 252–66. New York: Springer International Publishing.

———. 2022. "A Hierarchical Network-Oriented Analysis of User Participation in Misinformation Spread on WhatsApp." *Information Processing & Management* 59 (1). doi:0.48550/arXiv.2109.10462.

Nonyane, Mduduzi. 2021. "Enough Is Enough: Soweto Residents Commemorate June 16 Differently." *City Press*, June 16. www.news24.com.

Nyamnjoh, Francis B. 2010. "Racism, Ethnicity and the Media in Africa: Reflections Inspired by Studies of Xenophobia in Cameroon and South Africa." *Africa Spectrum* 45 (1): 57–93. doi:10.1177/000203971004500103.

Nyhan, Brendan, and Jason Reifler. 2010. "When Corrections Fail: The Persistence of Political Misperceptions." *Political Behaviour* 32: 303–30.

Obadare, Ebenezer, and Wendy Willems, eds. 2014. *Civic Agency in Africa: Arts of Resistance in the 21st Century.* Rochester, NY: James Currey.

Obodo, Chimere Arinze. 2022. "Free and Fair Election: A Sacrosanct Principle of Democracy?" *African Journal of Law and Human Rights* 6 (2), 192–98.

Ogbonnaya, Maurice. 2020. "#EndSARS Protest: More than a Call for Security Sector Reform in Nigeria?" https://kujenga-amani.ssrc.org/.

Ohme, Jakob, and Theo Araujo. 2022. "Digital Data Donations: A Quest for Best Practices." *Patterns* 3 (4). doi:10.1016/j.patter.2022.100467.

Okoro, Efeosasere, Benjamin Abara, Alex Umagba, Anyalewa Alan Ajonye, and Zayyad Isa. 2018. "A Hybrid Approach to Fake News Detection on Social Media." *Nigerian Journal of Technology* 37 (2): 454–62. doi:10.4314/njt.v37i2.22.

Okoro, Nnanyelugo, and Nathan Oguche Emmanuel. 2019. "Beyond Misinformation: Survival Alternatives for Nigerian Media in the 'Post-Truth' Era." *African Journalism Studies* 39 (4): 67–90. doi:10.1080/23743670.2018.1551810.

Olaniran, Samuel, and Stephanie Diepeveen. 2023. "Influencers and Incumbency: Digital Disinformation and Discontent in Nigeria's Presidential Elections." https://odi.org.

OMCT World Organisation against Torture. 2021. "Cameroon: Death Sentence Marks a New Turn in the Anglophone Crisis." www.omct.org.

Omidyar Network. 2022. "Keeping Misinformation, Hate, and Violence from Going Viral." *Omidyar Network*, January 11. https://medium.com.

Ong, Jonathan Corpus, and Jason Vincent A. Cabañes. 2018. *Architects of Networked Disinformation: Behind the Scenes of Troll Accounts and Fake News Production in the Philippines.* Communication Department Faculty Publication Series 74. Amherst: University of Massachusetts Amherst. www.newtontechfordev.com.

Operation Dudula. 2023. "Operation Dudula (@OperationDudula) / X." https://twitter.com/OperationDudula.

O'Sullivan, Tim. 1993. *Key Concepts in Communication and Cultural Studies*, 2nd ed. New York: Routledge.

Ovadya, Aviv. 2021. "On-Device Context: A Defense against Misinformation in the Encrypted WhatsApp," January 30, https://medium.com.

Oversight Board. 2022. "Oversight Board Publishes Policy Advisory Opinion on Meta's Cross-Check Program." https://oversightboard.com.

Oyebode, Oluwabunmi O., and Adeyemi Adegoju. 2017. "Appraisal Resources in Select WhatsApp Political Broadcast Messages in the 2015 Presidential Election Campaign in Nigeria." *Journal of Pan African Studies* 10 (10): 29–47.

Oyeleke, Sodiq. 2021. "NBS Report Shows FG Lying about Employment, Job Creation—PDP." *Punch*, March 16. https://punchng.com/.

Ozawa, Joao Vicente Seno, Samuel C. Woolley, Joseph Straubhaar, Martin J. Riedl, Katie Joseff, and Jacob Gursky. 2023. "How Disinformation on WhatsApp Went from Campaign Weapon to Governmental Propaganda in Brazil." *Social Media + Society* 9 (1). doi:10.1177/20563051231160632.

Palfrey, John. 2020. "The Ever-Increasing Surveillance State." *Georgetown Journal of International Affairs*, March 2. https://gjia.georgetown.edu.

Palmer, Lindsay. 2018. *Becoming the Story: War Correspondents since 9/11*. Champaign: University of Illinois Press.

Pang, Nicholas, and Yik Tung Woo. 2020. "What about WhatsApp? A Systematic Review of WhatsApp and Its Role in Civic and Political Engagement." *First Monday* 25 (January). doi:10.5210/fm.v25i12.10417.

Parks, Perry. 2022. "'Disastrous to Take a Single Note': Memory and Materiality in a Century of U.S. Journalism Textbooks." *Journalism Practice*. doi:10.1080/17512786.2022.2124435.

Parlar Dal, Emel, and Emre Erdoğan. 2021. *Küresel Siyasette Yeni Dezenformasyon Ekosistemini Anlamak*. Istandbul: EDAM.

Pascual, Manuel G. 2021. "Sudamérica Pone Firme a WhatsApp." *El País*, June 15. https://elpais.com.

Pasquetto, Irene V., Eaman Jahani, Shubham Atreja, and Matthew Baum. 2022. "Social Debunking of Misinformation on WhatsApp: The Case for Strong and In-Group Ties." In *Proceedings of the ACM on Human-Computer Interaction* 6 (CSCW1): 1–35. New York: Association for Computing Machinery. doi:10.1145/3512964.

Pasquetto, Irene V., Alberto F. Olivieri, Luca Tacchetti, Gianni Riotta, and Alessandra Spada. 2022. "Disinformation as Infrastructure: Making and Maintaining the QAnon Conspiracy on Italian Digital Media." In *Proceedings of the ACM on Human-Computer Interaction* 6 (CSCW1): 1–31. New York: Association for Computing Machinery. doi:10.1145/3512931.

Pelican, Michaela. 2022. *The Anglophone Conflict in Cameroon—Historical and Political Background*. ABI Working Paper 20. www.arnold-bergstraesser.de.

Pennycook, Gordon, and David G. Rand. 2021. "The Psychology of Fake News." *Trends in Cognitive Sciences* 25 (5): 388–402. doi:10.1016/j.tics.2021.02.007.

Perrigo, Billy. 2019. "How Volunteers for India's Ruling Party Are Using WhatsApp to Fuel Fake News ahead of Elections." *Time*, January 25.

Pew Research Center. 2023. "Public Trust in Government: 1958–2022." www.pewresearch.org.

Piaia, Victor, Eurico Matos, Tatiana Dourado, Polyana Barboza, and Sabrina Almeida. 2022. "Ethical Issues in WhatsApp Research: Notes on Political Communication Studies in Brazil." *Revue Française des Sciences de l'Information et de la Communication* 25. doi:10.4000/rfsic.13328.

Picardo, Jacobo, Sarah K. McKenzie, Sunny Collings, and Gabrielle Jenkin. 2020. "Suicide and Self-Harm Content on Instagram: A Systematic Scoping Review." *PLOS ONE* 15 (9). doi:10.1371/journal.pone.0238603.

Pindayi, Brian. 2017. "Social Media Uses and Effects: The Case of WhatsApp in Africa." In *Impacts of the Media on African Socio-Economic Development*, edited by Okorie Nelson, Babatunde Raphael Ojebuyi, and Abiodun Salawu, 34–51. Hershey: IGI Global.

Pinheiro-Machado, Rosana, and Tatiana Vargas-Maia. 2023. *The Rise of the Radical Right in the Global South*. New York: Routledge.

Pizzi, Michael. 2015. "Case of Vice Reporters Underlines Turkish Crackdown on Internet Freedom." *Al Jazeera English*, August 3. http://america.aljazeera.com.

Plantin, Jean-Christophe, and Aswin Punathambekar. 2019. "Digital Media Infrastructures: Pipes, Platforms, and Politics." *Media, Culture & Society* 41 (2): 163–74.

Platteau, Jean-Philippe. 1996. "Physical Infrastructure as a Constraint on Agricultural Growth: The Case of Sub-Saharan Africa." *Oxford Development Studies* 24 (3): 189–219. doi:10.1080/13600819608424113.

Pocyte, Agniete. 2019. "From Russia with Fear: The Presence of Emotion in Russian Disinformation Tweets." Masters diss., Charles University.

Pohjonen, Matti, and Sahana Udupa. 2017. "Extreme Speech Online: An Anthropological Critique of Hate Speech Debates." *International Journal of Communication* 11: 1173–91.

Pommerolle, Marie-Emmanuelle, and Hans De Marie Heungoup. 2017. "The 'Anglophone Crisis': A Tale of the Cameroonian Postcolony." *African Affairs* 116 (464): 526–38. doi:10.1093/afraf/adx021.

Porteous, Robyn. 2023. "Fighting for SA's Future in the Face of Fake News." *Daily Maverick*, April 25. www.dailymaverick.co.za.

Postal & Telecommunications Regulatory Authority of Zimbabwe (POTRAZ). 2022. *Sector Performance Report: 4th Quarter 2022*. Harare: POTRAZ. www.potraz.gov.zw.

PTI. 2018a. "BJP Declares Rs 1,027-Crore Income in FY18: Report." *Economic Times*, December 17. https://economictimes.indiatimes.com.

———. 2018b. "WhatsApp Appoints Grievance Officer for India." *Economic Times*, September 24. https://economictimes.indiatimes.com.

———. 2019. "Facebook Removes Nearly 700 Pages Linked to Congress Due to 'Inauthentic Behavior.'" *Economic Times*, April 1. https://economictimes.indiatimes.com.

Purdeková, Andrea. 2015. *Making Ubumwe: Power, State and Camps in Rwanda's Unity-Building Project*. New York: Berghahn Books.

———. 2016. "'Mundane Sights' of Power: The History of Social Monitoring and Its Subversion in Rwanda." *African Studies Review* 59 (2): 59–86.

Rebelo, Karen. 2017. "Child Kidnapping Rumours in India Being Spread with Syria Image, Pak Video." www.boomlive.in.

———. 2018a. "Inside WhatsApp's Battle against Misinformation in India." www
.poynter.org.

———. 2018b. "Pakistani Ad Agency behind Child Kidnapping Video Breaks Its Silence." www.boomlive.in.

Recureo, Raquel, Felipe Soares, and Otávio Vinhas. 2021. "Discursive Strategies for
Disinformation on WhatsApp and Twitter during the 2018 Brazilian Presidential
Election." *First Monday* 26 (1). doi:10.5210/fm.v26i1.10551.

Reich, Zvi, and Aviv Barnoy. 2016. *Reconstructing Production Practices through Interviewing*. London: SAGE Publications.

Reis, Julio C. S., Philipe Melo, Kiran Garimella, and Fabrício Benevenuto. 2020. "Can
WhatsApp Benefit from Debunked Fact-Checked Stories to Reduce Misinformation?"
Harvard Kennedy School Misinformation Review 1 (5). doi:10.37016/mr-2020–035.

Rensburg, Ronel. 2019. "State Capture and the Demise of Bell Pottinger." In *The Global
Public Relations Handbook: Theory, Research, and Practice*, 3rd ed., edited by Krish-
namurthy Sriramesh, and Dejan Verčič. New York: Routledge.

Resende, Gustavo, Philipe Melo, Julio C. S. Reis, Marisa Vasconcelos, Jussara M.
Almeida, and Fabrício Benevenuto. 2019a. "Analyzing Textual (Mis)Informa-
tion Shared in WhatsApp Groups." In *Proceedings of the 10th ACM Conference
on Web Science*, 225–34. New York: Association for Computing Machinery.
doi:10.1145/3292522.3326029.

Resende, Gustavo, Philipe Melo, Hugo Sousa, Johnnatan Messias, Marisa Vasconcelos,
Jussara Almeida, and Fabrício Benevenuto. 2019b. "(Mis)Information Dissemination in WhatsApp: Gathering, Analyzing and Countermeasures." In *Proceedings
of the World Wide Web Conference*, 818–28. New York: Association for Computing
Machinery. doi:10.1145/3308558.3313688.

Retis, Jessica. 2019. "Homogenizing Heterogeneity in Transnational Contexts: Latin
American Diasporas and the Media in the Global North." In *The Handbook of Dias-
poras, Media, and Culture*, edited by Jessica Retis, and Roza Tsagarousianou, 113–36.
Hoboken, NJ: John Wiley & Sons.

Reuters, and Hindustan Times. 2018. "'He Looked like a Terrorist': How a Drive in Karnataka Ended in Mob Lynching." *Hindustan Times*, July 30. www.hindustantimes.com.

Richardson, Heather Cox. 2020. *How the South Won the Civil War: Oligarchy, Democracy, and the Continuing Fight for the Soul of America*, illus. ed. Oxford: Oxford
University Press.

Richardson, Leonard. 2023. "Beautiful Soup." www.crummy.com.

Riedl, Martin J., Joao Vicente Seno Ozawa, Samuel C. Woolley, and Kiran Garimella.
2022. *Talking Politics on WhatsApp: A Survey of Cuban, Indian, and Mexican
American Diaspora Communities in the United States*. Austin, TX: Center for Media
Engagement.

Robards, Brady, and Siân Lincoln. 2017. "Uncovering Longitudinal Life Narratives:
Scrolling Back on Facebook." *Qualitative Research* 17 (6): 715–30.

Rocha, Graciliano. 2022. "Bolsonaro va Tras los Pasos de Trump." *Washington Post*,
November 17. www.washingtonpost.com.

Rodgers, Scott. 2015. "Foreign Objects? Web Content Management Systems, Journalistic Cultures and the Ontology of Software." *Journalism* 16 (1). doi:10.1177/1464884914545729.

Rodny-Gumede, Yiva. 2018. "Fake It till You Make It: The Role, Impact and Consequences of Fake News." In *Perspectives on Political Communication in Africa*, edited by Bruce Mutsvairo, and Beschara Karam, 203–19. Cham: Palgrave Macmillan.

Rodriguez-Amat, Joan Ramon, and Yulia Belinskaya. 2023. "Desflecando Telegram: Un Campo de Batalla en la Esfera Pública Desbordada Rusa." In *Comunicación Política, Discursos Conspiranoicos y Movilización Social*, edited by Alberto Carratalá, Gemma López-García, and Marta Iranzo-Cabrera. Valencia: Tirant Lo Blanch.

Rogers, Richard. 2020. "Deplatforming: Following Extreme Internet Celebrities to Telegram and Alternative Social Media." *European Journal of Communication* 35 (3): 213–29. doi:10.1177/0267323120922066.

Roper, Chris. 2022. "South Africa." In *Reuters Institute Digital News Report 2022*, edited by Nic Newman, Richard Fletcher, Craig T. Robertson, Kirsten Eddy, and Rasmus Kleis Nielsen. Oxford: Reuters Institute for the Study of Journalism.

Rossini, Patrícia, Jennifer Stromer-Galley, Erica Anita Baptista, and Vanessa Veiga de Oliveira. 2021. "Dysfunctional Information Sharing on WhatsApp and Facebook: The Role of Political Talk, Cross-Cutting Exposure and Social Corrections." *New Media & Society* 23 (8): 2430–51. doi:10.1177/1461444820928059.

Rousset, Thierry, Gavaza Maluleke, and Adam Mendelsohn. 2022. "The Dynamics of Racism, Antisemitism and Xenophobia on Social Media in South Africa." www .jhbholocaust.co.za.

Ryfe, David M., and Markus Kemmelmeier. 2011. "Quoting Practices, Path Dependency and the Birth of Modern Journalism." *Journalism Studies* 12 (1): 10–26. doi:10.1080/1 461670X.2010.511943.

Saha, Punyajoy, Binny Mathew, Kiran Garimella, and Animesh Mukherjee. 2021. "'Short Is the Road that Leads from Fear to Hate': Fear Speech in Indian WhatsApp Groups." In *Proceedings of the Web Conference 2021*, 1110–21. New York: Association for Computing Machinery. doi:10.1145/3442381.3450137.

Sahlins, Marshall. 2014. *What Kinship Is—And Is Not*. Chicago: University of Chicago Press.

Sahoo, Sananda. 2022. "Political Posters Reveal a Tension in WhatsApp Platform Design: An Analysis of Digital Images from India's 2019 Elections." *Television & New Media* 23 (8): 874–99. doi:10.1177/15274764211052997.

Saka, Erkan. 2018. "Social Media in Turkey as a Space for Political Battles: AKTrolls and Other Politically Motivated Trolling." *Middle East Critique* 27 (2): 161–77. doi:10 .1080/19436149.2018.1439271.

———. 2021a. "Networks of Political Trolling in Turkey after the Consolidation of Power Under the Presidency." In *Digital Hate: The Global Conjuncture of Extreme Speech*, edited by Sahana Udupa, Iginio Gagliardone, and Peter Hervik, 240–55. Bloomington: Indiana University Press.

———. 2021b. "Research on Disinformation about EU in Turkey." IPA/2020/422-363. Delegation of the European Union to Turkey.

Salazar, Gabriel. 2019. "El 'Reventón Social' en Chile: Una Mirada Histórica." *Nueva Sociedad*, October 24. https://nuso.org.

Salikov, Alexey. 2019. "Telegram as a Means of Political Communication and Its Use by Russia's Ruling Elite." *Politologija* 3 (95): 83–110.

Saliu, Hassan A., and Solomon I. Ifejika. 2017. "The Independent National Electoral Commission (INEC) and the 2011 Elections: A Non-Romantic View." *South East Political Science Review* 1 (1): 268–95.

Santini, Rose Marie, Giulia Tucci, Débora Salles, and Alda Rosana D. de Almeida. 2021. "Do You Believe in Fake after All? WhatsApp Disinformation Campaign during the Brazilian 2018 Presidential Election." In *Politics of Disinformation: The Influence of Fake News on the Public Sphere*, edited by Guillermo López-García, Dolors Palau-Sampio, Bella Palomo, Eva Campos-Domínguez, and Pere Masip, 49–66. Hoboken, NJ: John Wiley & Sons.

Santos, Boaventura de Sousa. 2012. "Public Sphere and Epistemologies of the South." *Africa Development* 37 (1): 43–67.

Santos, Marcelo, and Plataforma Telar. 2022. "Un Año de Desinformación en la Convención Constitucional." *Plataforma Telar*, July 13.

Sathe, Gopal. 2019. "How the BJP Automated Political Propaganda on WhatsApp." *HuffPost*, April 16. www.huffpost.com.

Satish, Balla. 2018. "How WhatsApp Helped Turn an Indian Village into a Lynch Mob." *BBC News*, July 18. www.bbc.com.

Sauer, Pjotr. 2022. "Russia Bans Facebook and Instagram under 'Extremism' Law." *The Guardian*, March 21, sec. World news. www.theguardian.com.

Sawchuk, Kim. 2000. "No Logo: Taking Aim at the Brand Bullies." Review of *No Logo: Taking Aim at the Brand Bullies*, by Naomi Klein. *Canadian Journal of Communication* 25 (4): 585–88. doi:10.22230/cjc.2000v25n4a1186.

Scherman, Andrés, Nicolle Etchegaray, Magdalena Browne, Diego Mazorra, and Hernando Rojas. 2022. "WhatsApp, Polarization, and Non-Conventional Political Participation: Chile and Colombia Before the Social Outbursts of 2019." *Media and Communication* 10 (4): 77–93. doi:10.17645/mac.v10i4.5817.

Schneider, Anne, and Helen Ingram. 1993. "Social Construction of Target Populations. Implications for Politics and Policy." *American Political Science Review* 87 (2): 334–47.

Schoon, Alette, Hayes Mawindi Mabweazara, Tanja Bosch, and Harry Dugmore. 2020. "Decolonising Digital Media Research Methods: Positioning African Digital Experiences as Epistemic Sites of Knowledge Production." *African Journalism Studies* 41 (4): 1–15.

Schrape, Jan-Felix. 2016. *Social Media, Mass Media and the "Public Sphere." Differentiation, Complementarity and Co-Existence*. SOI Discussion Paper 2016–01. doi:10.2139/ssrn.2858891.

Schreier, Margrit. 2012. *Qualitative Content Analysis in Practice*. Thousand Oaks: SAGE Publications.

Schumann, Kim, Roxana Willis, James Angove, and Caroline Mbinkar. 2023. *Cameroon Conflict Human Rights Report 2022/23: From January 2021 to April 2023*. Oxford: Oxford Human Rights Hub.

Semenzin, Silvia, and Lucia Bainotti. 2020. "The Use of Telegram for Non-Consensual Dissemination of Intimate Images: Gendered Affordances and the Construction of Masculinities." *Social Media + Society* 6 (4): 2056305120984453. doi:10.1177/2056305120984453.

Senado.cl. 2019. "Acuerdo Por la Paz Social y la Nueva Constitución." www.senado.cl.

Shapiro, Ivor, Colette Brin, Isabelle Bédard-Brûlé, and Kasia Mychajlowycz. 2013. "Verification as a Strategic Ritual: How Journalists Retrospectively Describe Processes for Ensuring Accuracy." *Journalism Practice* 7 (6): 657–73. doi:10.1080/17512786.2013.765638.

Shekhar, Divya. 2017. "In a Post-Truth World, 'Check4Spam' Can Help You Sift Fake News." *Economic Times*, February 22. https://economictimes.indiatimes.com.

Sheller, Mimi. 2015. "News Now: Interface, Ambience, Flow, and the Disruptive Spatio-Temporalities of Mobile News Media." *Journalism Studies* 16 (1): 12–26. doi:10.1080/1461670X.2014.890324.

Shilongo, Kristophina, Araba Sey, and Hanani Hlomani. 2022. *Responses to Information Disorders: What Can Governments Do?* Cape Town: Research ICT Africa.

Shmargad, Yotam, Stephen A. Rains, Kevin Coe, Kate Kenski, and Steven Bethard. 2024. "Detecting Antisocial Norms in Large-Scale Online Discussions." In *Social Processes of Online Hate*, edited by Joseph B. Walther and Ronald Rice, 220–49. New York: Routledge.

Simmon, Erica S., and Nicholas Rush Smith. 2019. "The Case for Comparative Ethnography." *Comparative Politics* 51 (3): 341–59.

Simone, AbdouMaliq. 2018. *Improvised Lives: Rhythms of Endurance in an Urban South*. Cambridge: Polity Press.

Sinclair, Betsy. 2012. *The Social Citizen: Peer Networks and Political Behavior*. Chicago: University of Chicago Press.

Singer, Barbara, Caitlin M. Walsh, Lucky Gondwe, Katie Reynolds, Emily Lawrence, and Alinafe Kasiya. 2023. "WhatsApp as a Medium to Collect Qualitative Data among Adolescents: Lessons Learned and Considerations for Future Use." *Gates Open Research* 4 (130). https://doi.org/10.12688/gatesopenres.13169.2.

Singh, Shivam Shankar. 2019. *How to Win an Indian Election: What Political Parties Don't Want You to Know*. London: Ebury Press.

Smith, Thandi, and William Bird. 2020. "Disinformation in a Time of Covid-19: Weekly Trends in South Africa." *Daily Maverick*, December 6. https://www.dailymaverick.co.za.

Soares, Felipe, Raquel Recuero, Taiane Volcan, Giane Fagundes, and Giéle Sodré. 2021. "Research Note: Bolsonaro's Firehose: How Covid-19 Disinformation on WhatsApp Was Used to Fight a Government Political Crisis in Brazil." *Harvard Kennedy School Misinformation Review* 2 (January). doi:10.37016/mr-2020-54.

Social Science Research Council. n.d. "Harnessing Influencers to Counter Misinformation: Scalable Solutions in the Global South." www.ssrc.org.

Solomon, Hussein, and Hitomi Kosaka. 2013. "Xenophobia in South Africa: Reflections, Narratives and Recommendations." *Southern African Peace and Security Studies* 2 (2): 5–30.

South Africa History Archive. 2002. "SAHA—South African History Archive—You Strike the Women, You Strike the Rock!" www.saha.org.za.

South African History Online. 2015. "Xenophobic Violence in Democratic South Africa Timeline." www.sahistory.org.za.

South African National Editors' Forum. 2021. "Disinformation » SANEF | Protecting Media Freedom." https://sanef.org.za.

Southwood, Russell. 2022. *Africa 2.0: Inside a Continent's Communications Revolution.* Manchester: Manchester University Press.

Squires, Catherine R. 2002. "Rethinking the Black Public Sphere: An Alternative Vocabulary for Multiple Public Spheres." *Communication Theory* 12 (4): 446–68. doi:10.1111/j.1468–2885.2002.tb00278.x.

Sridhar, Nakul, and Vidhi Choudhary. 2018. "Ensure Steps to Prevent Spread of Rumours, Government Warns WhatsApp." *Hindustan Times*, July 4. www.hindustantimes.com.

Srinivasan, Ramesh. 2006. "Indigenous, Ethnic and Cultural Articulations of New Media." *International Journal of Cultural Studies* 9 (4): 497–518. doi:10.1177/1367877906069899.

Srivastava, Mehul, and Tom Wilson. 2019. "Inside the WhatsApp Hack: How an Israeli Technology Was Used to Spy." *Financial Times*, October 30. www.ft.com.

Srivastava, Vivek, and Mayank Singh. 2021. "PoliWAM: An Exploration of a Large Scale Corpus of Political Discussions on WhatsApp Messenger." In *Proceedings of the 7th Workshop on Noisy User-Generated Text*, 120–30. doi:10.18653/v1/2021.wnut-1.15.

Staender, Anna, and Edda Humprecht. 2021. "Types (Disinformation)." *DOCA-Database of Variables for Content Analysis.*

Stalcup, Meg. 2016. "The Aesthetic Politics of Unfinished Media: New Media Activism in Brazil." *Visual Anthropology Review* 32 (2): 144–56. doi.org/10.1111/var.12106.

Star, Susan Leigh. 1999. "The Ethnography of Infrastructure." *American Behavioral Scientist* 43 (3): 377–91.

Star, Susan Leigh, and Karen Ruhleder. 1996. "Steps toward an Ecology of Infrastructure: Design and Access for Large Information Spaces." *Information Systems Research* 7 (1): 111–34. doi:10.1287/isre.7.1.111.

Starbird, Kate, Ahmer Arif, and Tom Wilson. 2019. "Disinformation as Collaborative Work: Surfacing the Participatory Nature of Strategic Information Operations." In *Proceedings of the ACM on Human-Computer Interaction* 3 (CSCW), 1–26. New York: Association for Computing Machinery.

Statista. 2022a. "Nigeria: Leading Social Media Platforms 2022." www.statista.com.

———. 2022b. "Share of Internet Users in Africa as of January 2022, by Country." www.statista.com.

———. 2022c. "Social Media Sites Most Used as News Sources U.S. 2022." www.statista.com.

Statistik Austria. 2023. *Migration & Integration 2023 Statistisches Jahrbuch—Hauptergebnisse*, edited by Tobias Thomas. Statistik Austria. https://www.statistik.at.

Staudacher, Sandra, and Andrea Kaiser-Grolimund. 2016. "WhatsApp in Ethnographic Research: Methodological Nemer on New Edges of the Field." *Basel Papers on Political Transformations* (10): 24–40.

Steiger, Miriah, Timir J. Bharucha, Sukrit Venkatagiri, Martin J. Riedl, and Matthew Lease. 2021. "The Psychological Well-Being of Content Moderators: The Emotional Labor of Commercial Moderation and Avenues for Improving Support." In *Proceedings of the 2021 CHI Conference on Human Factors in Computing Systems*, 1–14.

Stroud, Natalie Jomini. 2010. "Polarization and Partisan Selective Exposure." *Journal of Communication* 60 (3): 556–76. doi:10.1111/j.1460-2466.2010.01497.x.

Suiter, Jane. 2016. "Post-Truth Politics." *Political Insight* 7 (3): 25–7.

Sun, Wanning, and Haiqing Yu. 2020. "WeChatting the Australian Election: Mandarin-Speaking Migrants and the Teaching of New Citizenship Practices." *Social Media + Society* 6 (1). doi:10.1177/2056305120903441.

Sunstein, Cass R. 2017. *#Republic: Divided Democracy in the Age of Social Media*. Princeton: Princeton University Press.

Swire-Thompson, Briony, Joseph DeGutis, and David Lazer. 2020. "Searching for the Backfire Effect: Measurement and Design Considerations." *Journal of Applied Research in Memory and Cognition* 9 (3): 286–99.

Tande, Dibussi. 2011a. "Cameroon's Diaspora-Driven Cyberspace: A Platform for Civic Engagement or a Threat to the Biya Regime?" www.dibussi.com.

———. 2011b. "The Biya Regime and Cameroon's Diaspora-Driven Cyberspace (Part 2)." www.dibussi.com.

Tansey, Oisín. 2007. "Process Tracing and Elite Interviewing: A Case for Non-Probability Sampling." *PS: Political Science & Politics* 40 (4): 765–72.

Tardáguila, Cristina, Fabrício Benevenuto, and Pablo Ortellado. 2018. "Fake News Is Poisoning Brazilian Politics. WhatsApp Can Stop It." *New York Times*, October 17. https://www.nytimes.com.

Taussig, M. 2020. *Mastery of Non-Mastery in the Age of Meltdown*. Chicago: University of Chicago Press.

Teixeira, Jacqueline M. 2022. "Masculinidade e pentecostalismo como tecnologia neoliberal." *Contemporânea—Revista de Sociologia da UFSCAR* 12: 743–67.

Tenhunen, Sirpa. 2018. *A Village Goes Mobile: Telephony, Mediation and Social Change in Rural India*. New York: Oxford University Press.

Timcke, Scott. 2022. "WhatsApp in African Trade Networks: Professional Practice and Obtaining Attention in AfCFTA Policy Formation." *First Monday* 27 (10). doi:10.5210/fm.v27i10.11610.

Timcke, Scott, Liz Orembo, and Hanani Hlomani. 2023. *Information Disorders in Africa: An Annotated Bibliography of Selected Countries*. Cape Town: Research ICT Africa.

Timcke, Scott, Liz Orembo, Hanani Hlomani, and Theresa Schültken. 2023. *The Materials of Misinformation on the African Continent: Mid-Year Report, 2023*. Cape Town: Research ICT Africa.

Tontodimamma, Alice, Eugenia Nissi, Annalina Sarra, and Lara Fontanella. 2021. "Thirty Years of Research into Hate Speech: Topics of Interest and Their Evolution." *Scientometrics* 126 (1): 157–79. doi:10.1007/s11192-020-03737-6.

Treré, Emiliano. 2018. *Hybrid Media Activism: Ecologies, Imaginaries, Algorithms*. New York: Routledge.

———. 2020. "The Banality of WhatsApp: On the Everyday Politics of Backstage Activism in Mexico and Spain." *First Monday* 25 (January). doi:10.5210/fm.v25i12.10404.

Tripp, Aili Mari. 2004. "The Changing Face of Authoritarianism in Africa: The Case of Uganda." *Africa Today* 50 (3): 3–26.

TRT Akademi. 2020. "TRT Akademi İnfodemi Araştırması Raporu." https://trtakademi .net.

Tucker, Joshua A., Pablo Barberá, Cristian Vaccari, Alexandra Siegel, Sergey Sanovich, Denis Stukal, and Brendan Nyhan. 2018. *Social Media, Political Polarization, and Political Disinformation: A Review of the Scientific Literature*. Menlo Park, CA: Hewlett Foundation.

Tufekci, Zeynep. 2014. "Big Questions for Social Media Big Data: Representativeness, Validity and Other Methodological Pitfalls." In *Proceedings of the 8th International AAAI Conference on Weblogs and Social Media*, vol. 8, 505–14. Washington, DC: Association for the Advancement of Artificial Intelligence.

———. 2017. *Twitter and Tear Gas: The Power and Fragility of Networked Protest*. New Haven, CT: Yale University Press.

Tully, Melissa, Dani Madrid-Morales, Herman Wasserman, Gregory Gondwe, and Kioko Ireri. 2021. "Who Is Responsible for Stopping the Spread of Misinformation? Examining Audience Perceptions of Responsibilities and Responses in Six Sub-Saharan African Countries." *Digital Journalism* 10 (5): 679–97. doi:10.1080/21670811 .2021.1965491.

Tyagi, Nirvan, Ian Miers, and Thomas Ristenpart. 2019. "Traceback for End-to-End Encrypted Messaging." In *Proceedings of the 2019 ACM SIGSAC Conference on Computer and Communications Security*, 413–30. New York: Association for Computing Machinery.

Udupa, Sahana. 2015. "Internet Hindus." In *Handbook of Religion and the Asian City: Aspiration and Urbanization in the Twenty-First Century*, edited by Peter van der Veer. Oakland: University of California Press.

———. 2018a. "Enterprise Hindutva and Social Media in Urban India." *Contemporary South Asia* 26 (4): 453–67.

———. 2018b. "Gaali Cultures: The Politics of Abusive Exchange on Social Media." *New Media & Society* 20 (4): 1506–22. doi:10.1177/1461444817698776.

———. 2019. "India Needs a Fresh Strategy to Tackle Online Extreme Speech." *Economic & Political Weekly* 54 (4, January 26).

———. 2023. "Extreme Speech." In *Challenges and Perspectives of Hate Speech Research*, edited by Christian Strippel, Sünje Paasch-Colberg, Martin Emmer, and Joachim Trebbe, 233–48. Berlin: Digital Communication Research.

Udupa, Sahana, and Ethiraj Gabriel Dattatreyan. 2023. *Digital Unsettling: Decoloniality and Dispossession in the Age of Social Media.* New York: NYU Press.

Udupa, Sahana, Iginio Gagliardone, and Peter Hervik, eds. 2021. *Digital Hate: The Global Conjuncture of Extreme Speech.* Chicago: Indiana University Press.

Udupa, Sahana, and Max Kramer. 2023. "Multiple Interfaces." *American Ethnologist* 50 (2): 247–59. doi:10.1111/amet.13117.

Udupa, Sahana, Antonis Maronikolakis, and Axel Wisiorek. 2023. "Ethical Scaling for Content Moderation: Extreme Speech and the (In)significance of Artificial Intelligence." *Big Data & Society* 10 (1). doi:10.1177/20539517231172424.

Udupa, Sahana, and Matti Pohjonen. 2019. "Extreme Speech | Extreme Speech and Global Digital Cultures—Introduction." *International Journal of Communication* 13: 3049–67.

Umar, Rusydi, Imam Riadi, and Guntur Maulana Zamroni. 2018. "Mobile Forensic Tools Evaluation for Digital Crime Investigation." *International Journal on Advanced Science, Engineering and Information Technology* 8 (3): 949–55.

Ünal, Recep, and Alp Şahin Çiçeklioğlu. 2019. "The Function and Importance of Fact-Checking Organizations in the Era of Fake News: Teyit.org, an Example from Turkey." *Media Studies* 10 (19): 140–60.

UNESCO. 2023. "Safeguarding Freedom of Expression and Access to Information: Guidelines for a Multistakeholder Approach in the Context of Regulating Digital Platforms." In *Proceedings of the Internet for Trust Conference.* Paris: UNESCO.

United Kingdom. 2019. *Online Harms White Paper.* London: HM Government. https://www.gov.uk/government/consultations/online-harms-white-paper.

United Nations. 2022. "South Africa: UN Experts Condemn Xenophobic Violence and Racial Discrimination against Foreign Nationals." www.ohchr.org.

United Nations High Commissioner for Refugees. n.d. "Refugees from Ukraine Recorded across Europe." https://data.unhcr.org.

Unver, Akin. 2019. *Russian Disinformation Ecosystem in Turkey.* Istandbul: EDAM.

———. 2020. *Fact-Checkers and Fact-Checking in Turkey.* EDAM Research Reports. Istandbul: EDAM.

USENIX Enigma Conference. 2017. "USENIX Enigma 2017—How WhatsApp Reduced Spam while Launching End-to-End Encryption." YouTube video, posted March 6. https://youtu.be/LBTOKlrhKXk.

Valenzuela, Sebastián. 2013. "Unpacking the Use of Social Media for Protest Behavior: The Roles of Information, Opinion Expression, and Activism." *American Behavioral Scientist* 57 (7): 920–42. doi:10.1177/0002764213479375.

Valenzuela, Sebastián, Ingrid Bachmann, and Matías Bargsted. 2021. "The Personal Is the Political? What Do WhatsApp Users Share and How It Matters for News Knowledge, Polarization and Participation in Chile." *Digital Journalism* 9 (2): 155–75.

Valenzuela, Sebastián, Daniel Halpern, James E. Katz, and Juan Pablo Miranda. 2019. "The Paradox of Participation versus Misinformation: Social Media, Political

Engagement, and the Spread of Misinformation." *Digital Journalism*, 7 (6): 802–23. doi:10.1080/21670811.2019.1623701.

Valenzuela, Sebastián, Carlos Muñiz, and Marcelo Santos. 2022. "Social Media and Belief in Misinformation in Mexico: A Case of Maximal Panic, Minimal Effects?" *International Journal of Press/Politics*. doi:10.1177/19401612221088988.

Valeriani, Augusto, and Cristian Vaccari. 2017. "Political Talk on Mobile Instant Messaging Services: A Comparative Analysis of Germany, Italy, and the UK." *Information, Communication & Society* 21 (11): 1715–31. doi:10.1080/1369118X.2017.1350730.

Van Dijck, José, Thomas Poell, and Martijn de Waal. 2018. *The Platform Society: Public Values in a Connective World*. New York: Oxford University Press.

Vanlıoğlu, Muhammet. 2018. "Siber Propaganda ve Dezenformasyon: Kitle Kaynaklı Troll Birimleri." *Uluslararası Kriz ve Siyaset Araştırmaları Dergisi* 2 (1): 206–35.

Van Leeuwen, Theo. 2007. "Legitimation in Discourse and Communication." *Discourse & Communication* 1 (1): 91–112.

Vasconcelos, Marisa, Erica Pereira, Samuel Guimarães, Manoel Horta Ribeiro, Philipe Melo, and Fabrício Benevenuto. 2020. "Analyzing YouTube Videos Shared on WhatsApp in the Early COVID-19 Crisis." *Education* 6 (1): 4–93.

Vasudeva, Feeza, and Nicholas Barkdull. 2020. "WhatsApp in India? A Case Study of Social Media Related Lynchings." *Social Identities* 26 (5): 574–89. doi:10.1080/135046 30.2020.1782730.

Velasquez, Alcides, Andrea M. Quenette, and Hernando Rojas. 2021. "WhatsApp Political Expression and Political Participation: The Role of Ethnic Minorities' Group Solidarity and Political Talk Ethnic Heterogeneity." *International Journal of Communication* 15: 22.

Verstraeten, Hans. 1996. "The Media and the Transformation of the Public Sphere: A Contribution for a Critical Political Economy of the Public Sphere." *European Journal of Communication* 11 (3): 347–70. doi:10.1177/0267323196011003004.

Villi, Mikko, Tali Aharoni, Keren Tenenboim-Weinblatt, Pablo J. Boczkowski, Kaori Hayashi, Eugenia Mitchelstein, Akira Tanaka, and Neta Kligler-Vilenchik. 2022. "Taking a Break from News: A Five-Nation Study of News Avoidance in the Digital Era." *Digital Journalism* 10 (1): 148–64. doi:10.1080/21670811.2021.1904266.

Vosoughi, Soroush, Deb Roy, and Sinan Aral. 2018. "The Spread of True and False News Online." *Science* 359 (6380): 1146–51. doi:10.1126/science.aap9559.

Walther, Joseph, and Ronald E. Rice, eds. 2024. *Social Processes of Hate*. New York: Routledge.

Wang, Yang, Pedro Giovanni Leon, Kevin Scott, Xiaoxuan Chen, Alessandro Acquisti, and Lorrie Faith Cranor. 2013. "Privacy Nudges for Social Media: An Exploratory Facebook Study." In *Proceedings of the 22nd International Conference on World Wide Web*, 763–70. doi:10.1145/2487788.2488038.

Wardle, Claire, and Hossein Derakhshan. 2017. *Information Disorder: Toward an Interdisciplinary Framework for Research and Policy Making*. Council of Europe report DGI(2017)09. Akindipe Strasbourg: Council of Europe.

Warner, Michael. 2021. *Publics and Counterpublics*. Princeton: Princeton University Press.

Wasserman, Herman. 2020. Fake News from Africa: Panics, Politics and Paradigms. *Journalism* 21 (1): 3–16. doi:10.1177/1464884917746861.

Wasserman, Herman, and Dani Madrid-Morales. 2019. "An Exploratory Study of 'Fake News' and Media Trust in Kenya, Nigeria and South Africa." *African Journalism Studies* 40 (1): 107–23. doi:10.1080/23743670.2019.1627230.

———, eds. 2022. *Disinformation in the Global South.* Hoboken, NJ: Wiley-Blackwell.

———. (Forthcoming). "Cynical or Critical Media Consumers? Exploring the Misinformation Literacy Needs of South African Youth."

Wasserman, Herman, Dani Madrid-Morales, Admire Mare, Khulekani Ndlovu, Melissa Tully, Emeka Umejei, and Chikezie E. Uzuegbunam. 2019. *Audience Motivations for Sharing Dis- and Misinformation: A Comparative Study in Five Sub-Saharan African Countries.* https://cyber.harvard.edu.

Waters, Stephenson. 2018. "The Effects of Mass Surveillance on Journalists' Relations with Confidential Sources: A Constant Comparative Study." *Digital Journalism* 6 (10): 1294–313.

We Are Social. 2021. "Digital Report 2021: El Informe Sobre las Tendencias Digitales, Redes Sociales y Mobile." https://wearesocial.com.

WhatsApp. 2023. "Communities Now Available!" (blog), https://blog.whatsapp.com /communities-now-available.

———. n.d. "Message Privately." www.whatsapp.com.

WhatsApp Help Center. n.d. "Requirements for Broadcasting a Message." https://faq .whatsapp.com.

Wildemuth, Barbara M., ed. 2016. *Applications of Social Research Methods to Questions in Information and Library Science.* New York: Bloomsbury.

Williams, Philippa, Lipika Kamra, Pushpendra Johar, Fatma Matin Khan, Mukesh Kumar, and Ekta Oza. 2022. "No Room for Dissent: Domesticating WhatsApp, Digital Private Spaces, and Lived Democracy in India." *Antipode* 54 (1): 305–30. doi:10.1111/anti.12779.

World Bank. 2016. *World Development Report 2016: Digital Dividends.* Washington, DC: World Bank. doi:10.1596/978-1-4648-0671-1.

———. 2023. "Individuals Using the Internet (% of Population)—Sub-Saharan Africa." https://data.worldbank.org.

World Economic Forum. 2023. *Global Risks Report.* Cologne/Geneva, Switzerland. https://www.weforum.org.

World Health Organization. n.d. "Infodemic." www.who.int.

Yee, Kyra, Alice Schoenauer Sebag, Olivia Redfield, Emily Sheng, Matthias Eck, and Luca Belli. 2023. "A Keyword Based Approach to Understanding the Overpenalization of Marginalized Groups by English Marginal Abuse Models on Twitter." In *Proceedings of the 3rd Workshop on Trustworthy Natural Language Processing,* edited by Anaelia Ovalle, Kai-Wei Chang, Ninareh Mehrabi, Yada Pruksachatkun, Aram Galystan, Jwala Dhamala, Apurv Verma, Trista Cao, Anoop Kumar, and Rahul Gupta, 108–20. Stroudsburg: Association for Computational Linguistics. doi:10.18653/v1/2023.trustnlp-1.10.

Yustitia, Senja, and Panji Dwi Asharianto. 2020. "Misinformation and Disinformation of COVID-19 on Social Media in Indonesia." In *Proceeding of LPPM UPN "VETERAN" Yogyakarta Conference Series 2020–Political and Social Science Series*, 1: 51–65.

Zelizer, Barbie. 2007. "On 'Having Been There': 'Eyewitnessing' as a Journalistic Key Word." *Critical Studies in Media Communication* 24 (5): 408–28. doi:10.1080/07393180701694614.

Zhou, Yanmengqian, and Lijiang Shen. 2022. "Confirmation Bias and the Persistence of Misinformation on Climate Change." *Communication Research* 49 (4): 500–23.

ABOUT THE CONTRIBUTORS

MUSTAQUE AHAMAD is a professor in the School of Computer Science at the Georgia Institute of Technology and associate director of its Institute for Information Security and Privacy.

JOÃO GUILHERME BASTOS DOS SANTOS is director of Thematic Studies on Democracia em Xeque (DX) and researcher at the Instituto Nacional de Ciencia e Tecnologia para a Democracia Digital (INCT.DD) in Brazil.

YULIA BELINSKAYA is a researcher at the Institut for Creative\Media/Technologies, University of Applied Sciences, St. Pölten, Austria, and a lecturer at the University of Vienna, Austria.

TANJA BOSCH is Professor of Media Studies and Production at the University of Cape Town, South Africa.

SIMON CHAUCHARD is an associate professor in the Department of Social Sciences at the University Carlos 3 of Madrid, Spain.

CAYLEY CLIFFORD is the deputy chief editor of Africa Check.

JORGE ORTIZ FUENTES is currently studying toward a master's degree in computer science at the University of Chile and is a research student at the Millennium Institute Foundational Research on Data.

IGINIO GAGLIARDONE is an associate professor in media and communication at the University of the Witwatersrand, South Africa.

KIRAN GARIMELLA is an assistant professor in the School of Communication and Information at Rutgers University, United States.

JENCY JACOB is the managing editor of BOOM, one of India's first fact-checking newsrooms, where he leads a team of twenty fact checkers across India, Myanmar, and Bangladesh.

DANI MADRID-MORALES is a lecturer in journalism and global communication at the School of Journalism, Media and Communication, University of Sheffield (UK), where he co-leads the Research Disinformation Cluster.

ADMIRE MARE is an associate professor and head of the Department of Communication and Media Studies at the University of Johannesburg, South Africa.

NASIR MEMON is the interim dean of computer science, data science, and engineering at NYU Shanghai, a professor of computer science and engineering at the Tandon School of Engineering, and an affiliate faculty at the Computer Science Department in NYU's Courant Institute of Mathematical Sciences.

NICHOLAS MICALLEF is a lecturer at Swansea University, UK.

RUTH MOON is an assistant professor of media and public affairs at Louisiana State University, United States.

ALLEN MUNORIYARWA is a senior lecturer in the Department of Media Studies at the University of Botswana, Gaborone and a senior research fellow in the Department of Communication and Media at the University of Johannesburg, South Africa.

SAMUEL OLANIRAN is a postdoctoral fellow at the University of Johannesburg, South Africa.

CAROLINA PARREIRAS is a researcher at the Department of Anthropology at de University of São Paulo (USP), Brazil, and coordinator of the Ethnographic Laboratory for Technological and Digital Studies (LETEC—USP).

SAMEER PATIL is an associate professor in Kahlert School of Computing, University of Utah, United States.

JOAN RAMON RODRIGUEZ-AMAT is a principal lecturer and research and innovation lead in the Department of Media Arts and Communication at Sheffield Hallam University, UK.

ERKAN SAKA is a professor in the Department of New Media and Communication at Istanbul Bilgi University, Turkey.

MARCELO SANTOS is an associate professor at Universidad Diego Portales and an adjunct researcher at Millennium Nucleus for the Study of Media, Public Opinion, and Politics in Chile (MEPOP).

KIM SCHUMANN is a PhD candidate of social and cultural anthropology at the University of Cologne, Germany.

NKULULEKO SIBIYA is a PhD candidate in the Department of Media Studies at the University of the Witwatersrand, South Africa.

AMBER SINHA is an information fellow at Tech Policy Press, United States.

INGA KRISTINA TRAUTHIG is the head of research of the Propaganda Research Lab at the Center for Media Engagement at the University of Texas at Austin, United States.

SCOTT TIMCKE is a senior research associate at Research ICT Africa and a research associate in the Department of Journalism, Stellenbosch University.

ABOUT THE EDITORS

SAHANA UDUPA is Professor of Media Anthropology at the University of Munich (LMU) and Berkman Klein Fellow at Harvard University. She is the author of *Making News in Global India: Media, Publics, Politics*, coauthor of *Digital Unsetting: Decoloniality and Dispossession in the Age of Social Media*, and coeditor of *Digital Hate: The Global Conjuncture of Extreme Speech* and *Media as Politics in South Asia*.

HERMAN WASSERMAN is Professor of Journalism and Chair of the Department of Journalism at Stellenbosch University. He is the author of *Tabloid Journalism in South Africa*; *Media, Geopolitics, and Power: A View from the Global South*; and *The Ethics of Engagement: Media, Conflict and Democracy in Africa*, and coeditor of *Disinformation in the Global South*.

INDEX

Page numbers in italics indicate figures and tables.

Aam Aadmi Party (AAP), 76, 79, 81
abortion, 43, *45*
activism, political, 164
aesthetic politics, 70
affective turn, 261
Afghanistan, 148
Africa: coordinated harm in, 313–18;
    fact-checking in, 303–8; grassroots
    movements in, 7; pan-Africanism, 15,
    61; sub-Saharan, 3, 128, 130, 178, 185,
    313n1. *See also specific countries*
Africa Center for Strategic Studies, 54
Africa Check, 23, 24, 218–19, 221, 303–8
African National Congress (ANC), 50, 55,
    57, 59–60, 66
Ahidjo, Ahmadou, 112
AI. *See* artificial intelligence
AI4Dignity project, 85
Alexandra township, South Africa, 51
All Progressives Congress (APC), 192,
    193, 197
Almeida, Jussara, 246, 288
"alternative" facts, 90
alumni groups, 214
Amnesty International, 116, 120
Amungwa, Nicodemus Nde Ntso, 118
ANC. *See* African National Congress
Anderson, Chris, 166
Andhra Pradesh, India, 224
Android, 293
Anglophone Crisis, 3, 111–18, 120, 124–25
Anglophone users, 3, 16, 21
anonymization, 19, 270, 273, 280–82

anti-Black ideology, 1
anti-immigration, 1, 14–15
antisocial commenting, 2
AoIR. *See* Association of Internet
    Researchers
apartheid rule, nostalgia for, 58–61, *59*, *60*
APC. *See* All Progressives Congress
API. *See* application programming interface
API-based data gathering, 3
application programming interface (API),
    143, 305
approval group, 198
apruebo (approve) option, 92–93, 97,
    100–105
Araujo, Theo, 271, 279
arbitrary phone searches, 16
Argentina, 97, 145
Armenia, 145, 242
artificial intelligence (AI), 4, 281
Arun, Chinmayi, 236
Ashukem, Jean-Claude N., 116
Assam, India, 224
assimilation, 113
Association of Internet Researchers
    (AoIR), 147, 263
audience disengagement, from disinfor-
    mation, 129–34
Australian Broadcasting Corporation, 185
Austria, 17, 142, 145, 148, 153
authoritarianism, 175, 241; hybrid states,
    188; postcolonial, 129
authorization, 147
automation, 289–90, 297

Ayah Foundation, 118
Azerbaijan, 242

backfire effect, 307, 317
bad actors, 35
Balague, Guillem, 218
banal technology, 32
Bangalore, India, 70–71, 74–75, 84
Bangladesh, 85
banking, 151
Barbosa, Sérgio, 144, 250, 258, 263
bathrooms, unisex, 42, 47
BBC, 226
Beatrice (political prisoner), 121, 123
Beautiful Soup, 290–91
Belinskaya, Yulia, 17, 149
Bell Pottinger, 128
Ben-Hassine, Wafa, 251
Bharatiya Janata Party (BJP), 9, 71, 74, 76
bias: confirmation, 90; Eurocentric, 264;
    response, 90; social desirability, 181
Bihar, India, 224
Birnbaum, Larry, 259, 288, 297
BJP. See Bharatiya Janata Party
Black communities, in South Africa, 50
Black people, 58–59
black propaganda, 193
Black users, 161
Boko Haram, 114
Bolsa Família Program, 37
"Bolsonarista" groups, 14
Bolsonaro, Jair, 1, 10, 30, 36–37, 40–43,
    42n12, 66, 97
BOOM Fact Check, 23, 24, 309, 310
BoomLive, 229
Bosnia, 13
Bozkanat, Esra, 243
Brasil sem Miséria, 39
Braun, Virginia, 212
Brazil, 29, 69, 141, 246, 250–51, 256, 272n2,
    282; Bolsonaro in, 1, 10, 30, 36–37,
    40–43, 42n12, 66, 97; conclusion to,
    46–49; deep extreme speech in, 85;

elections in, 1, 8, 243, 247, 259; extreme
    speech and moralities in, 34–46; fact-
    checking in, 24; favelas in, 1, 7, 11, 14,
    30–35, 86; LGDP in, 269; method of
    study in, 32–34; São Paulo, 11, 14
bribe solicitation, 118
Brighter AI, 281
#BringBackOurGirls, 7, 190
British Southern Cameroons, 111
"broadcast" groups, 6
Buea, University of, 124
Buhari, Muhammadu, 193
Bursztyn, Victor S., 259, 288, 297
business users, 7

Cambridge Analytica, 194
Cameroon, 1–2, 3, 16, 21; Anglophone
    Crisis in, 112–13; British Southern, 111;
    conclusion, 124–25; Francophone, 111;
    phone searches in practice, 117–24;
    phone searches theorized in, 115–17;
    separatist movement in, 112, 120; so-
    cial media in, 113–15; Twitter in, 114
Cameroon Database of Atrocities, 3, 115
Cameroon Human Rights Council, 119
Cameroon Renaissance Movement, 121
canvassers, 198
Cape Peninsula University of Technology
    (CPUT), 131–32, 134–37
capitalism, 24, 317
Capitol riots (January 6), 65, 97, 106
Carlos 3 Madrid, University, 279n9
Carrefour, 39
cascade logic, 245
cease and desist letters, 227
Cele, Bheki, 58
censorship, 236, 253
Center for Strategic Studies, 4
Centre for Human Rights and Democ-
    racy in Africa (CHRDA), 118, 123
Cesarino, Letícia, 41, 42n12
Channels (feature), 312
Chari, Freeman, 218

chat encryption, 125

Check4Spam, 228

Checkpoint Tipline, 230

Chennamma, Kittoor Rani, 69

Chhattisgarh, India, 224

child abuse, 42n13

child kidnappings, 309–10

children, missing, 309

Chile, 1, 15, 88; conclusion, 105–7; constitutional referendum of, 91–92; content analysis in, 93–95, 94; datasets in, 92–93; method of study in, 92–95; results and discussion, 95–107; SNA and, 95; uncheckable content in, 89–91

China, 4, 54, 128, 138, 188, 310

Chin'ono, Hopewell, 218

ChiShona (language), 210

CHRDA. See Centre for Human Rights and Democracy in Africa

civic engagement, 247

civil society, 7, 23, 189

climate change, 90, 148

closed groups, 6, 9, 18, 20, 143, 207, 262

Cloud Data Loss Prevention API (DLP API), 273

Clubhouse, 72

coding schemes, 257

collaboration: community, 23; interdisciplinary, 22; verification by, 183–84

collaborative system, 218–19

collective social correction, 220

Colombia, 97, 199

colonialism, 129, 264

commodification, 318

communication: general communication group, 198; "rhythm" of, 194; telecommunication policy research, 316–17. See also political communication

community: Black, 50; collaboration, 23; diaspora, 165–67, 170; Hispanic, 163; marginalized, 159, 160; perceived intimacy diaspora, 167; role of, 152–53; sense of, 10

comparative ethnography, 316

Complexo, 31–34, 36–37, 40, 42–43, 49

confirmation bias, 90

Congo, 63

conspiracy theories, 106, 126, 243, 306

constitutional referendum, of Chile, 91–92

contagion dynamics, 103, 104

"contenders," 116

content analysis: in Chile, 93–95, 94; ECA, 56; qualitative, 257

content moderation, 18–19, 86, 316

content modernization, 314

context collapse, 245

Continent, The (magazine), 129

convenience, of WhatsApp, 161

convivial bonds, 2

coordinated harm, 24, 313–18

coordinated xenophobia attacks, 314n2

corruption, 41

Council of the EU, 154, 156

counternarratives, of identity formation, 61

counterpublics, 161, 162

counterterrorism. See Cameroon

COVID-19 pandemic, 29, 37, 53, 131–33, 241–44, 247, 306, 310; digital tools during, 179; fact-checking and, 129; health-related misinformation and, 214; journalism during, 179, 182

CPI on Pedophilia (2008), 42

CPUT. See Cape Peninsula University of Technology

Criminal Procedure Code, 115, 119

critical media theory, 160

critical thinking skills, 215

cross-disciplinary studies, 5

cross-platform dynamics, 245

cross-platform research, 106

crowdsourced fact-checking, 219–21

Cruz, Edgar Gómez, 7, 31, 69

Cuban Americans, 17, 161, 165, 169, 171

cultures: digital hate, 259; multiculturalism, 163; oral, 129

cyberterrorists, 114

Damara Nama (language), 210
dark surrealism, 42
Das, Veena, 32, 35
data access, 19–20
data charges, 89
data collection, 20–21, 211, 216, 248–49, 274, 286; challenges for, 268–71; characteristics and spread of misinformation, 288–89; conclusion, 299; device setup, 290–91; discussion and implications, 297–98; ethical considerations for, 291–92; findings, 292; group discovery and, 292–93; group maintenance and, 294–97; group membership and, 293–94; inappropriate content or behavior and, 295; initial exploratory investigation for, 291; large-scale longitudinal deployment and, 296–97; manual intervention and, 297–98; method, 290–92; other issues, 295–96; public information access and, 298; related work on, 287–90; restraint in amount, 283; tools for, 289–90
data costs, 61–62
data donation, 268, 269, 271
datafication of life, 29
data minimization, 19, 270
Data Protection Act, 211
datasets: in Chile, 92–93; May, 92–96, 105; referendum, 93, 95, 100–101
Dawai! (forum), 145
Death and Life (Klimt), 142, 148
decolonial research methods, 264
deep extreme speech, 11, 15, 21, 68–70; in Brazil, 85; conclusion, 86–87; digital campaigns and, 71–73; in Global South, 85; intrusive and intimate channels of distribution in, 73–76; simulating the social and, 76–80, 77, 78, 79; themes in, 80–86
deep fakes, 4, 70, 84
#DefyHateNow, 114

Delhi, India, 70
democracy, 175, 188, 241; crisis of, 160; inclusive, 17
deplatforming, 8
derogatory speech, 35
"deviants," 116
diaspora communities, 165–67
difference, social markers of, 49
"digital body of the king," 42n12
digital campaigns, deep extreme speech and, 71–73
digital capitalism, 24
digital dividends, 313
digital hate cultures, 259
digital inequalities, 49
digital literacy skills, 31, 215, 242
digital media, 30
digital networks, 2
Digital News Report (Reuters Institute for the Study of Journalism), 8, 165
digital payments, 7
digital skills, 31
digital technology, 127
digital tools, 179
disappearing messages, 9
discourse analysis, 257
disinformation. See specific topics
Disinformation in Africa 2024 (Center for Strategic Studies), 4
divisive messaging, politics of, 14–16
DLP API. See Cloud Data Loss Prevention API
Doğruluk Payı, 242
Donovan, Joan, 2, 195, 314
Duda, 33, 33n5, 43
Durov, Pavel, 155

East Germany, 8
ECA. See ethnographic content analysis
Economic Freedom Fighters (EFF), 63, 64
educational public service announcements, 234
EFF. See Economic Freedom Fighters

Electoral Commission, 128
electoral fraud, 98
electoral politics, 8
electricity breakdown, 178
emergency aid, 37
emotion, 199
encrypted messaging, 4, 5, 89
encrypted private messaging (EPM), 245
#EndSARS, 7, 190
end-to-end encryption (E2EE), 144, 204, 230–31, 234, 251, 271, 310, 316
English language news reports, 223
EPM. *See* encrypted private messaging
ERC. *See* European Research Council
ERC POLARCHATS, 272, 272n2, 282, 284
Eritrea, 169
Escobar, Pablo, 199
ESPN, 218
established news media, 129
estallido social, 91
ethical research, 33n5, 266
ethics of care, 265
ethnographic content analysis (ECA), 56
ethnographic sensibility, 33
ethnography, 22, 316
EU. *See* European Union
Eurocentric bias, 264
European Commission (2017), 86
European Research Council (ERC), 279n9
European Union (EU), 154, 228, 241, 246–47, 252, 269
the everyday, the ordinary versus, 32
exclusion, 171
experience friction, 232–35
Extensible Messaging and Presence Protocol (XMPP), 230
extreme speech. *See specific topics*
extremism law, in Russia, 7
eyewitnessing, 177–78, 182–83

fabrications, 305, 306
Facebook, 7, 72, 165, 204, 249–50, 252, 254, 315; algorithms on, 225; Anglophone Crisis on, 115, 120; in Cameroon, 114; deplatforming and, 8; devaluing, 134; echo chambers on, 69; hate speech on, 3; as main source of information, 165; Rousseff and, 36; in South Africa, 128–29; xenophobia on, 53
"Fact Ambassadors" project, 307
"fact-checkability," 90
fact-checking, 18, 23, 31, 40, 62, 178, 204–7; amplifiers of information, 219; in Brazil, 24; COVID-19 and, 129; crowdsourced, 219–21; in Global South, 88; in India, 309–12; informal, 12, 208, 217, 221; self-appointed fact-checkers, 217–19; speech regulation and, 228–30; xenophobia and, 53
fact-checking, in Africa, 303; lessons learned for, 307–8; trends and patterns of misinformation, 305–6; *What's Crap on WhatsApp?* and, 304–5
FactCheckZW, 219, 221
"factual" facts, 90
fake news, 29–31, 33, 36–38, *39*, 54, 303
falsehoods, 202–3, 212
family, destruction of, *47*
far-right content, 58
favelas, 1, 7, 11, 14, 30–35, 86
fear, 199–202
Feng, Kevin K. J., 149, 249
Ferreira, Carlos Henrique Gomes, 287, 288
field enumerators, 274
filter bubble, 233
firsthand experience, 151
focus groups, 260
football fandom, 214
Force Atlas 2, 100, 196
foreign actors, 3
Fort Hare, University of, 131
4Chan, 9
Francophone Cameroonians, 111
Francophone Republique du Cameroun, 112
Fraser, Nancy, 161, 163

fraud, 40, 96–98, 102
Free Basics, 7
freedom of the press, 123
freelance journalists, 184
free speech, 236
Free State, University of the, 131, 133, 136
friction, experience, 232–35

Gagliardone, Iginio, 3, 14
Gandhi, Rahul, 80
Garimella, Kiran, 20, 153, 156, 288, 297
gatekeeping, 208n1
gatewatching, 208, 220
gathering space, WhatsApp as, 184–85
Gauteng Province, 54
GDPR. *See* General Data Protections Regulation
gender ideology, 42
general communication group, 198
General Data Protection Law (LGDP) (Brazil), 269
General Data Protections Regulation (GDPR), 269, 270
George, Cherian, 117–18, 194
Georgia, 145
Gephi, 100, 196, 260
Germany, 8, 112, 227
Ghana, 76
Giddens, Anthony, 221
Global North, 5, 12, 264, 317
Global Risks Report, 3
Global South, 2–3, 9, 15, 21, 69, 107, 128–29, 243, 267–68; "broadcast" WhatsApp groups in, 6; deep extreme speech in, 85; digital inequalities in, 49; exclusionary politics in, 50; fact-checking in, 88; multimedia strategy in, 87; political communication in, 255–56; political messaging in, 18, 66; social and political mobilizations in, 5; surveillance in, 11; "technology of life" in, 7. *See also specific countries*
Global Witness, 53

Goal.com, 218
good actors, 35
Google, 128, 273
government (state) surveillance, 12, 115
Gowda, Nange, 68
Gowda, Uri, 68
Grand National Dialogue, 116
Granovetter, Mark S., 164, 214
grassroots movements, 7, 189
gray interventions, 12
group discovery, 286, 292–93
group maintenance, 287, 294–97
group membership, 286–87, 293–94
group messages, 225–26
Guinness World Record, 54
Gujarat, India, 224

Habermas, Jürgen, 162, 165
Haddad, Fernando, 35–36
harassment, 236
Harindranath, Ramaswami, 7, 31, 69
Hassan, Idayat, 192, 195
hate cultures, digital, 259
hate speech, 4, 7–10, 31, 126, 227, 234, 263; on Facebook, 3; Prevention and Combating of Hate Crime and the Hate Speech Bill, 52; in South Africa, 51–54
hate spin, 194
health-related misinformation, 214
Heineken, 39
hereditary social status, 162
Herero (language), 210
Hinduism, 80
Hindu nationalists, 2
Hispanic communities, 163
Hispanic users, 161
Hitchen, Jamie, 192
hoaxes, 305, 306
homophobia, 244
homosexual people, 118
"How to Destroy a Black Man," 61
Новый Венский Журнал (New Viennese Journal), 145

human infrastructure, 190, 191–93
human networks, in Nigeria, 197–98
human rights, 50, 91
Human Rights Watch, 115, 120
hybrid authoritarian states, 188

ICR. *See* intercoder reliability tests
ICT Households, 30, 30n4, 31
identity formation, counternarratives of, 61
ideology, 200–201
IFCN. *See* International Fact Checkers Network
ignorance of the crowd, 220
illegal immigrants, 55
IM. *See* instant messaging
immigration, 63, 146; anti-immigration, 1, 14–15; attacks on migrants, 50–51; illegal, 55; undocumented, 315
imperialism, 264, 317
improvisation, 221
inappropriate content or behavior, 287, 295
INC. *See* Indian National Congress
inclusive democracy, 17
Independent Media, 54
Independent National Electoral Commission (INEC), 194, 199
India, 8, 13–15, 22–24, 68–70, 138, 256, 272n2, 282; Assam, 224; BJP in, 9, 71, 74, 76; conclusion, 86–87; digital campaigns in, 71–73; fact-checking in, 309–12; intrusive and intimate channels of distribution in, 73–76; Personal Data Protection Bill, 269; simulating the social and, 76–80, *77, 78, 79*; themes in, 80–86. *See also* speech regulation
Indian Americans, 17, 161, 166
Indian National Congress (INC), 71, 74, 76
Indigenous People of Biafra (IPOB), 192
individual interviews, 181
Indonesia, 13, 85

INEC. *See* Independent National Electoral Commission
inflammatory content, 9
"infodemic," 306
"infopolitics," 169
informal fact-checking, 12, 208, 217, 221
informality, 206–7
informal user correction, 216–21
information disorder, 127, 139
information dissemination, 286
Information Technology (Intermediary Guidelines and Digital Media Ethics Code) Rules, 228
informed consent, 211, 263
infrastructural studies, 177
infrastructure: human, 190, 191–93; of journalism, 176–78; physical, 179. *See also* reporting infrastructure
Ingram, Helen, 115, 116
Instagram, 7, 43, 72, 114, 141, 170, 198, 243
instant messaging (IM), 231
instant messaging services, 8, 18
institutional trust, 201
interaction networks, 196
intercoder reliability tests (ICR), 93, 94
interdisciplinary collaboration, 22
intermediaries, 228
Intermediary Guidelines and Digital Media Ethics Code (Information Technology) Rules, 228
internal content moderation, in South Africa, 61–65, *62*
International Fact Checkers Network (IFCN), 91, 92, 229, 235
internet access, 32
Internet Intermediary Rules, 22
internet penetration rate, 113
Internet Relay Chat (IRC), 230–31
Internet Research (AoIR), 263
internet usage, 313n1
intertextuality, 245

interviews: individual, 181; qualitative, 180, 260; reconstruction, 180; scroll-back method, 143–44, 261; small-group, 181

intimacy, 33; lived encryptions and, 10–11; perceived intimacy diaspora community, 167

introduction friction, 233

IPOB. *See* Indigenous People of Biafra

Ipsos, 3

Iran, 138, 141, 149

IRC. *See* Internet Relay Chat

IsiNdebele (language), 210

IsiXhosa (language), 210

isiZulu (language), 210

Islamic fundamentalism, 193

Istanbul Convention, 244

IT Rules, 311

Ivette (pseudonym), 170

İyi Party (Turkey), 248

Jacob, Jency, 23, 24

"Jai Shree Ram," 80

January 6 Capitol riots, 65, 97, 106

Japan, 62

Jharkhand, India, 224

Jonathan, Goodluck, 193

journalism, 229; during COVID-19 pandemic, 179, 182; crowdsourcing and, 219–20; freelance, 184; infrastructure of, 176–78

Kaiser-Grolimund, Andrea, 246, 250

Karim, Abdul, 120

Karnataka, India, 74, 224

Kazakhstan, 145

Kenya, 2, 13, 85, 307

Kerala, India, 224

Khumalo, Mario, 52

Kigali, 182, 184

kindred terms, 32

kin relations, 84

Klimt, Gustav, 142, 148

knowledge systems, 176

Krippendorf's alpha, 94

Kutumb, 72

KwaZulu-Natal, University of, 131, 133, 134, 138

Labour Party (LP), 192

language: of legitimation, 148–51; local, 210

large-scale longitudinal deployment, 296–97

Larkin, Brian, 18, 191, 197

Law 2014/028, 116–17

law enforcement agencies, South African lack of trust for, 57–58

Legal Resources Centre, 53

legitimation, 148–51, 157

Leiden algorithm, 196

Leopold Museum, 148

"Letter from the Prime Minster," 74

Levenshtein distance, 93

Lewis, Rebecca, 193, 195

LGBTQI+ individuals, 118, 243

LGDP. *See* General Data Protection Law

like-mindedness, 226

literacy, media, 214–15, 242

lived encryptions, 6, 10–14

local languages, 210

"long tail" phenomenon, 80

LP. *See* Labour Party

Lula da Silva, 36, 37–38, 40–41, 43–46, *44*

lurking, 20

Lux, Nhlanhla, 55

lynchings, 224, 231

Madhya Pradesh, India, 224

Madrid-Morales, Dani, 7, 16, 30, 53, 130

Maharashtra, India, 224

Mahere, Fadzayi, 218

Mail & Guardian (newspaper), 129

Malawi, 63

Malta, Magno, 42n13

Malviya, Amit, 73

manipulation, 236

manual verification, 298
the margin, 35
marginalization, 113, 171
marginalized communities, 159, 160
Maros, Alexandre, 246, 288
Marwick, Alice E., 193, 195
mass messages, 37–38, 227, 235
Matrix, 232
MAXQDA, 181
May dataset, 92–96, 105
Mbaku, Gilbert, 118, 119, 122
media literacy, 214–15, 242
Media Monitoring Africa, 128, 129
media team, 198
Melo, Philipe, 288, 289, 297
message traceability, 230–32
messaging apps, 254
Meta, 5, 7, 198, 205, 250, 308. *See also*
    Facebook; Instagram
metadata, 231–32, 276
metamorphic sublimity, 42
Mexican Americans, 17, 161, 163,
    166–69
Mexico, 97
Mhlanga, Blessed, 218
microhistories, 31
migrants: attacks on, 50–51; undocu-
    mented, 315
Milan, Stefania, 144, 250, 256, 258, 263
military service, 150
military veterans, 58
Ministry of Electronics and Information
    Technology, 231
Ministry of Health (Turkey), 248
minorities, derogatory speech against, 35.
    See also specific minorities
misinformation. *See specific topics*
missing children, 309
mobile penetration, 89
mobile phones, 31, 178, 186
mobilization, 150, 226, 246
mob violence, 223–24, 236–37
Modi, Narendra, 80

Moj, 72
moral evaluation, 147
moralities, in Brazil, 34–46
moral panics, 34
Mozambique, 63
multiculturalism, 163
Mumbai, India, 70
mundane technology, 32
Muslims, 13, 80, 81, 82; women, 224
mythopoesis, 147

Nagorno-Karabakh conflict, 242
Nairobi, 184
Namibia, 204–5; conceptual frameworks
    for, 206–9; conclusion, 221–22;
    informal user correction in, 216–21;
    methods and materials for, 209–12;
    sharing dis/misinformation in,
    212–16
Namibia Fact Check, 218, 219, 221
"nano-influencers," 74
narrowcasting, 12
National Party (NP), 58
National Peace and New Constitution
    Agreement, 91
national polarization, 40
national security, 111
#Nationalshutdown, 63
Nemer, David, 32, 190
Nestlé, 37
Network Enforcement Act (Netzwerk-
    durchsetzungsgesetz, NetzDG),
    227
network societies, 164
New Delhi, India, 73
Newman, Nic, 89, 165
news gathering, 176
news media, established, 129
newsroom computers, 179
New Viennese Journal (Новый Венский
    Журнал), 145
Neyazi, Taberez Ahmed, 206, 214
Ng, Sheryl Wei Ting, 206, 214

Nigeria, 7, 8, 9, 69, 127, 189–90, 307; conclusion, 202–3; disinformation campaigns in, 193–95; fear in, 199–202; human infrastructure in, 191–93; human networks in, 197–98; method and data, 195–96, *196*; propaganda in, 193; versatility of WhatsApp in, 199–202
Nobre, Gabriel Peres, 262, 288
NodeXL, 260
non-mass media, 163
Nostradamus, 153
notices, 234
NP. *See* National Party
nudges, 234
NVivo, 166
Nyathi, Elvis, 51

Obama, Barack, 169
Obi, Peter Gregory, 192, 195, 202
OD. *See* Operation Dudula
Odisha, India, 224
Ohme, Jakob, 271, 279
Olaniran, Samuel, 7, 9
"older" diaspora, 169–70
Oliyath, Shammas, 228
OMCT World Organisation Against Torture, 116
Omidyar Network, 251
Online Harms White Paper, 233
online safety and trust, perceived, 155–56
Online Safety Bill (UK), 23
"open" social media, 88
Operation Dudula (OD), 1, 14, 50–52, 54–58, 60–67
OpIndia.com, 229
oppressed caste groups, 13
oral culture, 129
the ordinary, the everyday versus, 32
Oshiwambo (language), 210
"otherness," 256
Oxford, University of, 8
Oxford English Dictionary, 32

Pakistan, 23, 309, 310
pan-Africanism, 15, 61
Pang, Nicholas, 143, 164, 179
Parreiras, Carolina, 11, 14
partisan content, 9
Pasquetto, Irene V., 192
passwords, 123
patriotic themes, 80–81
PDP. *See* Peoples Democratic Party
pedophilia, 42, 42n13
Peoples Democratic Party (PDP), 192, 193
perceived intimacy diaspora community, 167
Personal Data Protection Bill (India), 269
Personal Information Protection and Electronic Documents Act (PIPEDA) (Canada), 269
Peru, 97
Pew Research Center, 161
PGP, 251
the Philippines, 85
phone searches: arbitrary, 16; avoiding, 118–20; deletions for, 123–24; in practice, 117–24; theorizing, 115–17
physical infrastructure, 179
Pinterest, 72
PIPEDA. *See* Personal Information Protection and Electronic Documents Act
platformization, 29
platform moderation, 19
platform society, 29n1
PNAD, 30
polarization, political, 164
police, lack of confidence in, 58
policy group, 198
political activism, 164
political communication, 254; conclusion to, 265–66; decolonizing research, 264–65; ethical considerations for, 262–64; in Global South, 255–56; research methods for, 256–61
political conversations, 168–69

political disinformation, in South Africa, 127–29, 134–38
political engagement, 247
political information, types of, *94, 96*
political misinformation asymmetry, 96–97, 102
political persecution, 247
political polarization, 164
political propaganda, 126
"political remittances," 169
political talk, 8, 17
POPIA. *See* Protection of Personal Information Act
populism, 241
postcolonial authoritarianism, 129
power centralization, 188
power outages, 178
presence, witnessing without, 182–83
the press, freedom of, 123
prevalence metrics, 3
Prevention and Combating of Hate Crime and the Hate Speech Bill, 52
privacy, 276; challenges, 269–71; protection, 280–82; right to, 119; on Telegram, 155
private groups, 20, 121, 195, 198, 262
problematic content, 267
"problematic speech," 225
process-tracing, 180
profit maximization, 318
propaganda: in Nigeria, 193; political, 126
Propaganda Research Lab, 166
Protection of Personal Information Act (POPIA), 211
protests, 113
provisionality, 206–7, 208
PSA. *See* Put South Africa First
pseudoliberals, 81
PT. *See* Workers Party
public awareness videos, 23
public groups, 90, 91, 262
public information, access to, 298

public service announcements, educational, 234
public shaming, 247
public sphere, 159, 162–64, 220
public trust, crisis of, 160, 164, 170
Put South Africa First (PSA), 52, 64, 314
Python, 258, 260, 290

QR codes, 274
qualitative content analysis, 257
qualitative interviewing, 180, 260
qualitative methodologies, 260–61
qualitative research, 255, 265

Radebe, Dan, 65
Rajasthan, India, 224
rationalization, 147, 149, 157
ration-critical debate, 162
ReCal, 94
rechazo (reject) option, 92–93, 97, *98, 99,* 100–105
reconstruction interviews, 180
referendum dataset, 93, 95, 100–101
refugee crisis, 244
refugee protection laws, 50
regressive regimes, 2
regulatory models, 22
Reis, Julio C. S., 24, 204, 288
religions, 43, 80
religious cross, *46*
reporting infrastructure, WhatsApp as, 175, 179; conclusion, 185–86; findings, 182–85; infrastructure of journalism and, 176–78; method for, 180–81, *181*
research: cross-platform, 106; decolonial methods, 264; ethical, 33n5, 266; qualitative, 255, 265; telecommunication policy, 316–17. *See also* social science research
Resende, Gustavo, 286, 288, 296
response bias, 90
Reuters Institute for the Study of Journalism, 8, 165

reverse news sharing mechanism, 167–69
"rhythm" of communication, 194
rights: human, 50, 91; to privacy, 119
right-wing extremists, 65
"Rise Up Africa" (speech), 59
risk, rhetoric of, 177
Robards, Brady, 144, 261
Rockefeller, John D., 305
Romano, Fabrizio, 218
Romero, Gerard, 218
Rome Statute, 318
Rossini, Patrícia, 8, 9, 159
Rousseff, Dilma, 36, 36n6, 41
Russia, 4, 17, 54, 128, 138, 142; community
    and, 152–53; conclusion and reflec-
    tions, 156–58; diaspora messaging in,
    144–46; extremism law in, 7; language
    of legitimation, 148–51; methodologi-
    cal and ethical challenges in, 143–44;
    methodology for, 146, 146–56, 154, 155;
    perceived online safety and trust in,
    155–56; sources for, 153–55, 154, 155;
    Twitter in, 141, 243
Russian Embassy, 154–55, 156
Russian Federation, 149
Rwanda, 18, 175, 178, 183–87

SAA. See systematic anonymization audit
safety, perceived, 155–56
"safety by design" framework, 233
Saka, Erkan, 11, 20
Samajvadi Party, 71
SANDF. See South African National
    Defence Force
São Paulo, Brazil, 11, 14
SAPS. See South African Police Services
Sarv Webs Private Limited, 229
Scherman, Andrés, 69, 89, 159, 164
Schneider, Anne, 115, 116
Schoon, Alette, 260, 261
Schumann, Kim, 3, 16, 21
Scofield, Gilberto, 85
scrollback interview method, 143–44, 261

self-appointed fact-checkers, 217–19
self-checking, volunteer fact-checkers,
    217–19
self-correction, 207
self-help, 221
self-reflexivity, 265
semipublic space, 165
Senegal, 307
sentiment analysis, 258
separatist movement, in Cameroon, 112, 120
Serbia, 145
ShareChat, 72
Shettima, Ahmed, 192
Shiv Sena (political party), 71
"shutdown" campaign poster, 63
Sibiya, Nkululeko, 3, 14
Signal, 4–5, 176, 181, 184, 186, 231
Singapore, 214, 255
Siziba, Gift Ostallos, 218
skills: critical thinking, 215; digital lit-
    eracy, 31, 215, 242
Sky TV, 218
small-group interviews, 181
smartphones, 32
SM Hoax Slayer, 228
"smoke and mirrors game," 42, 46
SNA. See social network analysis
snowball sampling, 92
Soares, Felipe, 9, 69, 247
social class, 31
social context, of infrastructure systems,
    177
social correction, 18, 207
social desirability bias, 181
social markers of difference, 49
social media. See specific platforms
social network analysis (SNA), 95, 259
social relations, 70
social science research, 267; advantages,
    279–83; conclusion, 283–85; data col-
    lection challenges for, 268–71; data
    protocol for, 273–79, 275, 276, 277;
    possible strategy for, 271–73; practical

challenges, 271; privacy-related challenges, 269–71; strong anonymization and privacy protection in, 280–82; unexpected findings in, 282–83

Solhekol, Kaveh, 218

Somalia, 63

Sotho (language), 210

South Africa, 1–2, 13, 18, 23, 126, 204–5, 305, 307; Alexandra township in, 51; apartheid rule in, 58–61, 59, 60; audience disengagement from disinformation, 129–34; Black communities in, 50; conceptual frameworks, 206–9; conclusion, 65–67, 139–40, 221–22; coordinated xenophobia attacks, 314n2; counternarratives of identity formation in, 61; domestic and foreign dimensions of political disinformation, 134–38; Facebook in, 128–29; hate speech in, 51–54; informal user correction in, 216–21; internal content moderation in, 61–65, 62; methodology for study in, 54–57; methods and materials for, 209–12; mis/disinformation in, 51–54, 212–16; political disinformation in, 127–29; results and findings for, 57–65; sharing dis/misinformation in, 212–16; xenophobia in, 314n2, 316

South Africa First Party, 52

South African National Defence Force (SANDF), 58

South African National Editors' Forum, 53

South African Police Services (SAPS), 58

Southern Cameroons National Council, 113

southern theory, 264

South Korea, 255

Soviet Union. See Russia

Soweto Parliament, 55

spam farms, 234

speech regulation, 223–24; conclusion, 235–37; experience friction and, 232–35; fact-checking and, 228–30; message traceability and, 230–32; platform features for, 225–26; platform measures for, 227–35; regulatory responses, 227–35

SQLite database, 21, 289–91, 292, 296

Star, Susan Leigh, 175, 176, 177

the state, 35

state capture, 128

State of the Nation Address 2023, 65

state (government) surveillance, 12, 115

Staudacher, Sandra, 246, 250

stigmatization, of stereotypical media frames, 163

stochastic violence, 24

storytelling, 147

strong ties, 214

student union protests, 113

sub-Saharan Africa, 3, 128, 130, 178, 185, 313n1

subversive platforms, 159–61; communicative and societal role of WhatsApp, 164–65; conclusion, 170–71; findings, 167–70; methodology for, 165–67; public sphere and, 162–64

super users, 79

Supreme Court, 232

surrealism, dark, 42

surveillance, 188; fear of, 120; in Global South, 11; government, 12, 115

Syria, 23, 251, 309

systematic anonymization audit (SAA), 281

Tamil Nadu, India, 223, 224

Taussig, M., 41–42

Tchiroma, Issa, 114

"technology of life," 7, 31, 32

Telangana, India, 224

telecommunication policy research, 316–17

Telegram, 4, 17, 141–42, 156–59, 176, 245–46, 249–52, 254; channel C on, 154; data access to, 19; January 6 attack and, 65; privacy on, 155; public channels on, 143–44; "Topics" on, 147

Tembisa decuplets, 54
Temer, Michel, 36n6
terrorism, 97, 106, 111–17, 122, 124
Texas, University of, 166
text mining, 258
Teyit.org, 242
Thailand, 145
tight-knit networks, 11
TikTok, 43, 53, 128, 134, 190, 308
Tinubu, Bola Ahmed, 192, 195, 197, 199, 201
"Tinubu the drug dealer" campaign, 196
Tippu Sultan, 68, 68n1
Tor Browser, 251
Toronto, University of, 115
traceability, of messages, 230–32
trade protests, 113
trend alerts, 9
Tripura, India, 224
Trump, Donald, 42, 97, 169
trust, 33; crisis of public, 160, 164, 170; institutional, 201; perceived, 155–56
Tswana (language), 210
Tufekci, Zeynep, 195, 236
Turkey, 20, 22, 85, 145, 241–43; conclusion, 251–53; ethnographic research on, 244–48; İyi Party in, 248; methodological challenges in, 248–51; Ministry of Health, 248
Turkish Air Force, 243
"turn-the-vote" mobilization, 36
Twitter (X), 7, 72, 85, 136, 176, 190–95, 254, 308, 314–15; API access and, 143; in Cameroon, 114; censorship on, 236; conspiracy theories on, 106; deplatforming and, 8; echo chambers on, 69; in Russia, 141, 243; Trump on, 42
Twitter Academic API, 195
Twitter Spaces, 65, 202
2G internet connectivity, 227
typing indicator, 235
Tyson, Gareth, 153, 156, 288, 297

UAE. See United Arab Emirates
Udupa, Sahana, 15, 21, 122, 194, 200, 259, 261, 265
UFH. See University of Fort Hare
UFS. See University of the Free State
Uganda, 12
UK. See United Kingdom
Ukraine, 137, 141, 145, 146
UKZN. See University of KwaZulu-Natal
#Umzansi4Africa, 63
uncertainty, 241
uncheckable content, 89–91
"Understanding Disinformation Ecosystem in Global Politics" (Parlar Dal and Erdoğan), 241
undocumented migrants, 315
UNESCO, 3, 13, 24, 317
unisex bathrooms, 42, 47
United Arab Emirates (UAE), 145
United Kingdom (UK), 22, 23
United Nations, 53
United States, 17, 22, 127, 130, 138, 159, 161, 178; crisis of democracy in, 160; diaspora communities in, 165–67, 170; January 6 Capitol riots, 65, 97, 106
University Carlos 3 Madrid, 279n9
University of Buea, 124
University of Fort Hare (UFH), 131
University of KwaZulu-Natal (UKZN), 131, 133, 134, 138
University of Oxford, 8
University of Texas, Austin, 166
University of the Free State (UFS), 131, 133, 136
University of Toronto, 115
upstream information, 245
urban warfare, 35
user correction, informal, 216–21
Uzbekistan, 141

vaccines, 131–32
Venezuela, 99n5
Venezuelan Americans, 171

verification, 178; by collaboration, 183–84; manual, 298
versatility, of WhatsApp, 199–202
veterans, military, 58
Vienna, Austria, 145, 148
violence, 13, 32, 115, 282; depictions of, 122; extreme speech acts and, 61, 81; against migrants, 50; mob, 223–24, 236–37; stochastic, 24; women and, 35; xenophobia and, 53
viral media reports, 197
viral messages, 22, 24, 107, 199, 205, 309–10
virtual gatherings, 184–85
visa requirements, 145
visual content, 273
voice notes, 203, 304–5
volunteers, 189; fact-checkers, 217–19
vulnerable populations, 130

warning labels, 205
Wasserman, Herman, 7, 16, 30, 53
"Wathint' Abafazi, Wathint' Imbokodo" (You strike the women, you strike the rock) (song), 64
weak ties, 214
WeChat, 159, 180
Wells Fargo, 233
West Bengal, India, 224
WhatsApp. See specific topics
WhatsApp Explorer, 256, 257, 272, 277, 279
WhatsApp Messenger, 206
"WhatsApp penetration," 74
WhatsApp University, 87
What's Crap on WhatsApp? (podcast), 23–24, 304–5
White monopoly capital, 128
White supremacy, 59, 61
White users, 161
Wilson, Tom, 187, 195

wisdom of the crowd, 220
Witzel, Wilson, 37
women, 313–14; Muslim, 224; violence and, 35
Women's March, 64
Woo, Yik Tung, 143, 164, 179, 247
Workers Party (PT), 11, 35, 36, 36n6, 37
World Economic Forum, 3
World Health Organization, 306

X. See Twitter
xenophobia, 13, 15, 51–52, 61, 316; coordinated attacks, 314n2; fact-checking and, 53
XMPP. See Extensible Messaging and Presence Protocol

young people. See South Africa
You strike the women, you strike the rock ("Wathint' Abafazi, Wathint' Imbokodo") (song), 64
youth gangs, 51
YouTube, 53, 159, 227, 243
Yuh, Acho Wilson, 119

ZapZap, 7
ZEP. See Zimbabwe Exemption Permits
zero rating, 30, 48, 89
Zimbabwe, 58, 62, 63, 178, 188, 204–5, 305; conceptual frameworks, 206–9; conclusion, 221–22; informal user correction in, 216–21; methods and materials for, 209–12; sharing dis/misinformation in, 212–16
Zimbabwe Exemption Permits (ZEP), 63, 65
ZimFact, 218, 219, 221
Zoom, 181
Zuma, Jacob, 128

www.ingramcontent.com/pod-product-compliance
Lightning Source LLC
Chambersburg PA
CBHW031137020426
42333CB00013B/423